The Urban Book Series

The Urban Book Series is a resource for urban studies and geography research worldwide. It provides a unique and innovative resource for the latest developments in the field, nurturing a comprehensive and encompassing publication venue for urban studies, urban geography, planning and regional development.

The series publishes peer-reviewed volumes related to urbanization, sustainability, urban environments, sustainable urbanism, governance, globalization, urban and sustainable development, spatial and area studies, urban management, transport systems, urban infrastructure, urban dynamics, green cities and urban landscapes. It also invites research which documents urbanization processes and urban dynamics on a national, regional and local level, welcoming case studies, as well as comparative and applied research.

The series will appeal to urbanists, geographers, planners, engineers, architects, policy makers, and to all of those interested in a wide-ranging overview of contemporary urban studies and innovations in the field. It accepts monographs, edited volumes and textbooks.

Indexed by Scopus.

Hossein Bahrainy · Ameneh Bakhtiar

Urban Design in Seismic-Prone Regions

Springer

Hossein Bahrainy
Department of Urban Design and Planning
University of Tehran
Tehran, Iran

Ameneh Bakhtiar
Department of Architecture
University of Tehran
Tehran, Iran

ISSN 2365-757X ISSN 2365-7588 (electronic)
The Urban Book Series
ISBN 978-3-031-08320-4 ISBN 978-3-031-08321-1 (eBook)
https://doi.org/10.1007/978-3-031-08321-1

This Springer imprint is published by the registered company Springer Nature Switzerland AG
The registered company address is: Gewerbestrasse 11, 6330 Cham, Switzerland

Contents

Chapter 1
Urban Design and Disasters

Abstract We can define urban design as the purposeful decisions and actions which aim at establishing functional and formal order in the physical environment. It is obvious that this definition implies some abstract and value-free elements such as the decisions and actions involved and the formal and functional orders sought. In fact, the nature, characteristics, and interpretations of these elements normally change from one context to another. The general purpose of urban design, as one of the relatively new field of knowledge, has been to establish order in the physical environment. Today's interpretation of order, however, has changed drastically. This is because the urban environment has gained unprecedented complexity. Lack of order is reflected in the form, as well as activities of the urban environments. Urban design techniques and methods can be used to make changes and modifications in the built environment in order to establish desired order in the urban form and activities. The ambition to control "disorder," "chaos," "pollution," "contagion," "deviation," "sickness," etc., through spatial and social control (which includes organization and planning) and through regulation and restriction policies can be traced throughout the history of urbanism and governs today's approaches to urban planning and design. Vulnerability and seismic risk analysis in urban planning and design intends to mitigate seismic risk in the city. Urban design as a legitimate discipline and practice today depends on its ability and capacity to deal with the current issues society is facing and offer appropriate solutions to those problems, to the extent plausible and within the framework of the urban design tools and techniques.

Keywords Urban design · Disasters · Seismic risk · Basic needs · Quality of life · Sustainable development · Lessons learned · Cultural heritage

1.1 Introduction

From the time of early settlements, man, in keeping with one of his basic natural needs, has tried to gain control over his built environment. This control, which was sought through various natural, ritual, and functional means, manifested itself in some kind of order in the configuration, form and pattern of settlements. Throughout history, these general goals of control and order have always been with man, but their

H. Bahrainy and A. Bakhtiar, *Urban Design in Seismic-Prone Regions*,
The Urban Book Series, https://doi.org/10.1007/978-3-031-08321-1_1

meanings and the ways of interpreting and achieving those undergone considerable evolutionary and revolutionary changes from one period to another. The general purpose of urban design, as one of the relatively new field of knowledge, has been to establish order in the physical environment. Today's interpretation of order, however, has changed drastically. This is because the urban environment has gained unprecedented complexity. Lack of order is reflected in the form, as well as activities of the urban environments. Urban design techniques and methods can be used to make changes and modifications in the built environment in order to establish desired order in the urban form and activities (Fig. 1.1).

Accordingly, we can define urban design as the purposeful decisions and actions which aim at establishing functional and formal order in the physical environment. It is obvious that this definition implies some abstract and value-free elements such as the decisions and actions involved and the formal and functional orders sought. In fact, the nature, characteristics, and interpretations of these elements normally change from one context to another. They can, therefore, be defined specifically only in a particular context or culture.

It is intrinsic to this definition that urban design deals primarily with the *urban physical environment* which is the container of urban activities and hence, that urban

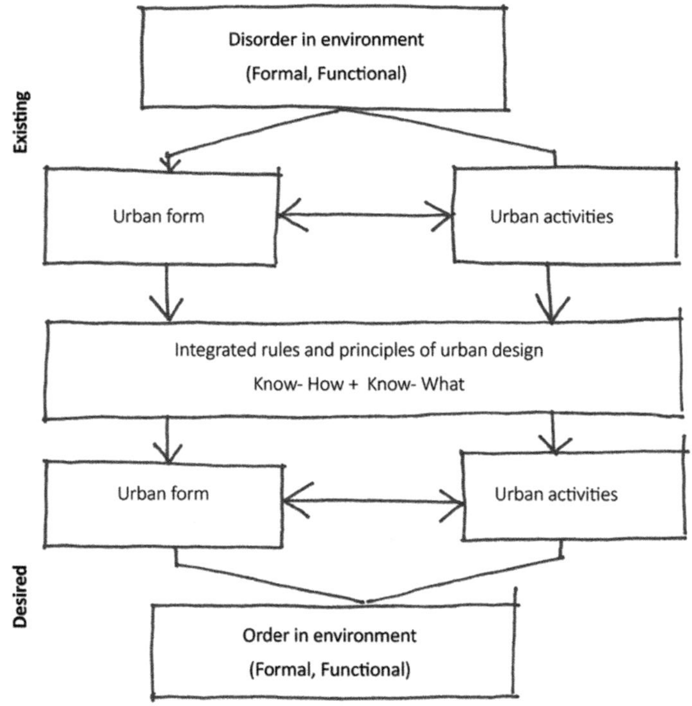

Fig. 1.1 Integrated rules are applied to the disordered environment, manipulating its elements to establish order (Bahrainy and Bakhtiar 2016)

activities, as the content, are also the concern of urban design; and that there are certain deliberate and purposeful decisions, actions and processes to be followed in order to achieve the desired goals. *Urban physical environment* in this definition implies urban form and all its attributes such as perceptual, visual, and aesthetic processes, and *urban activities* refer to the systems of activities which involve individuals, groups, and organizations in the urban environment. Urban design, therefore, deals with form as well function, and its purposes are efficiency and comfort, as well aesthetics (Bahrainy and Bakhtiar 2016; Banerjee 1992).

The ambition to control "disorder," "chaos," "pollution," "contagion," "deviation," "sickness," etc., through spatial and social control (which includes organization and planning) and through regulation and restriction policies can be traced throughout the history of urbanism and governs today's approaches to urban planning. Vulnerability and seismic risk analysis in urban planning and design intends to mitigate seismic risk in the city (Fig. 1.2).

Urban design as a legitimate discipline and practice today depends on its ability and capacity to deal with the current issues society is facing and offer appropriate solutions to the those problems, to the extent plausible and within the framework of the urban design tools and techniques.

The determinant factor of urban design, which makes it different from urban planning, is urban form. According to Alesch and Siembieda (2012), the importance of urban forms appropriate to the risks presented by hazards is especially critical in the case of rapid onset disasters such as near-field tsunamis affecting densely populated coastal locations. In cases like this, the vulnerable population has little time (typically only minutes or few hours) to make appropriate decisions about essential activities such as evacuation and sheltering, which in turn have a strong influence on the overall impact of the catastrophe. Moreover, in many occasions, these actions have to be autonomously conducted by the people, due to the cascading failure of emergency systems following a strong earthquake. This context of crisis can be alleviated by the urban form having characteristics that promote resilience in the form of support for rapid and effective response.

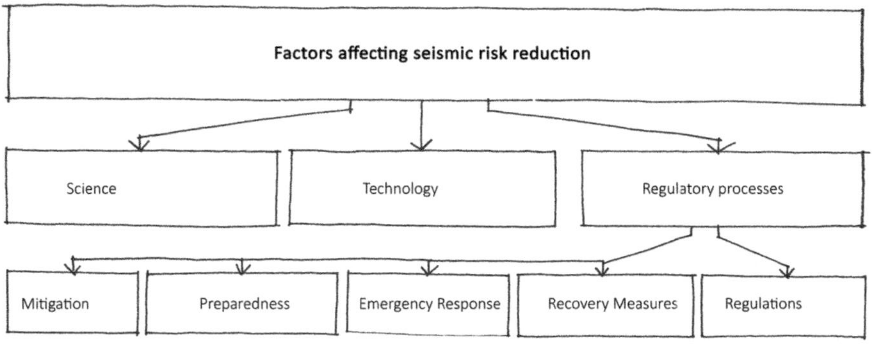

Fig. 1.2 Means and process of changing unacceptable risk into acceptable one

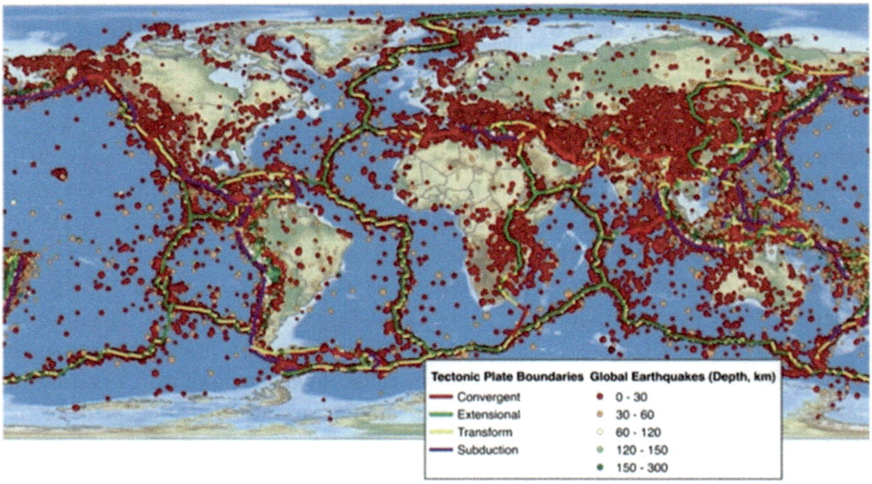

Fig. 1.3 Global distribution of earthquakes in the period from 1900 to 2014, and global plate boundaries (UNISDR 2017)

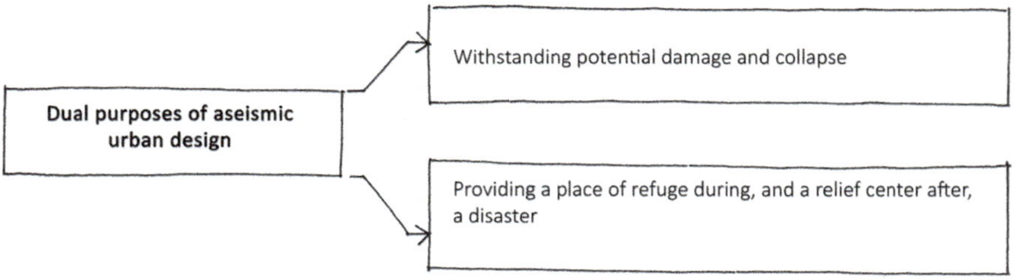

Fig. 1.4 Dual purpose of urban planning and design to mitigate disaster risk during and after a disaster

The global number of yearly reported natural disasters has quadrupled since the 1960s, reaching more than 400 events and 200,000 affected people per year in the 2010s (International Disaster Database 2011). Probably the most important factor is the increasing exposure of people to natural hazards, caused by human development patterns such as rapid urban growth and rising social inequalities, especially in the developing world. Moreover, as urbanization gathers pace, cities have become "hot spots" for disasters (Joerin and Shaw 2010; Wamsler 2014) (Figs. 1.3 and 1.4).

1.2 Human Basic Needs and Quality of Life

A study (Bahrainy and Tabibian 1998) on the quality of life indicators revealed that environmental safety is one of major components of the quality of life in urban areas and the third critical element of the basic needs of any human settlement. Protecting human settlements against natural disasters is, therefore, a major requirement for achieving sustainability and resiliency in our urbanizing world (Fig. 1.5).

Many factors have contributed to the evolution of cities, including demographic currents, cultural transitions, technological innovation, changes in values and social norms, and shifts in political structures. The goals of urban residents are convenience, efficiency, and economy, and urban planning was considered as a means of achieving these goals.

However, these goals are dependent upon the stability of daily life during peaceful times. Ensuring safety during emergencies has not been a prime consideration. Major cities where populations, resources, and capital are densely concentrated are highly susceptible when hit directly by a large-scale natural disaster. Urban planning and design should be required to devise and implement disaster prevention and land use in preparation for emergencies (Nishiyama 1993).

Figure 1.6 shows the factors to be considered in predictive modeling of earthquake ground motion. What is significant from the viewpoint of the subject of this book is the dominant role proper land use and city planning can play in the earthquake-prone regions to mitigate disaster and reduce loss of life (Fig. 1.6).

Since the built environment and urban structure provide the core framework for most human activity, it is crucial to develop an effective measure of resilience so it can withstand, and adapt to, the threats of natural and human-induced hazards (Bosher 2008).

Urban planners' and designers' mission is, therefore, to guide the urban development of an existing or new community while considering public and environmental welfare. Careful planning and design needs to take into account the factors that make it a convenient, healthful, efficient, and equitable place to live and work. At the same time, a good urban planner/designer will also take a sustainable, climate-informed, and eco-friendly approach that manages the environment, promotes longevity, and

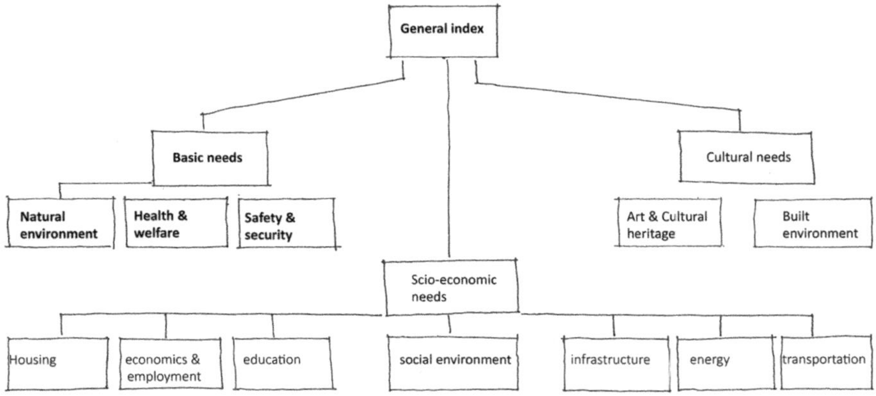

Fig. 1.5 Quality of life indicators and the determinant role of the basic needs (Bahrainy and Tabibian 1998)

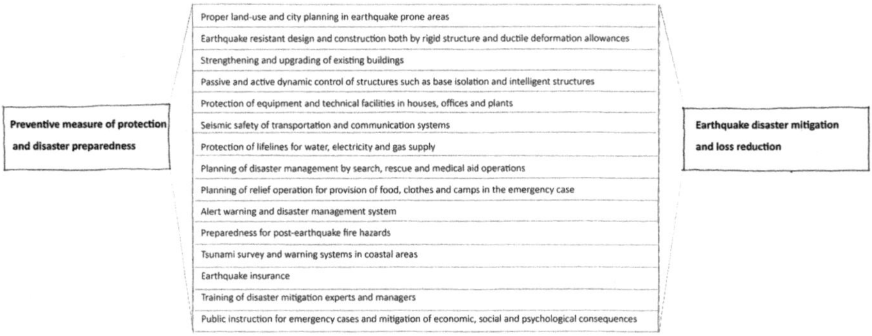

Fig. 1.6 Earthquake Prognostics Strategy: Preventive measures of protection and disaster response (Vogel and Brandes 1988)

reduces the risk of disasters (Jenkins 2021, Urban Planner Guide: 5 Tips for Urban Development | SimScale Blog. https://www.simscale.com/blog/2019/04/urban-planner-tips/).

So reducing natural disaster risks is becoming an increasingly significant goal of any planning/design activity in almost all countries throughout the world, but particularly in the seismic belt region of the globe. Therefore, today it may be claimed that reducing natural disasters, along with environmental protection, should be regarded as intrinsic and integral part of urban design and planning discipline, as well as practice.

Natural disasters cause the destruction and collapse of all services, activities, and functions of normal life in human settlements, and put a halt on their development trends. Disaster risk management includes three main stages: pre-disaster, during disaster, and post-disaster. The pre-disaster stage, which is the most relevant period in which urban planning and design activities may occur, includes prevention and preparedness. The disaster stage includes escape, refuge, rescue and relief. The post-disaster period focuses on rehabilitation and reconstruction. As it will be explained in the next chapters, urban design and planning can have an effective role in reducing disaster risk in each period, but particularly during the pre-disaster and post-disaster periods (Fig. 1.7).

Disaster mitigation may be broken down into active and passive activities. Active activities include hazard assessment, risk assessment and vulnerability assessment, which in turn will be divided into pre-disaster and post-disaster activities, both of which will further broken down into structural and non-structural elements. Non-structural elements will include legal, insurance, education, public awareness, organization, and early warning. Structural elements include engineered and non-engineered ones. For any specific area, it is necessary to accept the risk level and plan for preparedness to mitigate the risk and on its basis rescue, relief and rehabilitation. Urban design and planning can have an effective role in almost all stages and activities. Passive activities which are common in most of developing countries

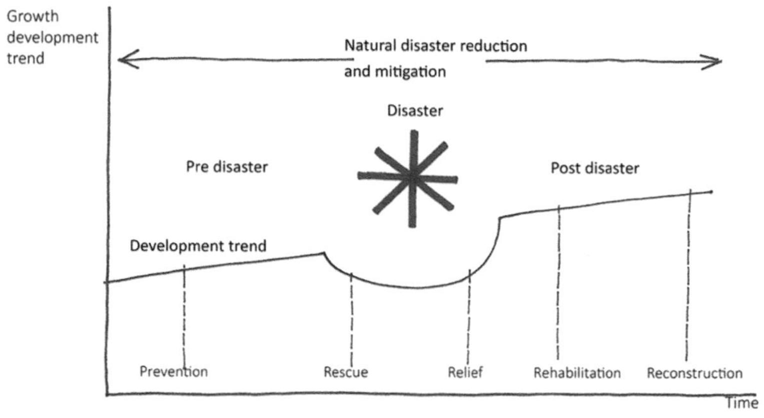

Fig. 1.7 Development trend and natural disasters

focus solely on post-disaster activities—rescue, relief, rehabilitation, and emergency, temporary and permanent settlements (Fig. 1.8).

1.3 Natural Disasters and Sustainable Development

Sustainable management of environment will help to the reduction of natural disasters. The measures taken to reduce or mitigate natural disasters will mostly be useful for the protection of environment.

It is clearly neither cost-effective or politically feasible to adopt measures to seismically strengthen the entire built environment nationwide. In some areas of the country, the most prudent strategy to manage the hazards associated with some structures is to simply leave them alone and let natural attrition take care of the problem over time. However, there are many structures in areas of high and moderate seismicity that are so important that they warrant special seismic safety programs and initiatives. Steps should be taken to make sure that throughout the nation critical facilities, communications networks, and other emergency-relevant resources can survive and remain operational following earthquakes and other disasters (Fig. 1.9).

Natural disasters are threatening sustainable development throughout the world. Most of the countries are facing serious problems because of these threats. Compared to only 30% in 1950, 54% of the world's population resided in urban areas as of 2014. This is predicted to increase to 66% by 2050. The number of individuals affected by natural disasters has also increased annually. It should be kept in mind that increasing vulnerability to these disasters is to a large extent due to our own activities. Most of the short-term strategies which are used under the term "development" have increased the adverse impacts of natural disasters and the vulnerability of communities. High rate of urbanization, environmental degradation, and industrial development in disaster-prone areas, increase natural disaster risks extensively. Deforestation, desertification,

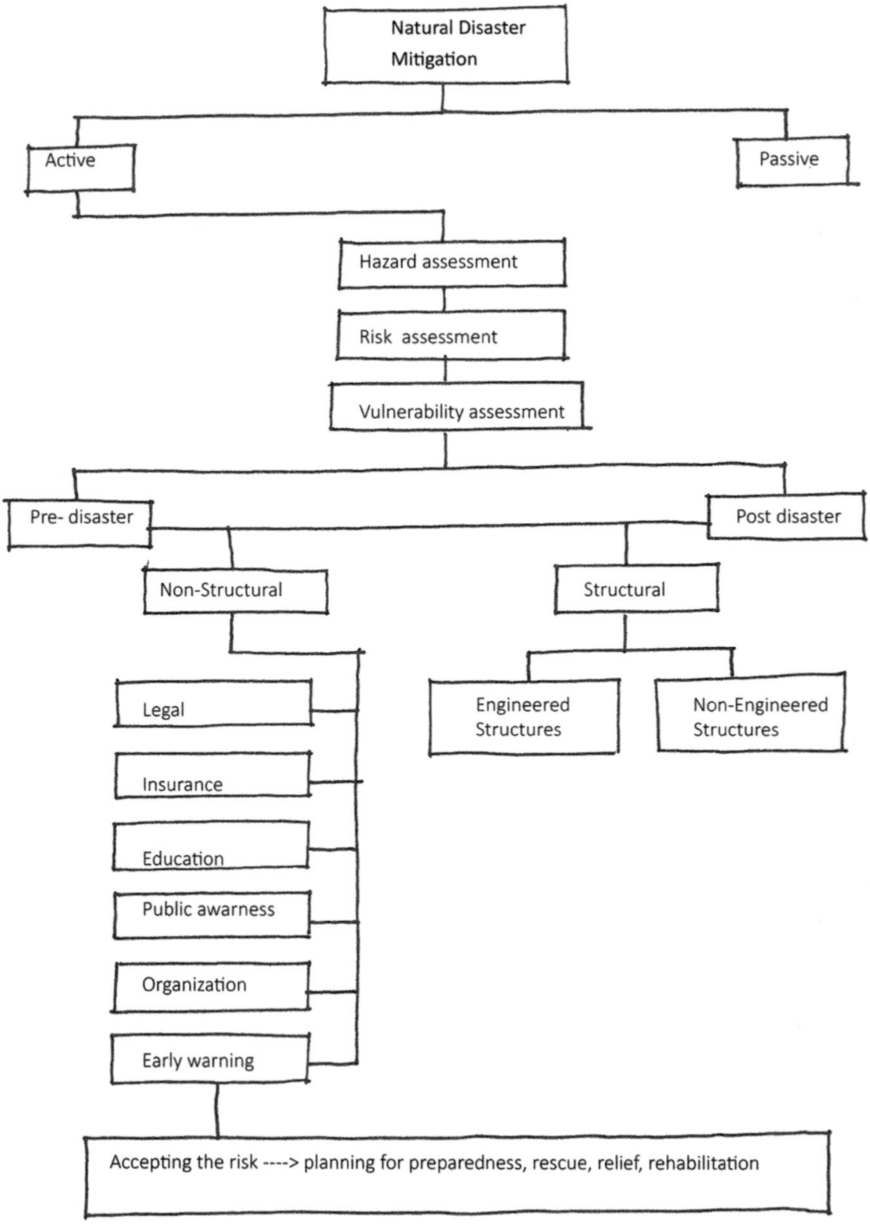

Fig. 1.8 Disaster management through active and passive methods (Bahrainy 2003a)

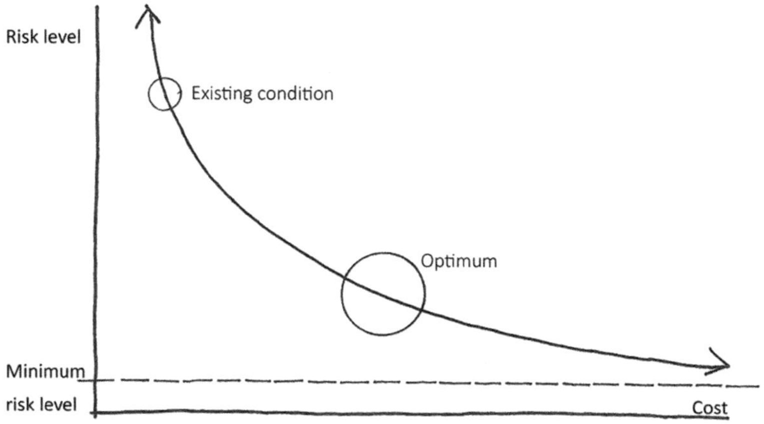

Fig. 1.9 Relationship between disaster costs and disaster risks

over use of pastures, weak management of irrigation and centralized agriculture are examples of land-use activities which cause certain natural disasters (IDNDR magazine 1996). Unplanned urbanization, migration, particularly of the rural poor to cities, often results in people living in slums with inadequate civic amenities and housing, thus exposing more people and assets to risks of loss in a disaster (UNISDR 2015).

Nobody can deny that the quality of urban environment depends on amenity, facility, health and safety. Among these the four environmental qualitative elements, amenity, facility and health of the urban environment greatly influence the extent to which we can enjoy our lives and activities each day. But we are apt to forget that safety from disasters is a basically important element of our urban environment. Rapid urbanization and development without consideration given to security against disasters take place every day in not only developed, but also developing countries (Nakabayashi 1993).

According to Nakabayashi (ibid, p. 2280, here are two objectives in disaster risk assessment. One is to clarify the subject of disaster prevention urban planning, which is to remake the urban structure and environment so that it is safer, and to be able to identify the target areas that are relatively more risky. The other is to estimate the demand for countermeasures against the damage caused from a disaster that is considered to be the next earthquake in each city. This damage estimation promotes the preparedness of countermeasures after earthquake disaster. The former is area vulnerability assessment of earthquake and the latter is damage estimation of earthquake.

However, a review of the literature shows that very few studies have dealt with the role urban design and planning can play in disaster reduction. What exist are all either too general or too specific—i.e., take on specific subjects such as transportation, land use, evacuation process, and temporary settlement, and not necessarily urban design and planning, which is the main subject of this book. Also the number of publications based on a more proactive rather than active attitude that include non-structural aspects and/or consider small-scale everyday disasters is very limited.

In recent years and particularly following the outbreak of COVID-19 pandemic the concept of "resilience," the capacity of human and physical systems to respond to extreme events, has become increasingly prominent in disaster research. The contemporary focus, therefore, has shifted to ensuring the capacity of the built environment to both resist and recover rapidly following a disaster event. It will therefore demand a paradigm shift in the way that built environment professionals integrate their activities and interact with the communities within which built assets reside (Futon 2020).

During the suburban era, we tried to solve most of our land-use planning problems by putting more space between people and buildings. But in cities, that is not possible. Instead of spacing our way out of problems, we have to design our way out. The threat of infectious disease is likely to ramp up urban design as a solution—perhaps, for example, by creating more separation in public spaces like restaurants and parks.

So, it is reasonable to assume we will begin to see small changes in urban design that separate people a little more and help make it easier to protect them.

Maybe the best analogy is the "defensible space" movement started by urban planner Oscar Newman (1972) during the nadir of urban life in the 1960s. At a time when parks and public housing projects were crime-ridden, Newman's view was that all space belonged to somebody. By giving residents a sense of ownership of these spaces, he contended, gangs wouldn't take them over. It was a revolutionary theory in urban design that changed everything—showing that, instead of fleeing cities, standing firm and re-designing can be a better solution.

Natural hazards and/or disasters have widely varying impacts on the built environment and related planning and design practices. Disasters can:

- Cause severe property damage and human loss.
- Have impact on city functions.
- Intensify urban hazards and create new ones.
- Increase urban inequalities.
- Create new challenges for future urban development.
- Create barriers to sustainable urban development.
- Have a negative impact on the resources invested in the built environment (Bosher 2008, 321).

It is only recent years that urban design and planning is considered to play an effective and determinant role in reducing seismic risk in urban areas. Several important development factors as well as seismicity and vulnerability of elements and land instability are among the items that should be studied in applying any changes and improvements of different areas of natural disaster reduction. What is critical is to fuse natural disaster reduction policies into urban plans.

Environmental deterioration will intensify the natural disaster impacts, so that would turn a normal thunder storm into a disaster. Intervention of river banks will intensify flooding or even create one.

Continuous growth of cities will lead to ecological devastation of surrounding areas and their eventual destruction. One of the urbanization results is that will increase the volume and speed of surface water which will lead to the intensity and speed of the floods.

The lowering and rising of ground water aquifer will cause subsidence, loss of soil resilience, unstable slopes and underground excavations. All these indicate reciprocal relation between the environment and natural disasters.

Sustainable management of environment will help to the reduction of natural disasters. The measures taken to reduce or mitigate natural disasters will mostly be useful for the protection of environment.

The impacts of natural disasters are increasing daily, because: more people and properties are affected, and also there will be more pressure on the environment. Land destruction will have numerous economic, social and ecological consequences, either directly or indirectly.

Natural disasters continue to pose serious threats to economic development. Precious development resources are lost when disaster wipes out the products of investment, with further loss of development resources when production of goods and services is reduced. Development activity is also negatively affected by disasters when ongoing programs have to be interrupted to allow a shift in resources from long-term programs to short-term recovery and emergency response programs.

Although disasters and particularly earthquakes have been continuously threatened human settlements throughout history in many parts of the world, no serious and comprehensive study have been carried out to show how urban design and planning as activities which focuses on the purposeful and deliberate shape giving of urban environment to achieve quality life for its residents (Bahrainy 1998).

1.4 Lessons Learned from Past Experiences

Several determinant events during the last three decades have contributed to the acceptance of the essential role urban design and planning which can play in reducing disaster risk in the built environment. *Ferdows* earthquake of 1968; *Manjil* earthquake of June 1990, the UN decade for disaster risk reduction (1990–2000), the Kobe earthquake of 1995, and Hurricane Katrina in 2005. The Manjil earthquake will be described in detail in Chap. 3, but here the critical goals and results of a research project carried out following this event entitled "Integrative approach to earthquake mitigation" will be summarized:

- *Develop and expand the knowledge for prediction, prevention, and mitigation of earthquake impacts.*
- *To mobilize and train national manpower in order to further motivate research in earthquake.*
- *Transfer of earthquake technology to the country.*
- *To develop guidelines, criteria, codes and regulations for planning, design and construction in seismic zones.*
- *To enhance public, professionals and authorities' awareness toward earthquake.*

- *To form a focal point for the continuation of research in earthquake.*
- *Production and collection of basic data to be used in research activities in earthquake.*
- *To translate theories in practice (actions).*
- *32 volumes of scientific and analytical reports on various aspects of earthquake.*
- *Of which 23 volumes are guidelines, criteria, models and recommendations for planning and design in seismic-prone areas* (Bahrainy (national director) 1992).

1.4.1 The Kobe Earthquake of 1995

On January 17, 1995, an earthquake rocked Kobe, Japan's sixth largest city, at 5:46 am, while most residents were asleep. The destruction from the 20-s, 7.2 magnitude (Japanese scale) tremor shocked the world. Kobe, which had been considered safe from earthquakes, found itself with overcrowded shelters, broken water mains, over 350 fires, inadequate supplies and jammed roadways. A large section of a central subway station caved in, the Kobe port was largely destroyed by liquefaction, and the Hanshin expressway collapsed in five places and rolled over (Mileti 1999).

The Kobe earthquake was the worst urban disaster to hit Japan since the great Kwanto quake of 1923. This earthquake showed that not only were the largely wooden-built older urban dwellings unable to withstand the shock but that some modern structures were also vulnerable. One of the great lessons learned was that the built environment remains at considerable risk despite the great technological advances that have been made in earthquake engineering (Whittow 1995).

The Kobe earthquake, in spite of its catastrophic impact on the city of Kobe and surrounding, for the first time brought the subject of urban planning and design as significant determinant factors which affect seismic risk in urban areas to the attention of engineers, planners, and decision makers. Major determinant factors, such as settlement location, topography, land uses, transportation network, density, urban form and structure, significantly affect seismic risk in urban areas (Menoni 2001).

The two earthquakes of Manjil (1992) and Kobe (1995) could be considered as a turning point for the role urban design and planning can effectively play in reducing seismic risk in urban areas. Before these events, civil engineers, and particularly earthquake engineers were making efforts to make buildings seismic resistance, which experience shows that it has not been quite successful. When modern structures did collapse, it seemed to be because the ground beneath them gave way. So the great lesson learned from these two experiences was that seismic resistance buildings do not necessarily make a city safer from earthquake. Kobe was the first significant disaster to strike a modern, developed metropolitan area, and it revealed the unique vulnerabilities of complex urban regions. So following these two events, it is realistic to point out that focus in earthquake risk reduction has shifted (may be as a paradigm shift) from engineering (building structure) to larger scale of urban design and planning (Bahrainy 1998) (see also chaps. 4, 5, and 6 of this book).

Fig. 1.10 Damaged (1995) and restored (1997) Port Island, Kita Park Ouay and Kobe bridge (Kobe Municipal Waterworks Bureau 1997)

As it was stated earlier, it is clearly neither cost-effective or politically feasible to adopt measures to seismically strengthen the entire built environment, especially in the old and deteriorated parts of city centers, where a high volume of buildings are severely vulnerable to earthquake. A general land-use plan, based on seismic microzonation may help to reduce vulnerability of these areas to some extent. However, there are some structures that are culturally and historically so important (such as *Arge Bam*) that they warrant special seismic safety programs and initiatives.

Implementation of the reconstruction activities following the Kobe earthquake indicates that for the first time the focus has shifted from seismic-resistant design of individual buildings to strengthening of infrastructure, improving accessibility network, appropriate distribution of land uses, open spaces and density (Fig. 1.10).

1.4.2 Hurricane Katrina

This experience, although a climatological hazard, was a large and destructive category 5 hurricane that flooded 85% of the city of New Orleans on August 29, 2005. It was believed that the cause of this disaster has been also design errors (Briaud et al. 2006). What is critically important from the view point of this book is that this event also placed the planning and design professions at the center of disaster prevention. Radical rethinking is the key indicator of a design moment. It is this process that we seek to describe the role urban design can play in disaster (seismic) mitigation.

Carmona (2021) suggests the substitution of the word "making" for "shaping." To him, urban design is not about making new places from scratch but is instead *always* about shaping places that already exist. These places may or may not already have been developed but will always be on, over or under an existing landscape, which—more often than not—will be part of an existing urban fabric. In this regard, we shape and reshape places over time (Carmona 2021).

There are also others (see, for example, Mileti 1999; Sennett 1990; Lynch 1972, 1984; Kostof 1991, 1992) who write about "design moments" as a period of time in which particular events occur that result in a process of urban restructuring that is physical, social and conceptual. A design moment can result from war, natural or man-made disasters, large-scale shifts in the mode of economic production, urban policies, severe demographic change, and biological events such as recent COVID-19. In this sense, a design moment is a critical juncture in the history of a city in which the most basic components of a city's character—its social fabric and urban form—are fundamentally altered (Fig. 1.11).

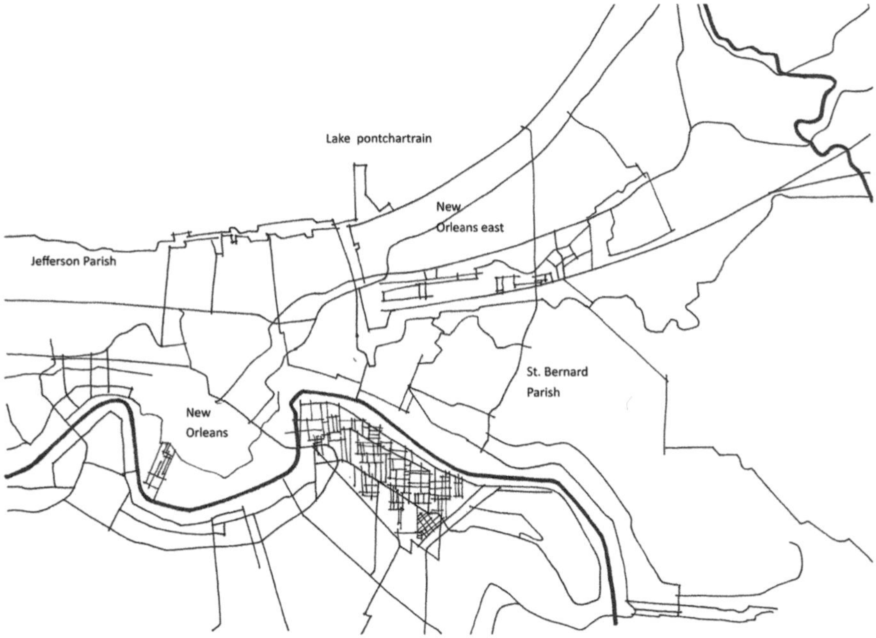

Fig. 1.11 Flood extent in New Orleans, Louisiana, on September 2, 2005. 2009 *Source* Map produced by J. Wagner; City of New Orleans, LA, GIS, 2006. 238 J. A. Wagner & N. Frisch (Wagner and Frisch)

1.5 The Role of Urban Design

It is only recently, and following learning critical lessons from past experiences, that urban design and planning are considered to have an effective and determinant role in reducing seismic risk in urban areas. Several important development factors, as well as seismicity and vulnerability of elements and land instability are among the items that should be studied in applying any changes and improvements of different areas of natural disaster reduction. What is critical is to fuse natural disaster reduction policies into urban plans.

Preventive measures of protection and disaster preparedness to mitigate earthquake disaster risk—i.e., life loss and property damage are many and diverse, from earthquake-resistant design and construction, strengthening and upgrading of existing buildings, passive and active dynamic control of structures, protection of equipment and technical facilities in houses, offices and plants, seismic safety of transportation and communication systems, protection of lifelines for water, electricity and gas supply, planning for disaster management by search, rescue and medical aid operations, planning of relief operations for provision of food, clothes and camps in the emergency case, alert warning and disaster management system, preparedness of post-earthquake fire hazards, tsunami survey and warning systems in coastal areas, earthquake insurance, training of disaster mitigation experts and managers, and public instructions for emergency cases and mitigation of economic, social and psychological consequences. But the most significant elements which can play determinant role in this regard are urban planning and design, which include all aspects, particularly issues of regional scale such as location analysis, topography, geology, transportation, and land use, but in the case of urban design it focuses on mainly on physical aspects, which includes: urban form and space, and urban activity systems (see Fig. 1.12).

In the case of urban design and planning disasters could affect the three substantive elements of these areas: urban activities, urban form and urban spaces (Table 1.1).

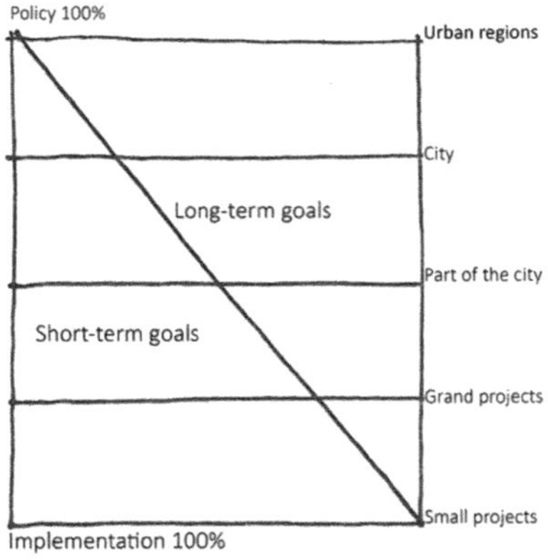

Fig. 1.12 Different urban design scales and their implementation tools and mechanisms (Bahrainy 2013)

Table 1.1 Risk assessment elements in urban design and planning

Activities	- Disruption of activities - Interruption of activities due to human losses and property damages - Economic and social damages due to interruption of activities - Mishaps in spatial and physical communication - Disruption of center-vicinity and macro–microcommunication
Form	- Collapse and chaos in physical form - Chaos and interruption of communication in settlements spatial organization - Collapse of "spatial hierarchy"
Space	- Surrounding buildings are damaged and the safe spaces in and around dwellings are lost - At the local level, the relation between dwellings and work places and nearby activities are broken - At the regional level the distribution and spatial organization of settlements are disrupted

Source Bahrainy (2013)

Preventive measures of protection and disaster preparedness include proper land use, city planning and design in the earthquake-prone areas, earthquake-resistant design and construction both by rigid structures and ductile deformation allowances, strengthening and upgrading of existing buildings, passive and active dynamic control of structures such as base isolation and intelligent structures, protection of equipment and technical facilities in houses, offices and plants, seismic safety of transportation and communication systems, protection of lifelines for water, electricity and gas supply, planning for disaster management by search, rescue and medical aid operations, planning of relief operations for provision of food, clothes and camps in the emergency case, alert warning and disaster management systems, preparedness for post-earthquake fire hazards, tsunami survey and warning systems in coastal area, earthquake insurance, training of disaster mitigation experts and managers, public instructions for emergency cases and mitigation of economic, social and psychological consequences. All these will eventually lead to earthquake disaster mitigation and loss reduction.

But as we will see in the following chapters, urban design and planning, aside from land-use planning can play a role in almost all the areas outlined above.

1.5.1 A Summary of Regional Actions

- Preparing seismic zonation map, which includes: Hazard assessment, risk assessment, vulnerability assessment. This is the basic information on its basis which different land uses could be assigned to appropriate zones depending on the vulnerability and sensitivity of uses and seismic risk of each zone.

- Formulating regional policies. By using smart growth concept and seismic zonation map, general policies may be devised to include: general land uses, transportation network, regional infrastructure, etc.
- Deciding on the spatial pattern and location of settlements on the basis of seismic risk zonation maps, and along with the regional infrastructure plan.
- Preparing various earthquake scenarios on the basis of intensity, location, type and time of the event.
- Formulating implementation guidelines and mechanisms.

1.5.2 Information Needed for the City Scale

- Seismology map
- Geology map
- Topography
- Land uses: sensitive, vulnerable and damaging.
- Age and deterioration rate.
- Construction techniques and quality
- Building materials
- Building form
- Urban texture (grain), open and green spaces.
- Density, building height
- Transportation networks (hierarchy, bridges, parking lots and structures, traffic volume.
- Urban infrastructure, utilities and facilities, water, sewage, electricity, gas, communication systems.
- Urban facilities(public health, education, fire department, security.

Topography and natural feature are determinant factor in risk level and eventually the vulnerability of different areas in a region. Any construction on the elevated areas with steep slopes are threatened; and as conditions, such as soft and watery soil permits landslide and liquefaction will occur.

Environmental deterioration will intensify the natural disaster impacts, so that would turn a normal thunder storm into a disaster. Intervention of river banks will intensify flooding or even create one.

Continuous growth of cities will lead to ecological devastation of surrounding areas and their eventual destruction. One of the urbanization results is that will increase the volume and speed of surface water which will lead to the intensity and speed of the floods.

The lowering and rising of ground water aquifer will cause subsidence, loss of soil resilience, unstable slopes and underground excavations. All these indicate reciprocal relation between the environment and natural disasters.

Since the built environment and urban infrastructure provide the core framework for most human activity, it is crucial to develop them with an effective measure of resilience so they can withstand, and adapt to, the threats of natural and human-induced hazards. These physical systems have traditionally been designed, built and maintained by the myriad professions. Therefore, designing a built environment that can cope with the impacts of disasters demands an in-depth understanding of the expertise and knowledge on avoiding and mitigating the effects of hazards in order to secure more sustainable future (Hamelin and Hauke 2005; Bosher 2007).

Recent natural and human-induced events have highlighted the fragility and vulnerability of the built environment to disasters. During the last few decades, the prevalence of disastrous events has stimulated a growth in theoretical developments in relation to the way in which disasters are avoided and managed. A paradigmatic shift has led to a focus on disaster preparedness, hazard mitigation and vulnerability reduction rather than disaster management and relief. In other words, there is a shift in the way disasters are being managed, i.e., moving away from the reactive attributes of disaster management toward the more proactive disaster risk management approach that should be mainstreamed into developmental initiatives. The discourse of resilience now resonates throughout the disciplines involved with the mitigation of disasters.

But what does urban design has to do with natural disasters? We have been wrestling with the above question in our applied planning and design research over the past several years. Our goal is to encourage scholarly reflection about our professional, pedagogical, political, and personal responses to the disasters and what lessons we can learn from the experiences for urban design. The Manjil (Iran) earthquake in June 1992, the 1995 Great Hanshin (Kobe) earthquake, and the 2007 Hurricane Katrina in New Orleans, to name just a few, were all significant to urban design practice and theory because they were the result of catastrophic design failure.

It is in the restructuring of urban space—or at least the potential reconfiguration of it—that we find the basic purpose and focus of the activities that encompass a design moment. Design moments always include reconceptualization of the city and in the case of wars, disasters and the intense disruption of existing cities, a design moment involves a response to physical destruction, urban ruins, and the dynamic of displacement and resettlement of former residents (Kostof 1992). The language of the design moment is revealing: organic growth versus planned districts, piecemeal versus comprehensive reconstruction; the ad hoc versus the orderly; top-down versus local control—the list of descriptors, while numerous, are often mere synonyms for a binary juxtaposition of the planned versus the unplanned response to urban design and rebuilding in the wake of disaster. The reality of the process is often far more nuanced than these "either/or" constructions of the design problem suggest.

It has been acknowledged that the way the built environment is delivered can itself lead to disasters, particularly in less developed nations where building codes and planning regulations may not be as well policed as they are in other nations (Ofori 2002). Also with socioeconomic progress, the built environment becomes more vulnerable as settlements become more reliant on their increasingly extended supply lines (Menoni 2001), and ever-ex-expanding and vital distribution networks of water, power, gas, and telecommunication systems. Moreover, constructed items represent most of every nation's savings. Therefore, the built environment is an investment that is worthy of protection.

Planners and designers to embark on their professional activities in the seismic-prone regions need seismic zonation. A rather long process should be followed to develop earthquake risk management which starts with preparing data bases and information which includes hazards, such as ground shaking, ground failure, surface faulting, tectonic deformation, tsunami runs up, and aftershocks. This information then lead to risk assessment, which includes hazard maps, inventory, vulnerability and location analysis. Now, a decision may be made as the level of acceptable versus unacceptable risk, on its basis which, seismic zonation map can be prepared. Micro-zonation is in fact the most useful tool for planners and designers who practice in seismic-prone regions. The next and last stage is risk management, which includes mitigation, preparedness, emergency response and recovery and reconstruction.

1.6 Risk Management Approach to the Preservation of Cultural Heritage

The 2016 handbook of the Canadian Conservation Institute (CCI) and ICCROM states that "cultural heritage tells the stories of the world's many peoples. The material part of that heritage, objects and sites, tells us of their activities, their perceptions, their skills, and their ideas. It is unique, irreplaceable and unfortunately, vulnerable" (ICCROM 2016, Bahrainy 2003a, b)

The risks threaten the cultural heritage vary from sudden and severe forces, such as earthquakes (Pompei, Arge Bam, …), floods, fire and armed conflicts (Iraq, Afghanistan, …) to gradual and cumulative processes (such as depreciation, corrosion, and chemical, physical, or biological degradation).

Examples of dangers to heritage assets at the different layers of enclosure: Strong earthquakes can cause the collapse of various heritage buildings and monuments in the affected area (seismic region).

Here, two cases of significant cultural heritage which were severely damaged due to earthquake will be briefly reviewed.

1.6.1 Ferdows *Earthquake*

The town of *Ferdows* is located in the southwest of the *Khorasan* Province. It is one of the historical cities in Iran, which due to its geographic and climatic conditions and the need for adjusting to the harsh climatic condition of desert holds unique characteristics in the urban structure and its relevant elements (Fig. 1.13). The catastrophic earthquake of 1968 in two consecutive days, led to the destruction of over 70% of buildings.

Fig. 1.13 View of the historical town of *Ferdows* before the 1968 earthquake

Earthquake shook the area twice, once in the afternoon of August 31, 1968, the epicenter which was *kakhak*. Several villages were completely destroyed and some 200 suffered extensive damage. The second one occurred just less than 20 h later, Saturday, September 1st to destroy whatever was damaged the day before. The first one had a magnitude of 7.3 Richter and the second one 6.3. As a result of these earthquakes over 10,000 people were killed in the affected area and 6000 were left homeless, but in the town of *Ferdows* only 300 people lost their life. The reason for the lower loss in the city was that just a day before strong earthquake occurred in *kakhak*, and also because the time of the earthquake was day time (UNDP-UNCH (Habitat) Project IRA 90//004 1994).

1.6.2 A 2003 Bam Earthquake

On Friday morning, December 26, 2003, at 1:56:56 Greenwich time (5:26:56 local time) a strong earthquake shook the city of *Bam* and surrounding areas in the province of Kerman, south of Iran. According to USGS, moment magnitude of 6.6 and a maximum Mercalli intensity of IX was recorded for this earthquake. The city of *Bam*, as well as many villages, was either completely destroyed or suffered serious damage. At the time of the earthquake the city had some 2,00,000 residents. The dominant building material was mud and sun-dried bricks, with an average age of 30 years. The most significant historical building in the area was *Arge (citadel) Bam*, which is said to be 2000 years old, and the world's largest adobe complex.

Fig. 1.14 (left) City of Bam leveled by the catastrophic earthquake of 1995; however, a few traditional dome-shaped structures remained untouched (photo by the authors). (right) Arge Bam, citadel as an international cultural heritage was severely damaged due to the Bam earthquake of 1995 (photo by the authors)

According to some estimates, over 70% of this complex was destroyed (Figs. 1.14) (BHRC 2003).

Bam's city texture had low density, but due to proximity of buildings and lack of adequate space between them, chain destruction occurred as a result. As with accessibility, arterials which play the role of city's main structure were not damaged and could play a critical role during early days immediately after the earthquake. Local access is divided into three groups: passages, which filled with building debris, and completely blocked; passages which were partially blocked, and passages which remained open and functional during and after the disaster.

Lack of due attention to construction of seismic resistance buildings and strengthening of the existing ones. The 2800 seismic guidelines, in spite of its critical role in improving construction quality in the country, has not been properly implemented.

One of the important lessons learned from the *Bam* earthquake deals with the reconstruction efforts. Owners of the damaged or ruined residential units preferred to stay in their own lot and set up temporary and later permanent units on the same site. This implies the rich meaning of settlement, ownership, and sense of belonging among the traditional residence of the city. To them, residing on the same site will also provide a sense security for the survivors (Fig. 1.15).

Other countries with historical and cultural heritage also made efforts to protect these valuable assets against natural disasters through available means. China, for example, will put an end to large-scale urban demolitions and preserve old historical buildings as part of a program to protect its cultural heritage.

China's rapid economic development has relied on breakneck rates of urban expansion and renewal, destroying entire neighborhoods and replacing thousands of old buildings with tower blocks and shopping malls.

The State Council, China's cabinet, said the aim of the new guidelines was to "systematically protect and pass on" the country's cultural heritage and "tell the stories of China and the Communist Party comprehensively and accurately" (Stanway 2021).

Fig. 1.15 (left) Traditional old fabric of the central section of *Kashan* in the hot-arid region of Iran as cultural heritage to be protected against earthquake disaster (photos by the authors). (right) The town of *Masooleh* in north of Iran as a cultural heritage which suffered minor damage due to Rubar-Manjil earthquake of June 1990 (photos by the authors)

The government of Canada has published a set of guideline regarding risk management of cultural heritage (ICCROM 2016). The five stages of control described in the book will be presented here.

1.6.3 The Five "Stages" of Control

In "A Guide to Risk Management of Cultural Heritage," a tool is introduced by which to develop risk treatment options. It includes five different *"stages"* to control the risks: Avoid, block, detect, respond, and recover (ICCROM 2016).

The five "stages of control" include both preventive and reactive measures to reduce risks to our heritage assets. Of course, prevention is much more important and more effective than reaction. However, good risk management always integrates these two types of action in order to achieve the best possible results.

1.7 Measures for Reduction of Earthquake Consequences and Mitigation of Seismic Risk

For the purpose of reduction of earthquake consequences and mitigation of seismic risk, short- and long-term studies and actions should be organized by responsible government authorities and professionals. The basic steps of these studies and actions are summarized for short-term and for long-term needs, respectively (see Chap. 3).

1.8 Taking Advantage of Disaster "Opportunity"

Although many researchers have argued that post-disaster recovery and reconstruction is an opportunity to rebuild with more resilience, residents may resist and attempt to recover their spaces quickly and return to the status quo.

1.8.1 *Maintaining and Adapting to Community Needs*

Disasters create a time-sensitive opportunity to plan for the future and attempt to become more resilient (Birkmann and von Teichman 2010). In terms of green spaces, this can apply to physical dimensions and collaborations. Especially in regard to recreational parks and their amenities, a redevelopment plan may consider future needs and physical adaptations. Rather than replacing as-is, this plan may allow it to meet new standards.

It is hoped that this book will stimulate interest and activities by professionals and governments to implement an integrated approach to disaster mitigation, pre-disaster planning, preparedness and prevention. This book, by using urban planning and design as effective tools intends to reduce earthquake consequences and the development of national preparedness and prevention programs.

1.9 Conclusion

Can urban design play any role in reducing seismic risk in the built environment? An accepted definition of urban design is the purposeful decisions and actions which aim at establishing functional and formal order in the physical environment. Today's interpretation of order, however, has changed drastically, because urban environment has gained unprecedented complexity and diversity. It is claimed that urban design as a legitimate discipline and practice depends on its ability and capacity to deal with the current issues society is facing and to offer appropriate solutions to the problems, to the extent plausible and within the framework of the urban design tools and techniques. Among the most serious problems, almost all societies are facing today, in one way or another, are natural disasters. The global number of yearly reported natural disasters has been exponentially increasing due to the fact that increasing number of people is exposed to natural disasters. Urban design techniques and methods can be used, as preventive measures, to make changes and modifications in the built environment in order to establish desired order in the urban form, space and activities and provide cities with safety which is one of major components of the quality of life in urban areas.

To learn from past experiences, a brief description is given for Kobe, *Ferdows* and *Bam* earthquakes (the *Manjil* earthquake will be described in more detail in Chap. 3), and also Hurricane Katrina, the cases which had significant impact on bringing the urban design and planning fields into the foreground of disaster management activities.

References

Alesch DJ, Siembieda W (2012) The role of the built environment in the recovery of cities and communities from extreme events. Int J Mass Emerg Disasters 30:197–211

Bahrainy H (national director) (1992) Natural Disaster Management. UNDP-UNCH (Habitat) Project IRA 90//004, Assistance in the implementation of Post earthquake Reconstruction Program, 8 vols

Bahrainy H (1998) Urban planning and design in a seismic-prone region—the case of Rasht in Northen Iran. J Urban Plan Dev 124–4:148–181

Bahrainy H (2003a) Natural disaster management in Iran during 1990's the need for a new structure. ASCE J Urban Plan Dev 129(3):140–160

Bahrainy H (2003b) Natural disasters and cultural heritage, the case of Iran. In: Proceedings of Natural disasters and cultural heritage, Chamberry, France

Bahrainy H (2013) Urban design process. University of Tehran Press (in Persian)

Bahrainy, Bakhtiar (2016) Toward an integrative theory of urban design. Springer

Bahrainy H, Tabibian M (1998) A model for evaluation of quality of life in urban areas. MOHIT SHENASI. J Environ Stud 24(21):41–56

Banerjee T (1992) Earthquakes, urban scale vulnerability and city design: some observations. School of Urban and Regional Planning, USC

BHRC, Building, and Housing Research Center (2003) Bam earthquake preliminary report, Ministry of Housing and Urban Development

Birkmann J, von Teichman K (2010) Integrating disaster risk reduction and climate change adaptation: key challenges—scales, knowledge, and norms. Sustain Sci 5:171–184. https://doi.org/10.1007/s11625-010-0108-y

Bosher LS (2007) Social and institutional elements disaster vulnerability: the case of South India. Academia Press, Bethesda, USA

Bosher LS (2008) Hazards and the built environment, attaining built-in-resilience. Rutledge

Briaud JL et al (2006) Investigation of the performance of the new orleans flood protection systems in Hurricane Katrina on August 29, 2005. CITRIS Center Report Number UCB/CITRIS-06/01. University of California, Berkeley, CA

Carmona M (2021) Public places urban spaces, the dimensions of urban design, 3rd edn. Rutledge

Database ID (2011) Disaster trends. Available at: http://www.emdat.be/disaster-trends. Accessed 20 Mar 2012

Futon W (2020) Perspective: COVID-19 and cities, health, housing, placemaking, urban planning

Hamelin J-P, Hauke B (2005) Focus areas: quality of life: towards a sustainable built environment. European Construction Technology Platform, Paris

Hein, C. (n.d.) *Toshikeikaku* and *Machizukuri* in Japanese Urban Planning, the reconstruction of inner city neighborhoods in Kobe. https://www.dijtokyo.org/doc/dij-jb_13-hein.pdf

ICCROM (2016) A guide to risk management of cultural heritage, Government of Canada, Canadian, Conservation Institute

IDNDR magazine (1996)

Jenkins M (2021) Urban planner guide: 5 tips for urban development | SimScale Blog. https://www.simscale.com/blog/2019/04/urban-planner-tips/

Joerin J, Shaw R (2010) Climate change adaptation and urban risk management. In: Shaw R, Pulhin J, Pereira J (eds) Climate change adaptation and disaster risk reduction: issues and challenges (community, environment and disaster risk management). Emerald, Bingley, pp 195–215

Kobe Municipal Waterworks Bureau (1997). 1000 days after The 1995 Hanshin-Awaji Great Earthquake, November
Kostof S (1991) The city shaped: urban patterns and meanings through history. Bulfinch, Little, Brown, Boston, MA
Kostof S (1992) The city assembled: the elements of urban form through history. Bulfinch, Little Brown, Boston, MA
Lynch K (1972) What time is this place? MIT Press, Cambridge, MA
Lynch K (1984) Good city form. MIT Press, Cambridge, MA
Menoni S (2001) Chain of damages and failures in a metropolitan environment: some observations on the Kobe earthquake in (1995). J Hazard Mater 86(1–3):101–119
Mileti DS (1999) Disasters by design, a reassessment of natural hazards in the United States. Joseph Henry Press, Washington, DC
Nakabayashi I (1993) Urban planning based on disaster risk assessment. In: Disaster management in metropolitan areas for the 21st century. Proceedings of the INDR, November, Nagoya, Japan
Newman O (1972) Defensible space: crime prevention through urban design. Macmillan, New York
Nishiyama Y (1993) Urbanization and natural disasters. In: Disaster management in metropolitan areas for the 21st Century. Proceedings of the INDR, November, Nagoya, Japan
Ofori G (2002) Construction industry development for disaster prevention and response. In: Proceedings of the international conference on post-disaster reconstruction: planning for reconstruction, May 2325, Montreal, Canada
Sennett R (1990) Conscience of the eye. W. W. Norton, New York, NY
Stanway D (2021) China vows to curb urban demolitions, preserve cultural heritage. Reuters, September
UNDP-UNCH (Habitat) Project IRA 90//004 (1994) Assistance in the implementation of post earthquake reconstruction program, analytical evaluation of reconstruction of three earthquake affected towns of *Ferdows*, *Tabas* and *Golbaf* in Iran, Volume one: Evaluation of reconstruction of the town of *Ferdows*, Report no. 56 (in Persian)
UNISDR (2017) Words into action guidelines: national disaster risk assessment. Hazard specific risk assessment 1. Earthquake Hazard and Risk Assessment
UNISDR (2015)
Vogel A, Brandes K (eds) (1988) Earthquake prognostics, hazard assessment, risk evaluation and damage prevention. In: Proceedings from the 2nd international seminar held in Berlin, June 24–27. Friedr. Vieweg & Sohn
Wagner JA, Frisch M (2009) New orleans and the design moment. J Urban Des 14(3):237–255
Wamsler C (2014) Cities, disaster risk and adaptation. Rutledge, New York
Whittow J (1995) Disaster impact and the built environment. Built Environ 21(2/3)

Chapter 2
Definitions and Concepts

Abstract This chapter will provide a brief dentition and description of the commonly used terms and concepts in disaster, and particularly seismic risk management. Life losses and material damages of disasters may be attributed to three groups of elements and their mutual interactions:

- Geophysical hazards, i.e., the physical environment which includes hazardous events.
- Climatologically hazards.
- Biological hazards.

These will affect the following elements:

- Humans systems or the social and demographic characteristics of the communities that experience them (such as: population, culture, technology, social and economic class, and political status).
- And building and construction system (such as: buildings, roads, bridges, infrastructure, and housing).

No doubt each of these elements is complex, but what is more critical is the interaction between them which leads to very complex meta-problems. Three main influences are at work. First, the earth's physical systems are constantly changing—witness the current warming of global climate. Scientists expect a warming climate to produce more dramatic meteorological events such as storms, floods, droughts, and extreme temperatures. Second, recent and projected changes in the demographic composition and distribution which means greater exposure to many hazards. Third, the built environment—public utilities, transportation systems, communication, and homes and office buildings—is growing in density, making the potential losses from natural forces larger. Geophysical hazards: begin from exterior or interior of earth and inflict life loss and damages.

Keyword Liquefaction · Landslide · Tsunami · Seismicity · Acceleration lines · Zonation · Fault · Disaster · Hazard · Mitigation · Vulnerability · Preparedness

---

The information used in this chapter is mostly taken from a research project entitled "UNDP-UNCHS (Habitat) Project IRA/90/004, Assistance in the Implementation of Post-earthquake Rehabilitation Program": Guidelines for earthquake disaster management 7 volumes.

H. Bahrainy and A. Bakhtiar, *Urban Design in Seismic-Prone Regions*,
The Urban Book Series, https://doi.org/10.1007/978-3-031-08321-1_2

Life losses and material damages of disasters may be attributed to three groups of elements and their mutual interactions:

- Geophysical hazards, i.e., the physical environment which includes hazardous events.
- Climatologically hazards.
- Biological hazards.

These will affect the following elements:

- Humans systems, or the social and demographic characteristics of the communities that experience them (such as: population, culture, technology, social and economic class, and political status).
- And building and construction system (such as: buildings, roads, bridges, infrastructure, and housing).

No doubt each of these elements is complex, but what is more critical is the interaction between them which leads to very complex meta-problems. Three main influences are at work. First, the earth's physical systems are constantly changing—witness the current warming of global climate. Scientists expect a warming climate to produce more dramatic meteorological events such as storms, floods, droughts, and extreme temperatures. Second, recent and projected changes in the demographic composition and distribution which means greater exposure to many hazards. Third, the built environment—public utilities, transportation systems, communication, and homes and office buildings—is growing in density, making the potential losses from natural forces larger (Mileti 1999). **Geophysical** hazards: begin from exterior or interior of earth and inflict life loss and damages.

Earthquake: a sudden movement or shaking of earth which occurs due to the release of the energy which has been gradually accumulated by the fault activities or volcanoes.

Land instability due to earthquake: landslides, subsidence, expansive soils: movement of soil mass, mud, and rock from the heights downward

Liquefaction: Liquefaction potential of sand deposits. Soil liquefaction has been a major cause of damage to earthworks and foundations in past earthquakes and poses a potential threat to the integrity of structures and installations during feature earthquakes. Thus, the assessment of liquefaction potential of the ground is one of the most important considerations in designing foundation and earth structures and, also for conducting seismic zoning work. The evaluation of liquefaction can be made with different levels of accuracy depending upon the availability of basic information on geology and geotechnical data.

Liquefaction is caused by higher level of underground water and soft sediment drift. In other words in the areas where earth is sandy and there is water among sand, any ground shaking (e.g., earthquake) would press watery sand which will lead to the flow of water or mud.

Landslide: Land-sliding is a type of ground failure during earthquakes which is associated with loss of human lives and properties. Oddly, however, this issue has not arrested much attention among engineers because many slides have taken place in remote and even inaccessible mountain areas and consequent impact on human activities has not been seriously recognized. However, sprawl of urbanization to the areas of hillsides and mountains has exposed local villages and towns to the potential menace caused by landslides during earthquakes.

Rockfalls: Rockfalls are still other type of geotechnical hazard brought about by an earthquake. The rockfalls often occur on steep slopes, mountain walls, and road cuts. The amount of rock masses involved is not generally so much, but because of its sudden occurrence during main shaking of an earthquake, it is generally impossible to elude the danger of falling rocks, if it occurs on the spot. Major issue of importance in the rockfall hazard is blockade of roads or highways.

Tsunamis: The waves created from sudden and severe vertical shock to the bottom of the ocean which may be happen due to landslide, earthquake, or volcanoes of the ocean bottom.

Seismicity: The relative frequency and distribution of earthquakes.

The **real acceleration lines** (which are based on a real earthquake in the region) or hypothetical (which are based on a probable earthquake scenario) show the degree of risk and vulnerability of different areas. These maps will be the basis of design and planning decisions in the seismic-prone regions. So the accuracy and precision of these maps will guarantee the quality and reliability of the design decisions.

Seismic waves: Sudden eruption in an epicenter causes waves which spread in all directions.

Seismic Parameters.

Seismic (earthquake) zonation map shows the areas with similar seismicity, shaking, and ground eruption. This map is a technical tool for planners on its basis which different land uses could be distributed. Zonation may be broken into micro and macro.

Foreshocks are minor shakings which occur in a place to show that the area is becoming active.

Main shock is the strongest shaking which is called the main earthquake.

Aftershocks are the shakings followed the main shaking and may continue for days, months, and even years.

Seismology is scientific study of earthquakes.

Foreshocks: Before a strong earthquake, several minor earthquakes might occur in an area, which indicates that the area is seismically active. However, this does not mean that foreshocks are prerequisite for all great earthquakes.

Main shock: It is the strongest shock during an earthquake.

Aftershock: It is the shocks following the main shock. These shocks may continue for a few days, weeks, months, or even years.

Types of earthquake: There are generally two types of earthquakes: (1) non-tectonic earthquakes which are caused by earth movement due to volcanic activities, and (2) the tectonic earthquakes, which are caused by fault movements.

Macro-seismic Epicenter

Mesoseismal: It is of or relating to the center of an area of earthquake disturbance.

Fault: It is a rupture on the earth surface which could extend deep low.

Strike-slip fault is a fault which is resulted from horizontal movement.

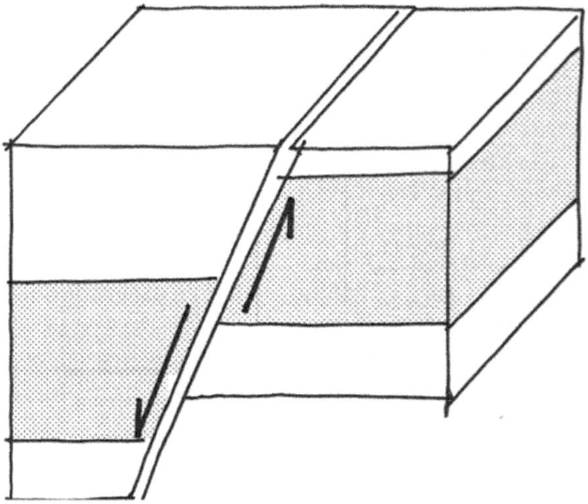

Fig. 2.1 Fault occurs when the land is pulled apart vertically, as shown here, or horizontally (Beer and Higgins 1999)

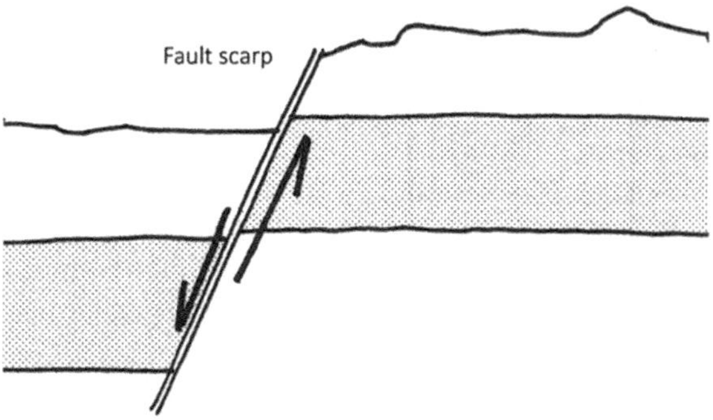

Fig. 2.2 Scarps can develop along a fault

Normal fault is a fault which is resulted from a vertical movement.

Capable (active) fault is a fault with the following characteristics:

- Occurrence of historical earthquakes somewhere along the fault.
- Revealing epicenters for new earthquakes along the fault.
- Different level or movement of surface which is not eliminated by erosion.
- Occurrence of mild local earthquakes of magnitude from one to four.

Active (capable) fault: faults with following characteristics are called active:

- Historical occurrence of earthquakes in certain point of a fault.
- Finding new epicenters throughout the fault.
- Difference in earth level or earth movement.
- Occurrence of local minor earthquakes along the fault (Figs. 2.1 and 2.2).

There is also another important type of waves: love, Rayleigh and P. waves (Fig. 2.3).

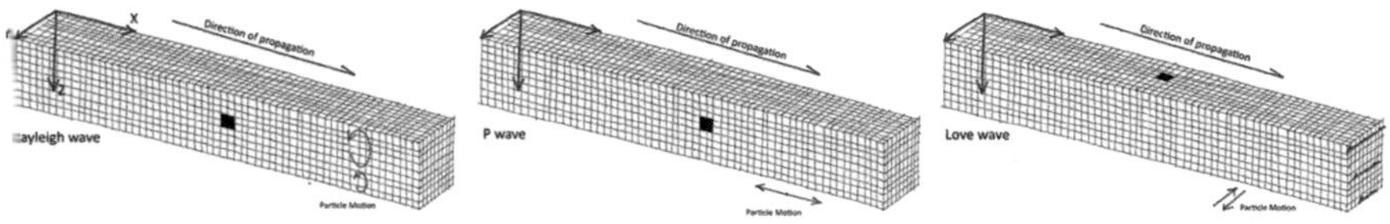

Fig. 2.3 Three types of fault movement: Rayleigh wave(1), P wave(2), and Love wave(3)

Faults may be further divided into two groups: body waves and surface waves.

Biological hazards: In the last months of 2019, the world witnessed an unprecedented event: the outbreak of COVID-19. It started as an epidemic in the city of Wuhan, China, and in a very short time spread to other parts of the world and turned into a global phenomenon as a pandemic. No other disaster, man-made or otherwise, in the history of mankind, has been able to affect every aspect of life in the planet. Scientists, politicians, leaders, … were quite shocked by their inabilities to grasp the phenomena and find solutions to put it under control. All technological advances proved useless. All societies, poor as well as rich, developing as well as developed, West and East, and all ages, genders, etc., faced the crises. The whole mankind fell in a stage of shock.

What is more critical is when some other disasters, such as earthquake, flooding, hurricane and tsunami, occur at the same time that society is dealing with COVID-19. This will lead to real catastrophe.

Disaster: Charles Fritz (Fritz 1961, p.655) defined disaster as:

> . .*an event, concentrated in time and space, in which a society, or a relatively self-sufficient subdivision of a society, undergoes severe danger and incurs such losses to its members and physical appurtenances that the social structure is disrupted and the fulfillment of all or some of the essential functions of the society is prevented.*

A disaster is defined by the United Nations International Strategy for Disaster Reduction (UN/ISDR 2004: Appendix 1) as:

A serious disruption of the functioning of a community or a society causing widespread human, material, economic or environmental losses which exceed the ability of the affected community or society to cope using its own resources. A disaster is a function of the risk process. It results from the combination of hazards, conditions of vulnerability and insufficient capacity or measures to reduce the potential negative consequences of risk.

But today opinions about what constitutes a disaster diverge. Latter-day theories include social constructionism, postmodernism, and conflict-based and political economy theories. Some view disaster as an analog for war, as social vulnerability, and as a condition that endangers uncertainty. Others define disaster as a breaking of the routines of social life in such a way that extraordinary measures are needed for survival. Horlick-Jones (1995) argues that disasters originate in the conditions of modern society and are disruptions in cultural expectations that cause a loss of faith in the institutions that are supposed to keep hazards under control (Mileti 1999).

The disaster vulnerability of an area or region is a function of a set of characteristics (physical elements and their functional links) pertaining to a developed chain of its human settlements. Therefore, the human settlement planning undertaken without due regard to the potential of natural hazards often risks increasing, if not the multiplied, impact of disasters. The development planning, and its physical realization and coordination through the process of spatial and/or master urban planning, influences where the population and the structures are located and concentrated, what materials and methods are used for construction, how topography is altered, what lifeline and utility systems are in place and are necessary to be developed, what emergency systems are developed and supported, what resources and reserves are maintained, what economic dependencies are fostered, etc.

Pre-earthquake emergency planning is the best way to ensure that the earthquake can be handled effectively. In particular, this is economically justified for countries of low economic strength that cannot handle the problem by prevention sector. A well-coordinated emergency response to an earthquake is likely to save many lives, greatly reduce the socioeconomic and political disruption of the population, as well as prevent a chain of secondary hazards escalating the magnitude of the disaster. Poor emergency response, accompanied with follow on (secondary) disasters can multiply the effects, human casualty and economic loss toll of an earthquake.

One of the major lessons learned from the earthquake that struck Kobe, and reinforced lessons learned from Northridge, was that failures of infrastructure systems can cause the most disruption to the social and economic system. Research on earthquake-resistant design was originally focused on buildings, but over the past twenty-five years, it has become increasingly concerned with infrastructure systems, particularly transportation facilities, water supply and distribution systems, gas and electric distribution, and telephone and other communications. Infrastructure systems are extremely vulnerable and their failure can result in failures of other lifelines and losses from disruption of activities that are much greater than the cost to repair damage. Mitigation measures can reduce risks at costs that are relatively modest (Eguchi 1997, 18).

In the case of urban design and planning disasters could affect the three substantive elements of these areas: activities, form and spaces (Table 2.1).

Hazards: Hazards are potential natural risks, severe phenomena but with a low frequency, originated in meteorology or geology and biology which when target humans and human settlements and infrastructure cause extensive losses and damages.

Hazard identification: Identification of potentially damaging physical events, phenomenon or human activity that may cause the loss of life or injury, property damage, social and economic disruption or environmental degradation.

Earthquake Damage Inspection: Earthquake damage and usability classification, after moderate or large-scale damaging earthquakes should be performed based on a uniformly established methodology within the country or wider region in order to create a uniform basis for assessment of physical damage and estimation of economic losses. Uniform data sets can thereby be constructed as a means to assess possible future earthquake effects.

Earthquake damage inspection as developed within this manual as based on a uniform methodology established within Working Group D of the UNDP/UNIDO

Table 2.1 Risk assessment elements in urban design and planning

Activities	• Disruption of activities • Interruption of activities due to human losses and property damages • Economic and social damages due to interruption of activities • Mishaps in spatial and physical communication • Disruption of center-vicinity and macro–microcommunication
Form	• Collapse and chaos in physical form • Chaos and interruption of communication in settlements spatial organization • Collapse of "spatial hierarchy"
Space	• Surrounding buildings are damaged and the safe spaces in and around dwellings are lost • At the local level, the relation between dwellings and work places and nearby activities are broken • At the regional level the distribution and spatial organization of settlements are disrupted

project "Building construction under seismic conditions in Balkan Region" and accepted for application by Balkan countries. This methodology for earthquake damage and usability classification is synthesized in the earthquake damage inspection Form and developed on the basis of the experience gathered in earthquake damage and usability, classification in past earthquakes.

The methodology on earthquake damage and usability classification is directly connected with the earthquake damage inspection Form enclosed on the next page. Explanations are presented on the back page of the inspection form. These are basic instruction materials for the inspection teams in order to perform damage and usability classification in uniform manner.

Response: Refers to the actions taken immediately before, during, and after a disaster occurs to save lives, minimize damage to property, and enhance the effectiveness of recovery.

2.1 Earthquake Damage and Usability Classification

Nature of damaging earthquake hazards

Earthquake damage to buildings, structures and utilities results from different types of seismic hazards. The main hazards posed by earthquakes may be summarized as follows:

1. Ground shaking of different severity.
2. Differential ground settlement, landslides and mudslides, soil liquefaction, ground lurching, and avalanches.
3. Ground displacements within fault zones.
4. Floods from dam.
5. Fires resulting from earthquakes.

All these hazards have occurred in past earthquakes with the dominant influence of ground shaking and also hazards associated with soil instabilities. By far, ground shaking has been the most damaging as it causes buildings and structures to collapse partially or totally and produces damage far from the epicentral area: Ground shaking affects the soil and foundation beneath structures, and much structural damage in earthquakes is a consequence of ground failure arid differential ground settlements. Sometimes, the ground will lurch, particularly along roadsides, culverts, river banks, and in low lying areas. Ground shaking can also initiate devastating rock and mudslides, which themselves can produce some of greatest disasters ever experienced from seismic causes. A very common hazard in earthquakes is the liquefaction of sandy soil, particularly in river valleys and coastal regions. During earthquake shaking, fine grained soil and sands, saturated by water, take on a liquid character owing to alternations in shearing stress. Water saturated sands are so widespread, particularly in flat areas where populations tends to concentrate, that soil liquefaction and resulting damage to buildings and structures have been observed in almost every damaging earthquake. Significantly, soil liquefaction effects are very frequently associated with comparatively low accelerations of ground shaking. A much more restricted hazard comes from the surface rupture within geological fault zones. Buildings that straddle fault displacements may be critically wrenched. Elimination of this hazard is more difficult in practice and depends upon adequate building codes and the availability of special geological fault maps.

Other earthquake hazards involve water and fire. Due to undersea faulting, gigantic sea waves (tsunamis) may rush up along the coastline and devastate coastal properties. Floods from sudden failure of dams in earthquakes as an ever present danger that could create enormous destructive effects sometime larger than those produced by the ground shaking itself. Fires are potential secondary effects in modern urbanized areas with the presence of chemical industry, and oil and gas supplies. Ground shaking could cause breakage of pipelines, failure of oil and gas tanks, and damage to chemical industries. Explosions, release of toxic chemicals and even fires in neighborhoods or entire towns could result (Tokyo Earthquake 1923; Niigata Earthquake 1964).

These earthquake hazards are reviewed so that earthquake damage inspection teams may differentiate the influences of different earthquake hazards on damaging effects.

Hazard assessment: Earthquake hazard assessment enables the likelihood of ground shaking across a region to be calculated, which is a fundamental component in earthquake risk assessment or hazard mapping for design codes. Figure 2.4, schematically shows the factors to be considered in predictive modeling of earthquake ground motion.

Risk: The probability of occurring a disaster. Risk is fundamentally a product of hazard and vulnerability.

Disaster mitigation planning: While the natural hazards are integral features of the environment, human casualty and the extent of damages are to a significant degree a function of decisions made, or not made, in the process of human settlement development. In most developing countries, a little attention has been given to natural disasters potential in national and regional (sub-national) economic and

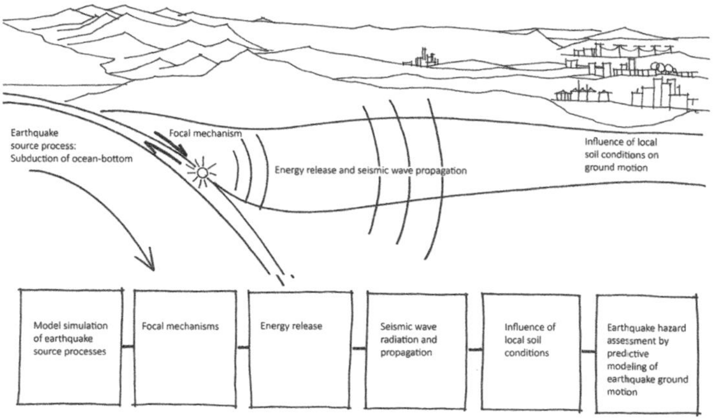

Fig. 2.4 Earthquake Prognostics Strategy: Deterministic approach to earthquake hazard assessment by predictive modeling of site-independent strong ground motion (Vogel and Klaus 1988)

spatial planning. Consequently, the response of national (and international) community has been to focus primarily on post-disaster activities (relief and rehabilitation) rather than to disaster prevention and preparedness.

The disaster susceptibility of a given region, from the viewpoint of human settlement, refers to the likelihood of human, physical and economic loss attributable to natural hazards. The disaster vulnerability of an area or region is a function of a set of characteristics (physical elements and their functional links) pertaining to a developed chain of its human settlements.

Therefore, the human settlement planning undertaken without due regard to the potential of natural hazards often risks increasing, if not the multiplied, impact of disasters. The development planning, and its physical realization and coordination through the process of spatial and/or master urban planning, influences where the population and the structures are located and concentrated, what materials and methods are used for construction, how topography is altered, what lifeline and utility systems are in place and are necessary to be developed, what emergency systems are developed and supported, what resources and reserves are maintained, what economic dependencies are fostered, etc.

Tools for hazard mitigation: Of the many techniques for coping with natural hazards and disasters that people have attempted to use, five categories have proven to be the most useful for minimizing losses and reducing social and economic disruption: Land-use planning, building codes, insurance, engineering, and warnings. Each is supported by its own body of research, disciplines, experts, and government and private-sector management structure.

It is believed that no single approach to bringing hazard mitigation into existence shows more promise at this time than increased use of sound and equitable land-use management. By planning for and managing land use to accomplish sustainable hazards mitigation, disasters can be reduced to a scale that can be borne by governments, communities, individuals, and businesses exposed to them.

Exposure: The extent of the people, properties, and the like being exposed to certain disaster event.

Vulnerability: The life loss and property damage occurred due to an event. In essence, the concept of vulnerability cannot be separated from the social and cultural conditions under which it exists. For any person or entity, vulnerability to a given threat is a holistic phenomenon that cannot be divided up into components.

Resilience: The extent one can endure and sustain an unexpected event, bearing capacity. Vulnerability assessment (Bosher, 2008).

Criteria for determination of earthquake parameters for planning and design: Earthquake disaster management could be defined as a complex process composed of three directly interrelated phases: disaster prevention, disaster preparedness and emergency relief, and reconstruction and rehabilitation. The structure of all three phases involves a variety of skills and different organizations all of which must work in complete accord due to high degree of interdependency between the various components. What is not achieved to upgrade systems to disasters by preventive measures has to be carefully planned to be covered by disaster preparedness and emergency relief as well as reconstruction and rehabilitation phase.

The ultimate disaster prevention is only ideally attainable; however, the disaster mitigation is always practically possible. The disaster prevention is a long-term expensive measure and can be afforded on priority basis by a few developed countries, only. On contrary, the disaster (emergency) preparedness can substantially reduce the impact of the disasters at a cost-efficient base that may be affordable by developing countries also.

While prevention consists of long-term measures undertaken in the scientific, technical, and economic fields, and designated to avoid or reduce significantly the level of expected loss, the mitigation leans solely on planning and managing disaster related activities.

Recovery and rehabilitation: Decisions and actions taken after a disaster with a view to restoring or improving the pre-disaster living conditions of the stricken community, while encouraging and facilitating necessary adjustments to reduce disaster risk. Recovery (rehabilitation and reconstruction) affords an opportunity to develop and apply disaster risk reduction measures.

Mitigation, mitigative adaptations, or hazard mitigation (structural, infrastructural, and non-structural); structural mitigation such as the strengthening of buildings and infrastructure exposed to hazards (via building codes, engineering designs and construction practices, etc.). **Non-structural mitigation** includes directing new development away from known hazard locations through land-use plans and regulations, relocating existing developments to safer areas and maintaining protective features of the natural environment (such as sand dunes, forests and vegetated areas that can absorb and reduce hazard impacts). The construction sector can play an important role in the structural elements of mitigation, while developers and urban planners and designers should be able to positively influence the non-structural elements of mitigation.

Preparedness planning: Activities and measures taken in advance to ensure effective response to the impact of hazards, including the issuance of timely and effective early warnings and the temporary evacuation of people and property from threatened locations.

Damage assessment

So put some of these definitions in a simple way:

Risk, vulnerability, exposure and capacity of humans to respond to extreme events.

Or:

Hazard + Context = Vulnerability = life & property losses.

Strengthening and retrofitting: The concepts of strengthening and retrofitting are often improperly associated. The structure could be upgraded without needing any strengthening (increase of strength). Stiffness-only, strength-only or ductility-only intervention can be carried out on a single element as part of a selective retrofit approach.

Climatologically hazards: These hazards include drought and dust storm, floods, fog, hail, wildfires and heat, hurricane and tropical storm, snow and extreme cold, wind, ice, sleet, avalanche, lightning, and tornadoes. But since this book takes on seismic issues, which is classified under geological hazards, these hazards are not discussed in this book.

Intensity: Earthquake intensity is based on observations and impacts on structures and humans. In fact, the intensity or severity of an earthquake reflects the force of an earthquake in certain locale.

Intensity scale: There are two scales to measure the intensity of an earthquake: MM1 (Mercali) which starts from 1 and goes to 12; and MSK.

Macro-seismic epicenter: Where the most severe damages in engineered buildings, earth relocation, rockfall, etc., take place. This is the point on earth surface that vertically leads to hypocenter deep in the earth. This point has the highest intensity and one move away from this point the shaking will reduce.

Focal depth: It is the perpendicular distance between epicenter, on the earth's surface, and the macro-seismic epicenter. Higher depth means lower intensity on the surface.

Earthquake scenarios: To predict appropriate scenario earthquakes for vulnerability and seismic risk analysis on regional and urban level, assessment of damageability of buildings and structures and formulation of requirements for building regulations and seismic design codes, several levels of seismic hazard should be implemented based on probabilistic hazard analyses and simulation of well documented destructive earthquakes in the past, including widely developed strong motion networks and dense array records. These earthquake scenarios could be predicted based on seismic zoning studies on the national and regional level, seismic microzoning investigations of urban areas, specific site investigations of important buildings and structures and use of recorded or simulated strong earthquake ground motions.

Risk analysis: Risk means the probability of an event or condition occurring. It encompasses risk assessment and risk management. The two challenges in risk analysis are how to handle uncertainty in knowledge about risk and determining how safe is safe enough (Kunreuther and Slovic 1996).

Exposure is the measure of people, property, or other interests that would be subject to a given risk.

Vulnerability is the measure of the capacity to weather, resist, or recover from the impacts of a hazard in the long term, as well as short term.

Intensity: The intensity of an earthquake depends on its impacts on structures and humans in a specific location.

Intensity scale: Scales are developed to measure the intensity of an earthquake, two of these scales are MM1 and MSK, which were presented by Italian scientist Mercalli in 1902, and ranges from I to XII.

Isoseismal map: When the points where the intensity is the same are connected the result is isoseismal and map is called isoseismal map.

Earthquake magnitude is a measurement of the energy it produces, not what people or the constructed environment experience during the event. It is a quantitative measurement of the extent and area of the earthquake and it is shown in recording equipments.

Magnitude scale: It is the most common scale to measure earthquakes and ranges from 2, the lowest to 8.9, the highest.

Hypocenter: It is the point depth under the earth where actual earthquake takes place.

Peak velocity describes aspects of the motion—how fast the ground is moving.

Peak acceleration means how quickly the speed of the ground is changing.

Frequency implies energy is released in waves and the waves vibrate at different frequencies just as sound waves do.

Duration: How long the shaking lasts. Three factors primarily dictate what people feel (and objects endure) in an earthquake: magnitude, distance from the fault, and local soil conditions.

Recovery: The term recovery has been used interchangeably with reconstruction, restoration, rehabilitation, and post-disaster redevelopment. Regardless of the term used, the meaning has historically implied putting a disaster-stricken community back together.

Preparedness: The purpose of preparedness is to anticipate problems in disasters so that ways can be devised to address the problems effectively and so that the resources needed for an effective response are in place beforehand. Preparedness includes such activities as formulating, testing, and exercising disaster plans; providing training for disaster responders and the general public; and communicating with the public and others about disaster vulnerability and what to do to reduce it.

Emergency settlement, temporary settlement, and reconstruction: Immediately after a disaster, there is need for a safe and easily accessible place where injured may find refuge and receive first aid, while others calm down and wait for transferring to a temporary settlement. Since staying in this place might take a while, basic needs, such as health, education, sanitary, water, food, and electricity facilities should be provided. Reconstruction of the damaged areas, or the building a new settlement, requires extensive study to find the right location, construction techniques, financial resources, as well as legal considerations (Bahrainy and Akhoundi 2001).

Sustainable hazards mitigation has six essential components: environmental quality, quality of life, disaster resiliency, economic vitality, inter- and intra-generational equity, and a participatory process.

Guidelines for earthquake disaster management. Guidelines for earthquake disaster management, covers the needs of government officials, planners, economists, architects, engineers, and scientists, with the purpose of planning for seismic risk reduction and the practical implementation of methodologies and findings in earthquake affected regions.

While the natural hazards are integral features of the environment, human casualty and the extent of damages are to a significant degree a function of decisions made, or not made, in the process of human settlement development. In most developing countries, a little attention has been given to natural disasters potential in national and regional (sub-national) economic and spatial planning. Consequently, the response of national (and international) community has been to focus primarily on post-disaster activities (relief and rehabilitation) rather than to disaster prevention and preparedness.

Secondary damages: Following an earthquake several destructive events might occur, such as:

- Fire.
- Explosions.
- Flooding.
- Subsidence.
- Landslide.
- Psychological trauma of shake.
- Disorientation.
- Uncertainty.
- Grief.

The most significant of those events are subsidence and expansive soils (Figs. 2.5, 2.6 and 2.7).

Isoseismal map of the *Rudbar -Manjil* Earthquake (Fig. 2.8), Macro-seismic intensity and the isoseismal map of the *Bam* earthquake (Fig. 2.9), *Manjil* Earthquake of 20 June, 1990, Recorded at *Abbar* (Fig. 2.10) and the empirical vulnerability functions of damaged and collapsed buildings due to Manjil earthquake (Fig. 2.11) are as follows:

Based on the data of *Manjil* earthquake of June 20, 1990.

Evacuation is used in a broad sense to describe the movement of some or all the building occupants to a place of acceptable safety in the event of fire or any other disaster.

Issues and problems brought to public attention in recent years with respect to conventional means of egress include:

- The limited capacity of stairs for occupant egress.
- The limited capacity of stairs for emergency responder access during occupant evacuation.

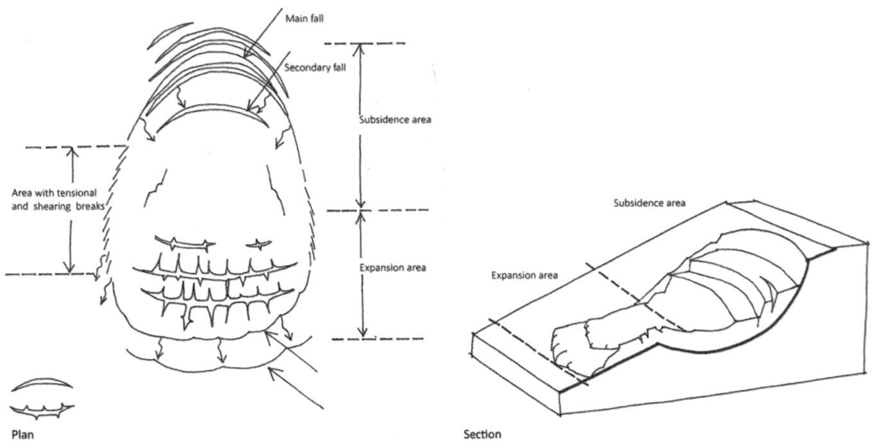

Fig. 2.5 Plan and section of a landslide

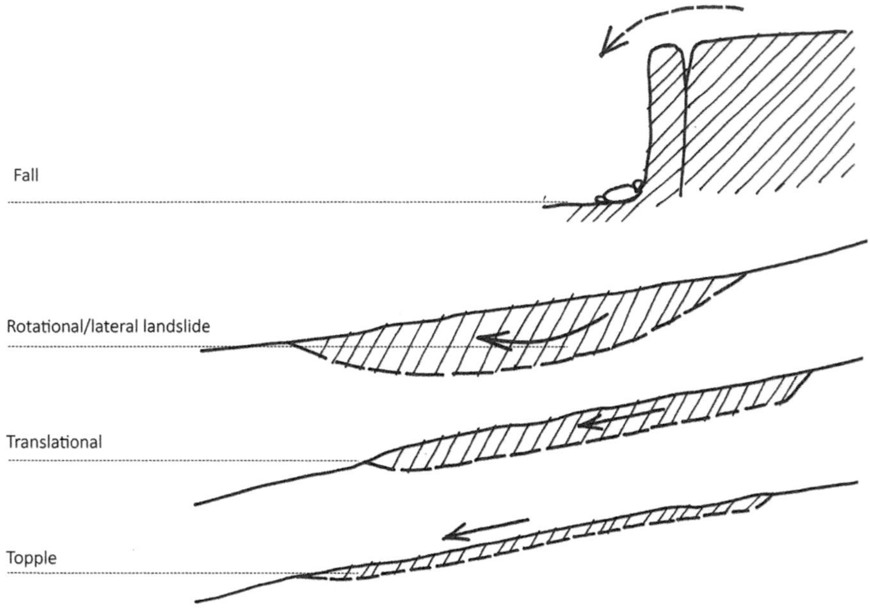

Fig. 2.6 Four different types of landslide

- The inherent limitations of evacuating persons with mobility impairments.
- The lack of alternatives when a single event compromises stairs and, where used in emergency situations, elevators.

In the event of fire, evacuation will be from the floor on which the emergency has occurred and the two floors immediately above and below the emergency floor to a

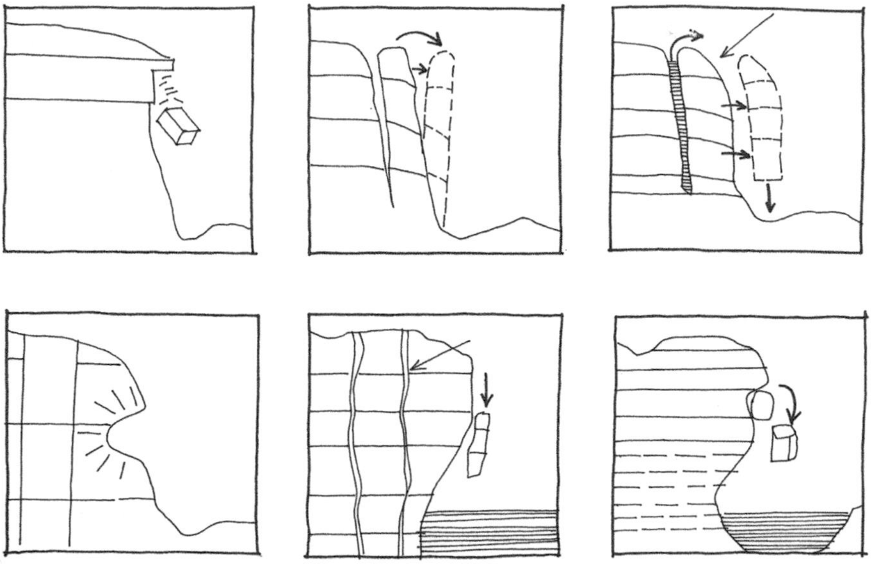

Fig. 2.7 Six different types of rockfalls

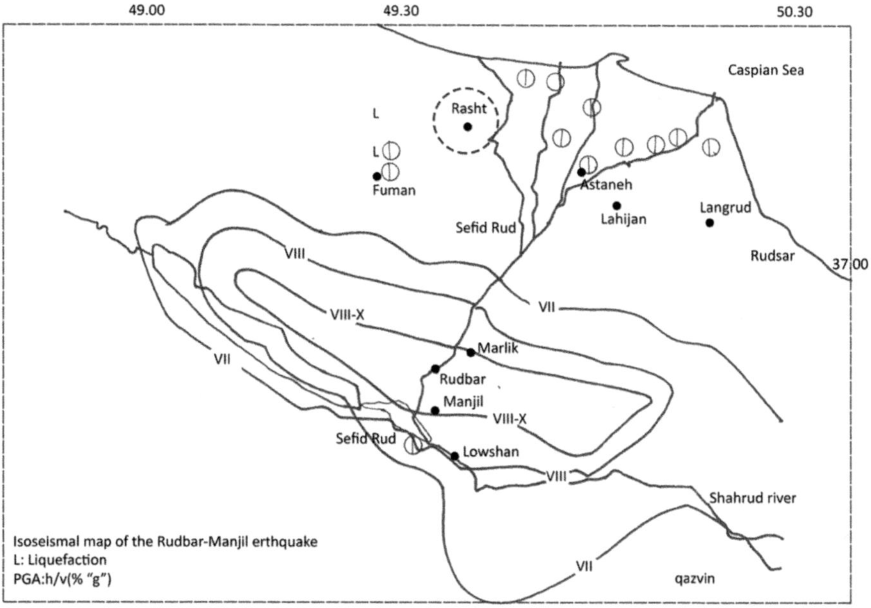

Fig. 2.8 Isoseismal map of the *Rudbar -Manjil* earthquake

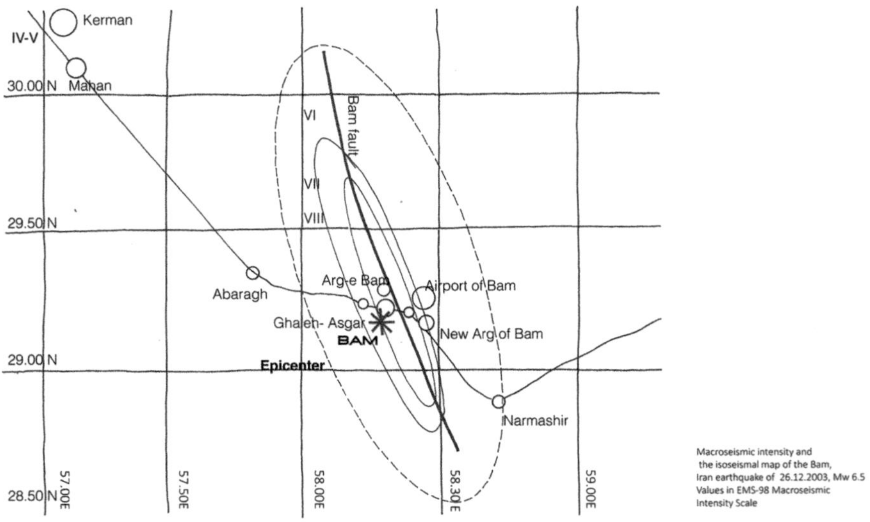

Fig. 2.9 Macro-seismic intensity and the isoseismal map of the Bam earthquake, 26.12.2003

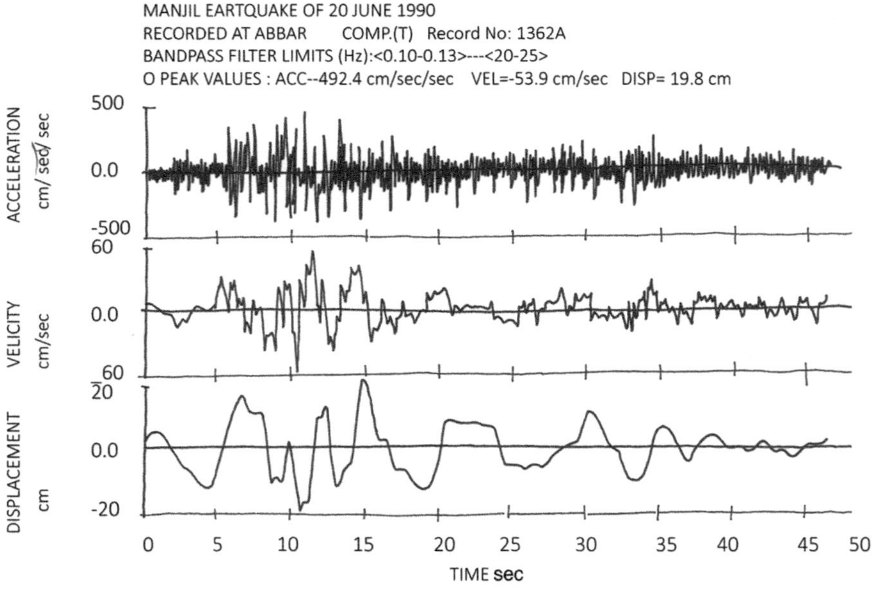

Fig. 2.10 Manjil Earthquake of June 20, 1990, Recorded Abbar

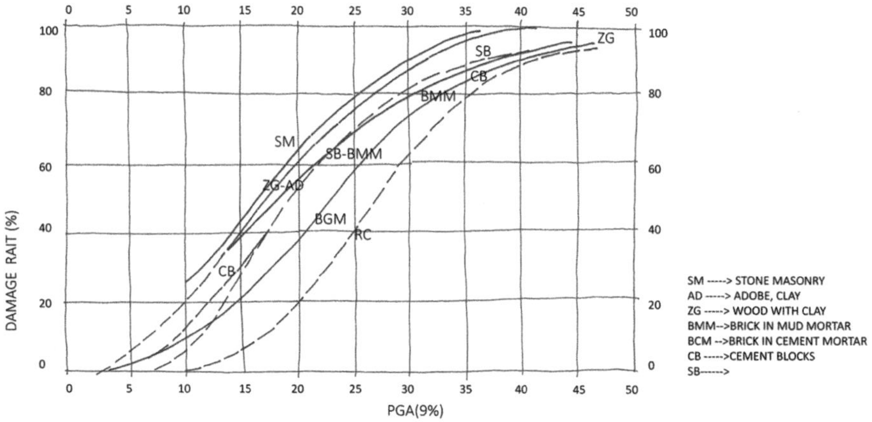

Fig. 2.11 Empirical vulnerability functions of damaged and collapsed buildings

safe point above or below the critical area. Evacuation should be accomplished by way of fire stairwells.

Evacuation in high-rise buildings

Fire escape/ Fire staircase. A fire escape is a special kind of emergency exit, usually mounted to the outside of a building or occasionally inside but separate from the main areas of the building. It consists of a number of horizontal platforms, one at each story of a building, with ladders or stairs connecting them. It provides a method of escape in the event of a fire or other emergency that makes the stairwells inside a building inaccessible (Fig. 2.12).

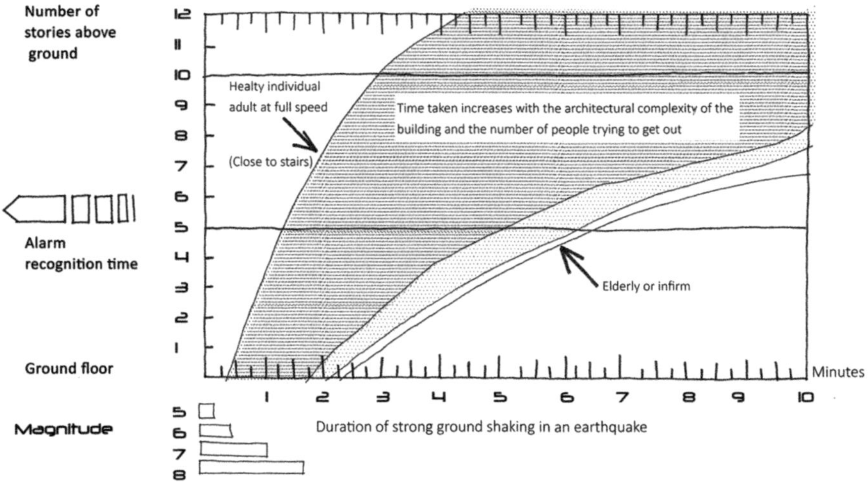

Fig. 2.12 Evacuation times from multi-story buildings

Vulnerability of building form due to fire: This vulnerability is determined on the basis of following criteria and ranges from very high to very low:

Higher vertical and horizontal density will increase the possibility of fire spreading.

The more open space, the lower the fire spreading.

The more height difference in adjunct buildings the more possibility of fire spreading from lower buildings to the higher ones.

The more the openings, the higher the possibility of fire spreading (Fig. 2.13) (https://independent.academia.edu/reshmarameshan?swp=tc-au-17494346).

An understanding of the potential damage probable earthquakes is of fundamental importance to determining expected levels of the environment exposure, assessment of seismic risk and determining the most appropriate measures and initiatives for planning of earthquake protection and reduction of earthquake damage. More recently, there have been significant attempts to describe the potential effects of earthquakes and present their impact through scenarios on regional and urban level in order to provide appropriate basis for reduction of seismic risk incorporated in development planning and earthquake disaster preparedness organization.

Seismic zoning maps: objectives and scope of seismic zoning.

Seismic zoning is a process under which subdivision of a region, or country into areas or zones that have relatively similar exposure to earthquake associated effects (primarily ground shaking) is made. The aim of zoning is to estimate the location, recurrence interval, and relative severity of future seismic events in an area so that the potential hazards can be assessed and the effects can be mitigated or avoided. Results of seismic zoning are commonly present in seismic zoning map where the studied effects are mapped and presented in order to provide a technical basis for engineering as well as social and political decisions.

To predict appropriate **scenario** earthquakes for vulnerability and seismic risk analysis on regional and urban level, assessment of damageability of buildings and

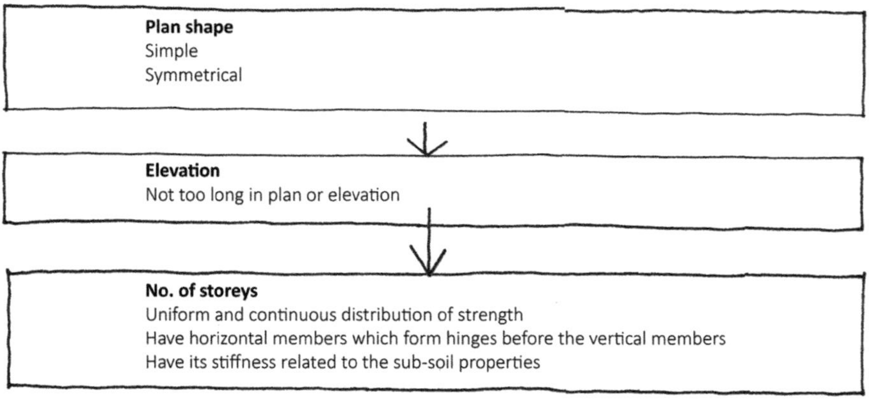

Fig. 2.13 Factors influencing safe and swift evacuation from high-rise buildings.

structures and formulation of requirements for building regulations and seismic design codes, several levels of seismic hazard should be implemented based on probabilistic hazard analyses and simulation of well-documented destructive earthquakes in the past, including widely developed strong motion networks and dense array records.

Earthquake disaster management could be defined as a complex process composed of three directly interrelated phases: disaster prevention, disaster preparedness and emergency relief, and reconstruction and rehabilitation.

Land instability represented by **land-sliding** and **liquefaction** has been recognized as a major case of distress caused by earthquake in recent times. The ground failure reemphasizes the importance of land-sliding and liquefaction as major cause of hazards associated with earthquakes.

It is well known that the hardness or softness of the soils or rocks forming the ground or slopes is a dominant factor influencing the instability of lands during earthquakes.

Soil liquefaction has been a major cause of damage to earthworks and foundations in past earthquakes and poses a potential threat to the integrity of structures and installations during feature earthquakes. Thus, the assessment of liquefaction potential of the ground is one of the most important considerations in designing foundation and earth structures and, also for conducting seismic zoning work. The evaluation of liquefaction can be made with different levels of accuracy depending upon the availability of basic information on geology and geotechnical data.

Generally speaking, there are two types of land-sliding induced by earthquakes. One is the sliding which occurs all of a sudden simultaneously with seismic shaking. This type of collapse is induced by the inertia force due to seismic shaking and has a tendency to occur over steep slopes at the shoulder of hilltops or near the top of mountains. When the place of such a topographical prominence is covered by layers of disintegrated rocks or transported soils, the chances of destruction are very high by landslides in that it occurs all of a sudden and it involves a large amount of soil or rock masses. Sometimes, the slide turns into a catastrophic debris flow, if the sliding mass contains a large amount of water.

The second type of land-sliding is the one which occurs in a quiet period of time several hours or several days after an earthquake. This type of landslide is sometimes a ground movement due to reactivation of old dormant landslide. Ground water is generally associated with the provocation of such a slide. It occurs slowly and signs of land movement are always detected well in advance, and hence, casualty is generally small. But the loss of properties and damage to public facilities are great because of a large amount of soils involved in the slide.

Rockfalls are still other type of geotechnical hazard brought about by an earthquake. The rockfalls often occur on steep slopes, mountain walls, and road cuts. The amount of rock masses involved is not generally so much, but because of its sudden occurrence during main shaking of an earthquake, it is generally impossible to elude the danger of falling rocks, if it occurs on the spot. Major issue of importance in the rockfall hazard is blockade of roads or highways, paralyzing the traffic which is often fatal for relief and rescue operation. The physical mechanism for occurrence of rockfalls is poorly understood, because of too many factors or conditions involved in individual cases of rock collapse. It is likely that highly fractured rocks or overhung rock masses tend to fall and crumble. If there is heavy rainfall preceding earthquakes, water tends to seep into fissures or cleavages, thereby loosing the stability of rock masses.

Prevention versus mitigation. The ultimate disaster **prevention** is only ideally attainable; however, the **disaster mitigation** is always practically possible. The disaster prevention is a long-term expensive measure and can be afforded on priority basis by a few developed countries, only. On contrary, the disaster (emergency) preparedness can substantially reduce the impact of the disasters at a cost-efficient base that may be affordable by developing countries also.

While prevention consists of long-term measures undertaken in the scientific, technical, and economic fields, and designated to avoid or reduce significantly the level of expected loss, the mitigation leans solely on planning and managing disaster related activities. However, the common denominator for either disaster prevention or disaster mitigation are adequate and reliable estimators on disaster agents whose adverse impacts should be minimized by disaster prevention measures and activities or amortized by disaster—mitigation (relief) operations. For this, an adequate research base and reliable quantification on disaster agents are required in order to draw out an overall and strategy approach to the general problems associated with disasters (disaster planning), and define the specific tactics to be used in particular contingencies associated with actual or expected cases of disaster prevention, mitigation, response, or recovery (disaster management) (Wolfe, 1981).

2.2 Conclusion

To help designers easily understand the techniques and methods used to reduce seismic risk in urban areas, brief dentition and description of the commonly used terms and concepts in disaster, and particularly seismic risk management, are given in this chapter. Life losses and material damages of disasters may be attributed to three groups of elements and their mutual interactions: geophysical hazards, climatologically hazards and biological hazards. These hazards will affect the following elements in human settlements: humans systems, and building and construction systems. Some of the significant terms explained here in this chapter are: liquefaction, landslide, rockfalls, tsunamis, acceleration lines, seismic zonation map, mesoseismal, faults, active faults, disaster, hazards, disaster mitigation, exposure, vulnerability, preparedness, strengthening and retrofitting, epicenter, earthquake scenarios, intensity, intensity scale, isoseismal map, peak velocity, guidelines for earthquake disaster management, secondary damages, evacuation, and prevention.

References

Bahrainy H, Akhoundi AA (2001) Reconstruction management of disaster-affected areas, the experiences of reconstruction of housing in the earthquake affected areas of the Guilan and Zanjan provinces, Iran, 1990–1992. University of Tehran Press (in Persian)
Beer AN, Higgins C (1999) Environmental Planning for Site Development, A manual for sustainable local planning and design. E & FN SPON, New York
Bosher L (ed) (2008) Hazards and built environment, attaining built-in resilience. Taylor and Francis
Eguchi RT (1997) Mitigating risks to lifeline systems through natural hazard reduction and design. In: Economic consequences of earthquakes: preparing for the unexpected, pp 111–123
Fritz C (1961) Disaster. In: Merton RK, Nisbet RA (eds) Contemporary social problems. Harcourt Press, New York
Horlick-Jones T (1995) Modern disasters as outrage and betrayal. Int J Mass Emerg Disasters 13(3):305–315. https://independent.academia.edu/reshmarameshan?swp=tc-au-17494346
Kunreuther H, Slovic P (eds) (1996) Challenges in risk assessment and risk management. Ann Am Acad Polit Soc Sci 545:1–200
Mileti DS (1999) Disaster by design. A reassessment of natural hazards in the United States. Joseph Henry Press, Washington DC
NCEER-SP-0001, National centre for earthquake engineering research, Buffalo, New York
UN/ISDR (2004) Appendix 1
UNDP UNCHS (Habitat) Project IRA/90/004 Assistance in the Implementation of Post-earthquake Rehabilitation Program, Housing Foundation of Islamic Revolution, I.R. Iran
Vogel A, Brandes K (eds) (1988) Earthquake prognostics, hazard assessment, risk evaluation and damage prevention. In: Proceedings from the 2nd International Seminar held in Berlin, June 24–27, 1986. Friedr. Vieweg & Sohn
Wolfe MR (1981) Urban scale vulnerability: focus down from city regions to community clusters and building typologies. Joint USA-Italy workshop on earthquake disaster mitigation, Oct 12–16, Rome, Italy

Chapter 3
Manjil Earthquake of June 20, 1990, The Lessons Learned

Abstract The June 20, 1990, *Manjil* Earthquake in northwest Iran was the most damaging event to date in the country. That produced the worst earthquake disaster in the history of Iran after the *Tabas*-e-*Golshan* earthquake of September 16, 1978, of similar magnitude and intensity in the narrow epicentral area. Considering the damaging effects of one of the most devastating earthquakes in Iran, Manjil–Rudbar Earthquake, with magnitude Ms = 7.4 and estimated epicentral intensity of IX - X degrees of MM Scale, caused 13,911 deaths, 36,693 people treated for injuries, and over 8000 people badly injured taken for hospitalization in Tehran. Damage and losses to the built environment were extremely high, estimated at 4.77 billion dollars in the densely populated region of the provinces of *Gilan*, *Zanjan,* and Eastern *Azerbaijan*, located in the western part of Alborz imbricated mountain belt of northwestern Iran. The real acceleration lines (which are based on a real earthquake in the region) or hypothetical (which are based on a probable earthquake scenario) show the degree of risk and vulnerability of different areas. These maps will be the basis of design and planning decisions in the seismic-prone regions. So the accuracy and precision of these maps will guarantee on the quality and reliability of the design decisions. The June 20, 1990, *Manjil* Earthquake was strongly felt over an area of 600,000 square kilometers, including Tehran, Tabriz, and several other major cities in Iran. Area of damaging effects in the provinces of *Gilan*, *Zanjan,* and Eastern *Azerbaijan*, with ground acceleration larger than 10% g, was estimated to 49,574 square kilometers affecting 3152 villages and 45 towns and cities in 24 districts with 98 rural districts in the province of Gilan as well as 16 districts with 90 rural districts in the province of *Zanjan*. More than 214,000 residential units, 1329 school buildings, over 300 health units as well as 82 medical centers and hospitals, a large number of agricultural land and facilities, religious and administrative units, service centers including 68 factories were destroyed or heavily damaged. Over 500,000 families were left homeless, and 178 village locations were abandoned due to landslides and other ground instabilities. In the epicentral region with ground acceleration larger than 30% g with an area of about 8000 square kilometers, 383 villages and the towns of *Manjil*, *Rudbar*, *Loshan*, *Jirandeh,* and *Abbar*, most of the buildings were destroyed or damaged beyond repair. More than 67 landslides and many rockfalls were triggered in the wider epicentral region, causing destruction and damages to the transportation systems, water supply, electricity, and other lifelines of regional

H. Bahrainy and A. Bakhtiar, *Urban Design in Seismic-Prone Regions*,
The Urban Book Series, https://doi.org/10.1007/978-3-031-08321-1_3

and urban infrastructure. One of the most dramatic landslides triggered by Manjil Earthquake was the *Fatalak* landslide, where a huge mass of about one million cubic meters slide down the slope suddenly and all houses of the *Fatalak* Village were buried with 137 habitants. Lessons learned from this experience and a set of guidelines formulated on its basis will be presented at the end of this chapter.

Keywords *Manjil* earthquake · Spatial–physical conditions · *Rudbar* · *Loshan* · Infrastructure · Damage distribution · *Rasht* · Landslides · Liquefaction · Deficiencies

3.1 Introduction

As it was mentioned before, Iran is a seismically active country and buildings in most parts of the country may be subjected to ground shaking. Until recently, however, most buildings in Iran were not specifically designed and constructed to resist earthquake loading. As a result, performance of Iranian buildings during earthquakes has been poor, especially during earthquakes such as the 1990 *Manjil* event.

The urbanization in the region can be traced to earlier than prevalence of Islam, especially in *Rudbar*, Rasht, and *Siahkal*, though not as it is defined today, *Siahkal* was a commerce center for mountain dwellers of *Daylaman*. *Bazaar*, then was defined as the transacting center between mountains and plains, which latter aided as important factor in urban development. *Sangar* was also center for regional market and during last few decades due to population increase has acquired the urban status. Despite the fact that *Rudbar* remained a capital for various rulers (in local governments), till 1966 was not considered as a town. From this date, by joining of many historical villages, (that were separated by obstacles), this place acquired urban status. *Manjil*, which 1931, due to the construction of *Sefidrood* Dam and, Tehran–Rasht highway, started a rapid growth, finally in the year 1976 acquiring urban status. Till 1976 *Loshan* was a rural nude, with a long history of settlement. During 1976–1986 period by merger of many villages, it acquired an urban status.

Rasht did not enjoy much importance, till *Safavid* dynasty, and only due to congenial climatic condition was a village suitable for local livelihood, in the *Gilan* plain. But, since then it was selected as the capital, thus becoming the crossroads of various commerce and business interactions, from Bandar *Anzali* to other parts of the country.

Due to a short history of urbanization, the formation and development of many towns in the region were dictated by political and official needs, that is why except Rasht. The rest of towns are devoid of structural characteristics of a city, which created various problems regarding their development.

3.2 Spatial–Physical Conditions

The towns of the region are situated at the belt connecting central Iranian plateau with the Caspian coasts. Moreover, Rasht is also situated at the interstate belt connecting all the important towns in a chain formation.

Situation of the towns in the region, considering the natural conditions dictating the location and functioning of these towns, has resulted in linear situation of *Loshan*, *Manjil*, *Rudbar*, and *Rostamabad* towns which are an extension of Tehran–Rasht highway. Thus, due to this, Rasht has acquired a central position, which has been an aiding factor in the development of its functioning and premises.

Rudbar, *Loshan Manjil*, and *Rostamabad* are closely situated, (*Loshan* to *Manjil* 28 km, *Manjil* to *Rudbar* 6 km, and *Rudbar* to *Rostamabad* 15 km), and moreover, these towns, through other connecting roads are joined to the other parts of the *Tarom Sofla*, *Tarom Olya*, *Anarloo*, and *Rahmatabad*. *Rudbar* more or less is in a dead-end situation, though through road it is connected to one of the villages in *Rahmatabad* rural district, the location, and access to this road restrains direct access to it. From the services and urban infrastructure, there is a wide disparity between Rasht and other towns of the region.

3.3 Spatial Structure, Regional Distribution of Settlements

Regional distribution of settlements in the evaluated area has historically followed spatial characteristics. From the spatial structural point of view, these characteristics can be divided into environmental, social, and economics, as described:

Environmental characteristics:

1. Geographical variations.
2. Prevalence of various geographical features such as plains, foothills, and mountains.
3. Temperate climatic conditions in plains, foothills, and cold climatic conditions in mountains.
4. Perennial water resources.
5. Access routes and their role in settlement patterns.
6. Region's situation with respect to its neighbors.
7. Vegetation cover.
8. Forest areas, especially in foothills and mountains.
9. Agricultural lands in plains.

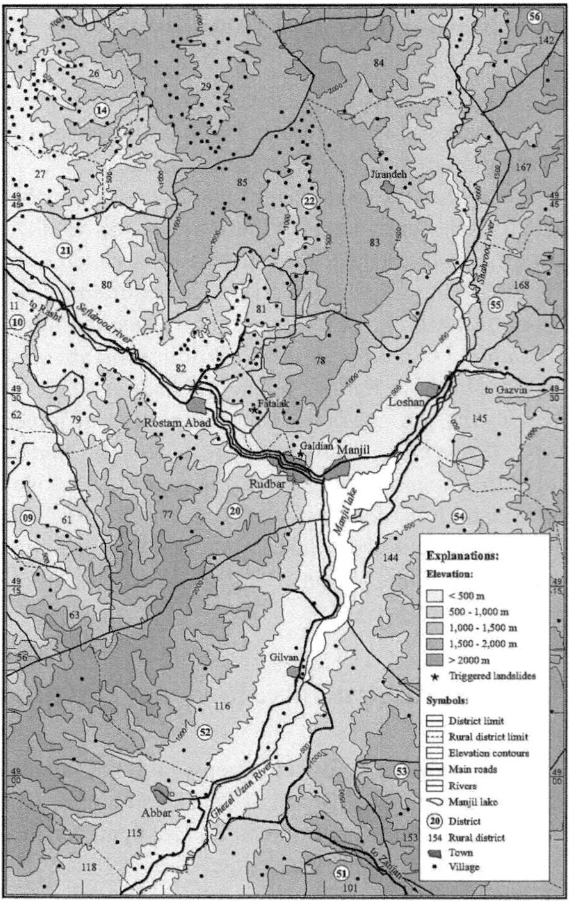

Fig. 3.1 Topographic map of Rudbar, Manjil, and Loshan region strongly affected by June 20, 1990, Manjil earthquake

3.4 Regional Topography

Topography and natural feature are determinant factors in risk level and eventually the vulnerability of different areas in a region. Any construction on the elevated areas with steep slopes is threatened; and as conditions, such as soft and watery soil permits landslide and liquefaction will occur (Fig. 3.1).

3.5 A Summary of the Social Characteristics of the Area

1. Long history of settlements.
2. Tribal culture.
3. Religious and cultural variations.

4. High population density in plains and lower density in mountains. Total population of the evaluated region is about 768,548 persons, out of whom 405,798 persons are in urban areas, and 362,750 are rural. Except the city of Rasht, that with 359,388 population occupies a middle position in the urban classification. The rest of the towns with lesser population number, are classified under the category of small cities. This aspect has had to a heterogeneity in service distribution between the cities. Moreover, the lack of big urban centers in Shahrood and Tarom valleys had forced, the residing population, to meet their needs from the neighboring urban centers, such as Zanjan and Qazvin.
5. Wide differences in settlement population in number of plains in comparison with mountains. In general, out of 855 villages of the region, 397 villages (46.4%) are in mountainous area; 269 villages (31.5%) in foothills, and 189 villages (22.1%) in plains, thus most of the villages of the region are situated in mountains and foothills (Taleb 1997).

3.6 A Summary of Economic Characteristics

1. Dominance of agro-economics in plains.
2. Importance of industrial economy in industrial belts and centers.
3. Dominance of herding economy in mountains.
4. Economic variation of different parts of the region (Taleb and Alizadeh 1997).

Based on statistical data of the World Bank, the International Bank for Reconstruction and Development and the World Almanac and book of facts 1994, gross national product (GNP) for I.R. of Iran was 116,040 million US dollars, based on local currencies which are deflated to constant 1994 local currency values before conversion to US dollars equivalents. For the estimated economic losses due to June 20, 1990, Manjil Earthquake at 4.11% of GNP of 116 billion US dollars, economic losses could be estimated in total to 4.77 billion US dollars (Petrovski 1998).

3.7 Regional Infrastructures

Over all, out of six towns of the region, only the city of Rasht and town of Sangar are situated at plains and have a mild slope, while Rudbar, Loshan, Siahkal, and Manjil are situated at foothills and have sharp slopes of higher degrees. Loshan, Manjil, and Rudbar, due to their location at natural passes and national transit system, have strategic importance. The city of Rasht also due to its situation at plains and national transit route is of great importance.

Water and power supply facilities have created congenial condition for infrastructural development of the region. Despite the existence of such facilities, the proper infrastructures to meet the rural settlement requirements have not been established.

From the transportation viewpoint, the region is still far from the satisfactory. As the main economic activity of the region is agriculture, thus the access roads are of great importance.

At the times of disaster, the continuous functioning of the water and power supply is most important for aid and rescue operations. The proper supply of clean water can reduce the casualties and prevent epidemics. Proper road and transportation network systems are also critical factors in the time of disaster, as an important lesson learned from the June 20, 1990, Manjil Earthquake.

3.8 Transportation Network

Like most of other difficulties, the region under assessment, faces acute problems in transportation network, the most important of which are as follows:

Airport: The sole airport of the region is situated in the city of Rasht, which is national and small. During the earthquake due to the meager capacity of it, rescue aids could not be flown into the region in bulks required.

Rail roads: The region has no railroad network. The nearest railroad network is Tehran Zanjan.

Inter-city conveyances: In between main villages, located near main roads, and urban centers, there are vehicle transportation services. These vehicles are primarily minibuses, followed by personal cars, which act as private carriers. The access program is two minibuses per day, starting early morning from villages to cities and afternoons reverse. The prevailing situation causes great distress, which in the emergency times acquires menacing prospect.

3.9 General Characteristics of June 20, 1990, *Manjil* Earthquake

The June 20, 1990, *Manjil* Earthquake in northwest Iran was the most damaging event to date in the country. That produced the worst earthquake disaster in the history of Iran after the *Tabas*-e-*Golshan* earthquake of September 16, 1978, of similar magnitude and intensity in the narrow epicentral area.

Considering the damaging effects of one of the most devastating earthquakes in Iran, Manjil–Rudbar Earthquake, with magnitude Ms = 7.4 and estimated epicentral intensity of IX–X degrees of MM Scale, caused 13,911 deaths, 36,693 people treated for injuries, and over 8000 people badly injured taken for hospitalization in Tehran. Damage and losses to the built environment were extremely high, estimated at 4.77 billion dollars in the densely populated region of the provinces of *Gilan*, *Zanjan,* and Eastern *Azerbaijan,* located in the western part of Alborz imbricated mountain belt of northwestern Iran.

The real acceleration lines (which are based on a real earthquake in the region) or hypothetical (which are based on a probable earthquake scenario) show the degree of risk and vulnerability of different areas. These maps will be the basis of design and planning decisions in the seismic-prone regions. So the accuracy and precision of these maps will guarantee the quality and reliability of the design decisions (Fig. 3.2).

The main event of June 20, 1990, was followed by a sequence of 5 strong aftershocks with magnitude Ms = 4.4–5.3 causing additional damage and landslides (Figs. 3.3 and 3.4).

The June 20, 1990, *Manjil* Earthquake was strongly felt over an area of 600,000 square kilometers, including Tehran, Tabriz, and several other major cities in Iran. Area of damaging effects in the provinces of *Gilan*, *Zanjan,* and Eastern *Azerbaijan*, with ground acceleration larger than 10% g, was estimated to 49,574 square kilometers affecting 3152 villages and 45 towns and cities in 24 districts with 98 rural districts in the province of Gilan as well as 16 districts with 90 rural districts in the province of *Zanjan*. More than 214,000 residential units, 1329 school buildings, over 300 health units as well as 82 medical centers and hospitals, a large number of agricultural land and facilities, religious and administrative units, service centers

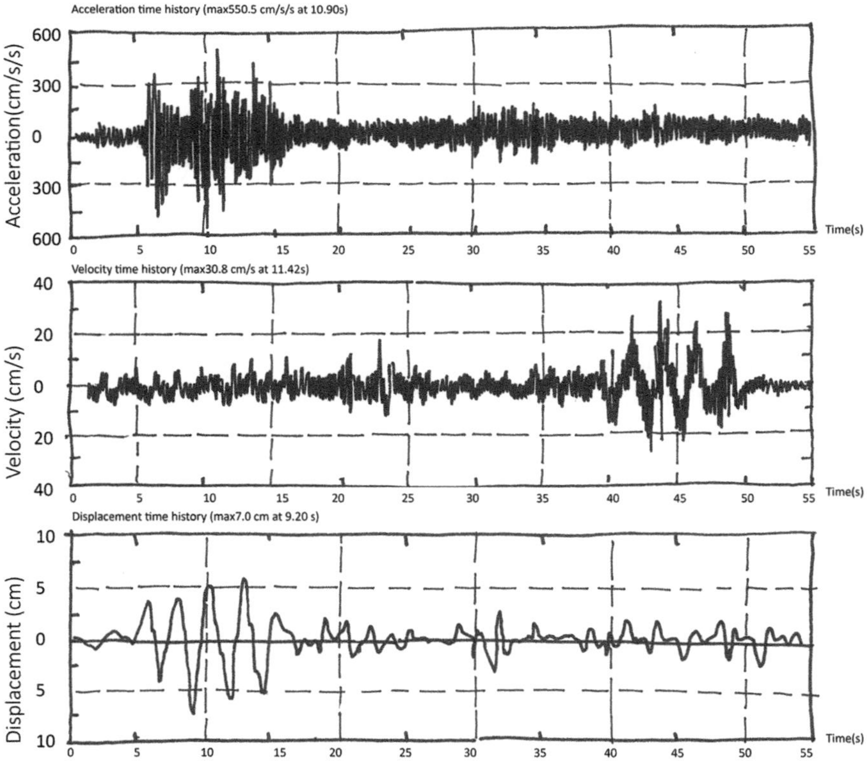

Fig. 3.2 Acceleration, velocity, and displacement time histories June 20, 1990, Manjil Earthquake, Record Abbar

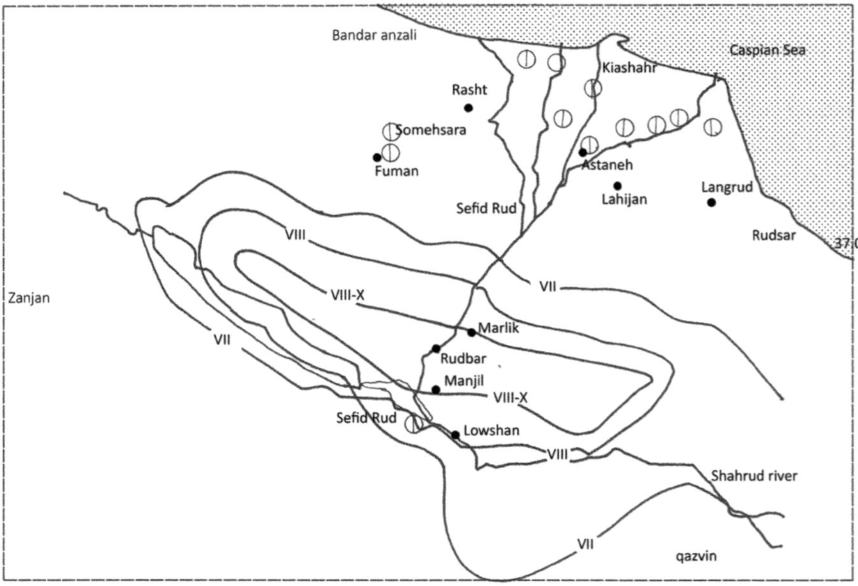

Fig. 3.3 Isoseismal map of Manjil Earthquake of June 20, 1990 (after Berberian and Qorashi 1991) Maximum ground accelerations

including 68 factories were destroyed or heavily damaged. Over 500,000 families were left homeless, and 178 village locations were abandoned due to landslides and other ground instabilities.

In the epicentral region with ground acceleration larger than 30% g with an area of about 8000 square kilometers, 383 villages and the towns of *Manjil*, *Rudbar*, *Loshan*, *Jirandeh,* and *Abbar*, most of the buildings were destroyed or damaged beyond repair. More than 67 landslides and many rockfalls were triggered in the wider epicentral region, causing destruction and damages to the transportation systems, water supply, electricity and other lifelines of regional and urban infrastructure. One of the most dramatic landslides triggered by Manjil Earthquake was the *Fatalak* landslide, where a huge mass of about one million cubic meters slide down the slope suddenly and all houses of the *Fatalak* Village were buried with 137 habitants. Only 14 persons were able to clue the disaster.

Based on the detailed earthquake damage classification and damage cost estimation of residential, educational, health, religious, administrative and other public buildings as well as industrial and services buildings, transportation and communication system, water, electricity, gas and oil supply systems and other lifelines of regional and urban infrastructure, economic losses were determined by sectors separately for the province of Gilan and Zanjan and summarized for the earthquake-affected region (Taleb and Alizadeh 1996) (Fig. 3.5).

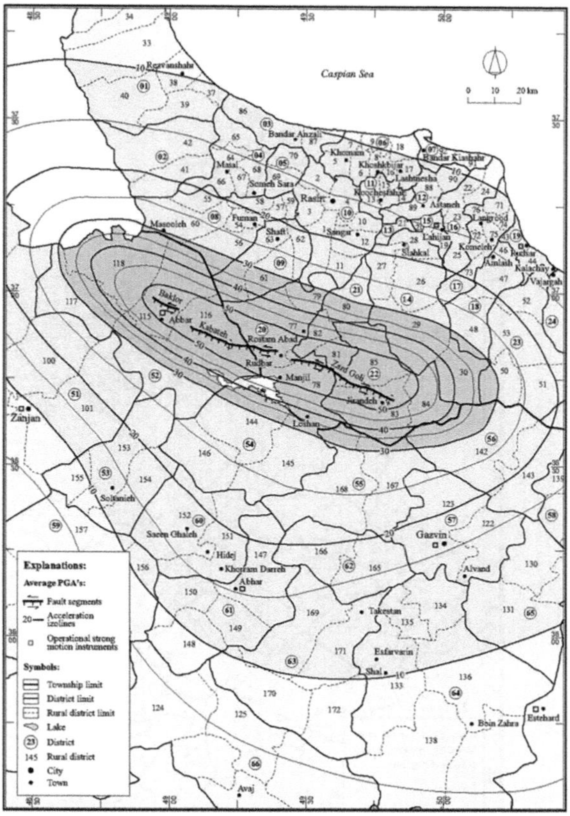

Fig. 3.4 Average maximum ground acceleration distribution due to June 20, 1990, Manjil Earthquake in the area of regional studies in Gilan and Zanjan provinces (in % of g). Based on recorded surface ground notions, attenuation, and damage distribution analysis

3.10 Rural Housing

Rural houses in Gilan Province: Timber-framed houses were more resilient than masonry structures, but many of the older and weaker timber houses collapsed. Tiles and heavier roofing materials were shaken from the sloped roofs, and many frames were severely distorted (Bahrainy et al. 1997) (Fig. 3.6).

The mountain villages of Zanjan Province: Some houses are built with a high level of craftsmanship and skill, even using the limited range of local materials. Such houses had little damage. Here adobe walls taper toward their top improving their stability, regular maintenance, and mud plastering prevents deterioration. The house in the center has the traditional horizontal timber reinforcement in the adobe walls. Rubble alone masonry houses on the left are nearly built with stone wedged together and smaller stone used toward the top (Fig. 3.7).

The mountain villages of Zanjan Province: Houses are made from a range of materials including rubble masonry, adobe masonry, fired brick, and some structural use of timber. House forms are similar and flat roofs predominant (Figs. 3.8 and 3.9).

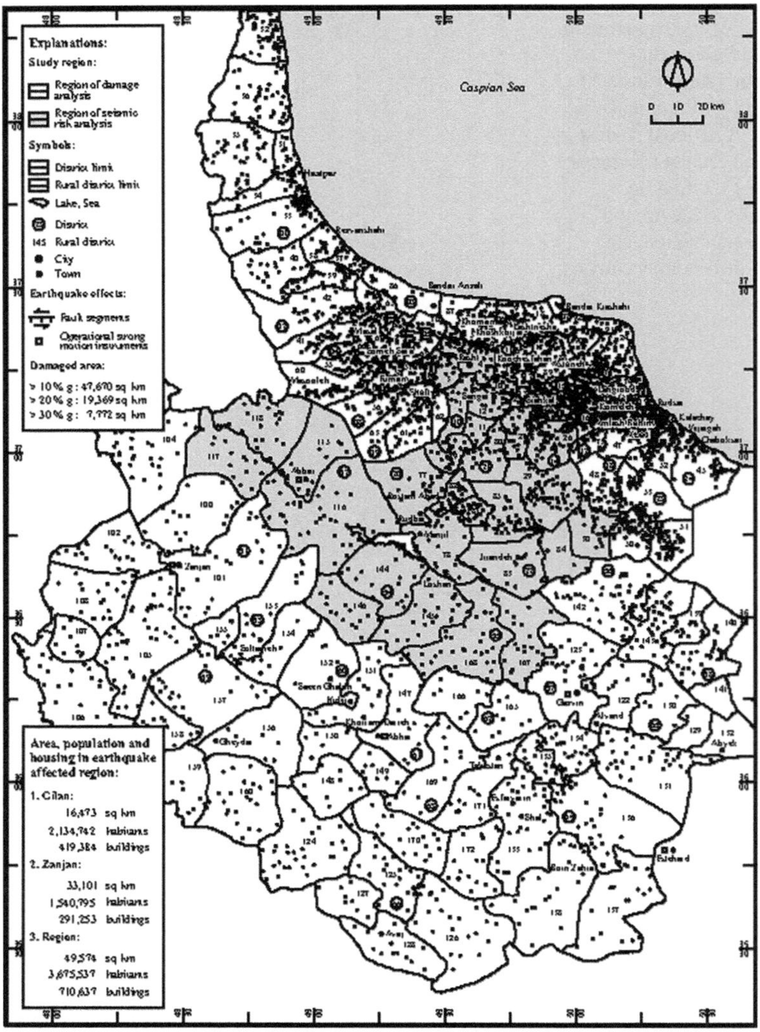

Fig. 3.5 Map of the region for damage distribution analysis and selected region for Seismic risk analysis and development planning

Damage of residential building in Rudbar, with intensive cracking and separation of the brick masonry infill walls of the steel frame structure due to disintegrated behavior of the flexible steel frame structure and baffle brick masonry (Fig. 3.10).

Failure of brittle brick infill masonry and intensive lateral of deformations on the second floor of steel frame structure of two-story building. First floor infill walls are less damaged (Figs. 3.11, 3.12, 3.13, and 3.14).

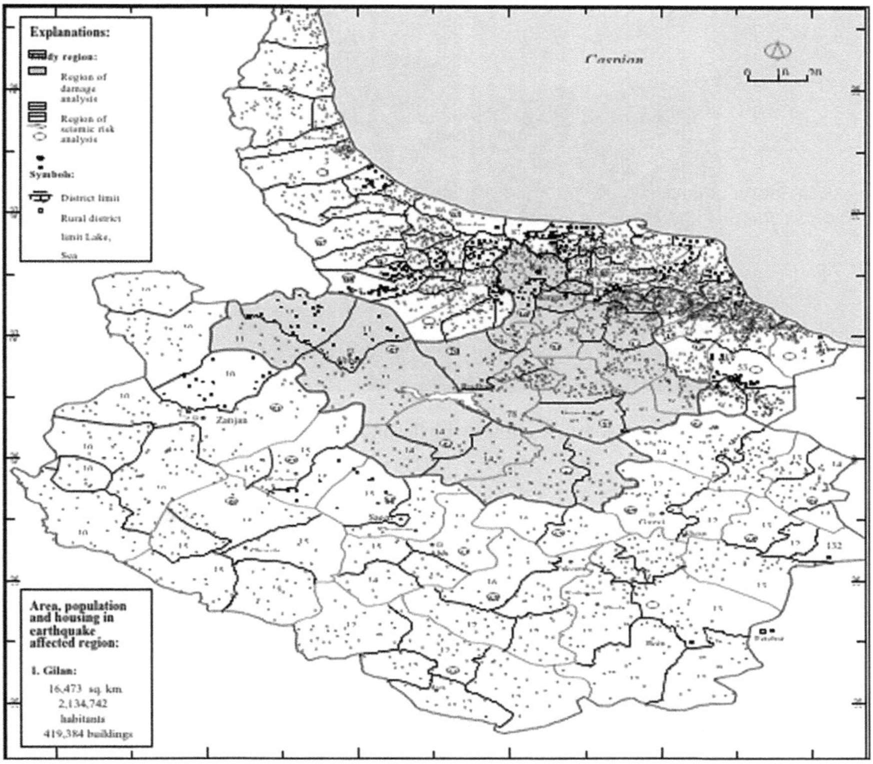

Fig. 3.6 Damage distribution of villages in rural districts of Gilan and Zanjan provinces, affected by June 20, 1990, Manjil Earthquake

Fig. 3.7 Timber-framed dwellings with sloped roofs in rural districts of Gilan Province (Coburn et al. 1990)

Fig. 3.8 Adobe and brick masonry dwellings with flat and heavy roofs in the rural district of Zanjan Province

Fig. 3.9 Upper mountain village in Zanjan Province. Dominantly of adobe dwellings with heavy flat roofs

3.11 The Area Affected by the June 20, 1990 Earthquake

The area is located in the mountainous region of the High Alborz-*Talesh* mountains, southwest of the Caspian Sea, that borders the northern side of the Oezel Owzan-Shahrud depressions. Strong earthquakes were frequently described striking the *Alborz* seismic belt in Northern Iran, but no evidence of active surface faulting has been reported for this region, and none of the historic earthquakes in the *Alborz* could be positively assigned to known active faults.

The mainshock of June 20, 1990, was followed by numerous aftershock activities. The largest aftershock have occurred on June 21 at 09:02:14 GMT and had a magnitude of Ms = 5.3, Mb = 5.8. Ln the following 12 h four aftershocks of

Fig. 3.10 Intensive cracking and dislocation of masonry infill walls and steel frame building in Rudbar

Fig. 3.11 Damaged steel frame two-story building with brittle brick masonry walls in Rudbar

magnitudes over 5.0 had occurred. They caused additional damage, casualties, and new landsliding and rockfalling. It caused 13,911 deaths, 36,693 people treated for injuries, and over 8000 people badly injured taken for hospitalization in Tehran. Damage and losses to the built environment were extremely high, estimated at 4.77 billion dollars in the densely populated region of the provinces of Gilan, Zanjan, and Eastern Azerbaijan, located in the western part of Alborz imbricated mountain belt of northwestern Iran. The main event of June 20, 1990, was followed by a sequence of 5 strong aftershocks with magnitude Ms = 4.4–5.3 causing additional damage and landslides.

Parts of the aftershock sequence of the June 20, 1990, earthquake were recorded by the Institute of Geophysics of the Tehran University, Atomic Energy Organization of I

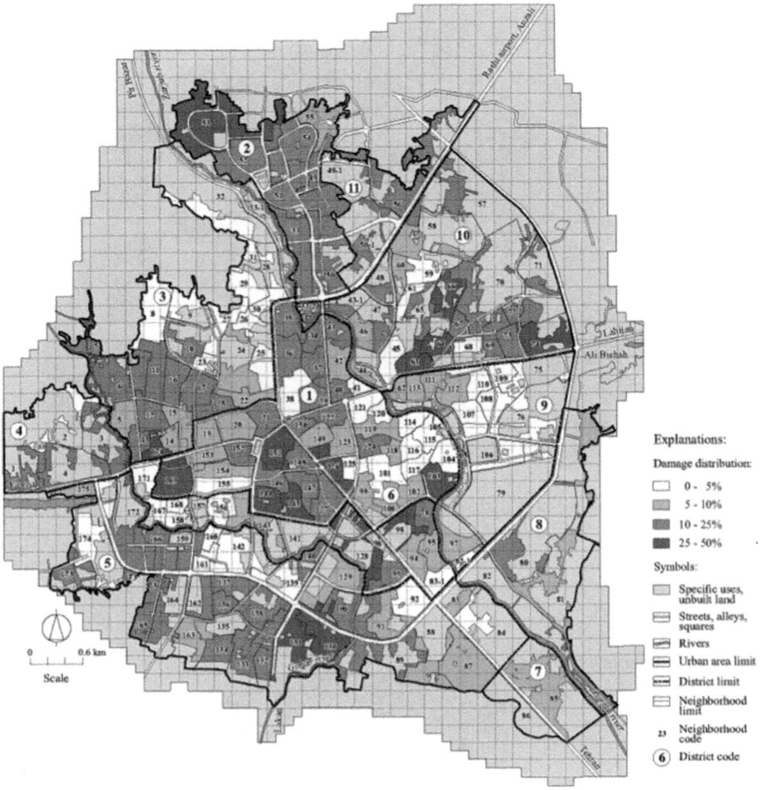

Fig. 3.12 Distribution of damaged dwellings in the neighborhoods of the city of *Rasht* due to June 20, 1990, *Manjil* Earthquake

ran, and research groups from China, Japan, and Russia using portable seismographs. The results of these studies were not available to be included in this report.

The Alborz imbricate mountain belt constitutes a broad arch of parallel folds, reverse faults, and nappes forming major imbricate structure which border the southern edge of the South Caspian Depression. They face the Depression on the north, and to the south grade into the plateau of Central Iran. Several surficial and non-tectonic features were developed along the steep slopes of the High Alborz, which made a great of difficulty to pinpoint the causative fault of the June 20, 1990, earthquake.

All mapped major faults, such as Manjil, Kelishom, Poshtkuh, and Deylaman (Berberian and Qorashi 1991), were carefully checked in the field, and no evidence of reactivation was found. Evidence of reactivation was found for some minor faults north of Abbar, as well as for a section of the Manjil thrust (west of Sefid Rud Dam) which moved due to the gravity and slided for about 30 cm down-slope of the Alborz imbricate mountain belt. Based on these field observations, Berberian and Qorashi (1991) (Fig. 3.2) considered an en-echelon strand consisting of three

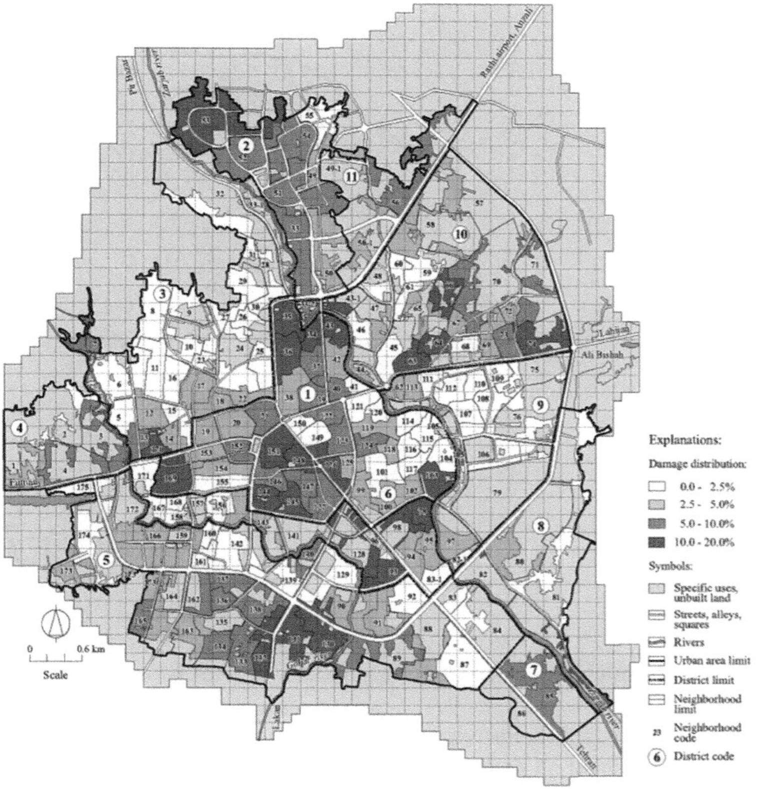

Fig. 3.13 Distribution of heavy and severe damaged Dwellings in the city of Rasht

segments stretching in N95E to N120E direction (Baklor fault-northwest, Kabateh fault-central, and Zard Goli fault-southeast) as a complex geological feature that caused the June 20, 1990, earthquake. This conclusion is supported by field observation of maximum displacements of about 60 cm in horizontal and of about 95 cm in vertical direction along the southeastern (Zard Goli) and central (Kabateh) segment.

In the same region, an E–W surface faulting parallel to already mapped Rudbar fault, passing about 300 m north of the Sefid Rud Dam and extending from Gilvan in the west to Pakdeh in the east is mapped by Moinfar and Naderzadeh, 1990 as a causative fault. Along with this 80 km long fault structure, observed are right-lateral displacements of about 20 cm and vertical motion of about 50 cm.

The most dominant tectonic structure in the region is the Manjil fault. Based on neotectonic field investigations (Jancevski and Qorashi 1998) outlined this fault as a tectonic border between the uplifting Alborz mountain block and the Manjil depression. Although there is no hard evidence about the movement during the June 20 earthquake, except for the section of the Manjil thrust west of Sefid Rud Dam which moved for about 30 cm downslope of the Alborz imbricate mountain belt,

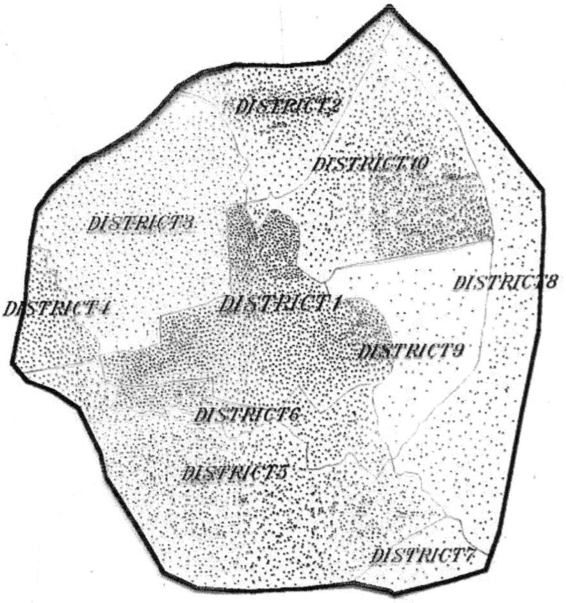

Fig. 3.14 Total damaged buildings in ten districts of the city of Rasht due to 1990 *Manjil* earthquake (Bahrainy 1992)

this fault was considered as a carrier of the June 20, 1990, seismicity (Jancevski and Qorashi 1998).

Although there is still controversy about the cause of the June 20, 1990, earthquake, it is obvious that all listed tectonic structures are close to each other and are in a very narrow strip within the epicentral region. However, due to the above-listed facts, the existing controversy is not affecting significantly studies for seismic hazard assessment and formulation of consistent earthquake scenarios applicable in seismic risk analysis and earthquake protection planning.

3.11.1 Landslides

The landsliding was another type of ground failure during earthquakes which is associated with loss of human lives and properties, Oddly, however, this issue has not arrested much attention among engineers because many slides have taken place in remote and even inaccessible mountain areas and consequent impact on human activities has not been seriously recognized, However, sprawl of urbanization to the areas of hillsides and mountains has exposed local villages and towns to the potential menace caused by landslides during earthquakes.

One of the most dramatic landslides triggered by *Manjil* Earthquake was the *Fatalak* landslide, where a huge mass of about one million cubic meters slide down the slope suddenly, and all houses of the *Fatalak* Village were buried with 137 habitants

(Fig. 3.17). The other major landslide that occurred by the *Manjil* earthquake was *Galdian* landslide (Fig. 3.15 and 3.16).

Generally speaking, there are two types of landsliding induced by earthquakes, one is the sliding which occurs all of a sudden simultaneously with seismic shaking. This type of collapse is induced by the inertia force due to seismic shaking and has a tendency to occur over steep slopes at the shoulder of hilltops or near the top of mountains. When the place of such a topographical prominence is covered by layers of disintegrated rocks or transported soils, the chances of destruction are very high by landslides in that it occurs all of a sudden and it involves a large amount of soil or rock masses. Sometimes, the slide turns into a catastrophic debris flow, if the sliding mass contains a large amount of water. At the time of the *Manjil* earthquake, this type of landslides occurred in the village of Fatalak, Tootkabon and also in the mountain area north of Backlor. Several hundred residents were buried in the sliding rubble, and loss of properties was really appalling.

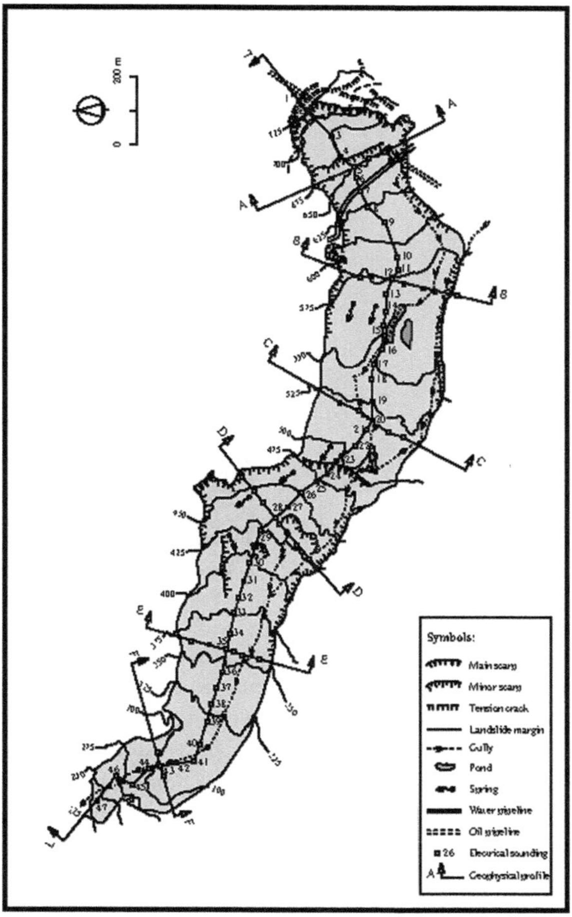

Fig. 3.15 Topographical map of the *Galdian* landslide

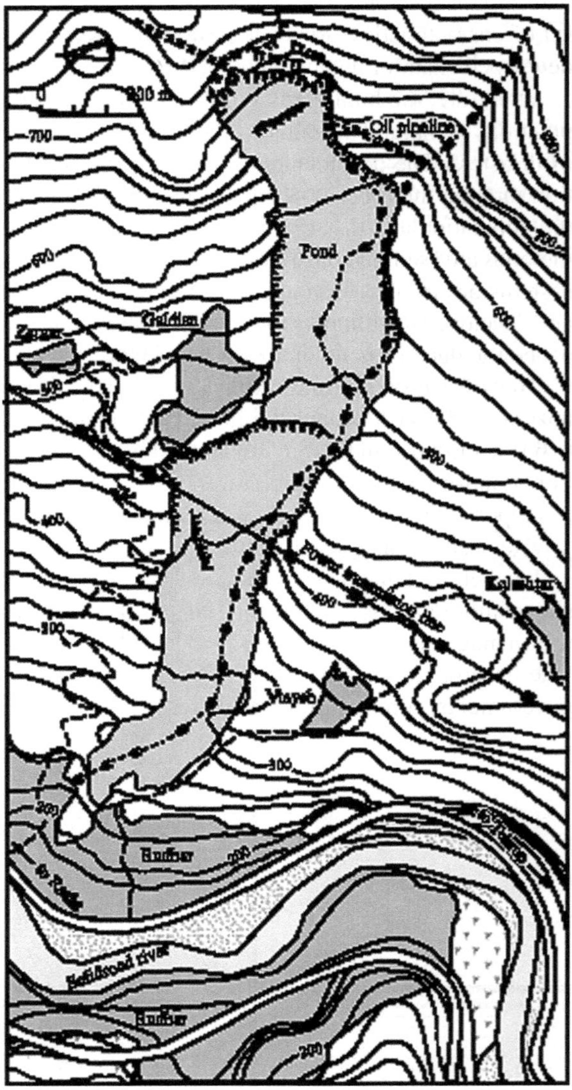

Fig. 3.16 General topographic map of the location of the *Galdian* Landslide

The second type of landsliding is the one which occurs in a quiet period of time several hours or several days after an earthquake. This type of landslide is sometimes a ground movement due to the reactivation of old dormant landslide. Groundwater is generally associated with the provocation of such a slide. It occurs slowly and signs of land movement are always detected well in advance and hence casualty is generally small. But the loss of properties and damage to public facilities are great because of a large amount of soils involved in the slide.

In some cases, a severe shaking during earthquakes is likely to change the groundwater conditions probably due to the small cracks or fissures produced in soils or

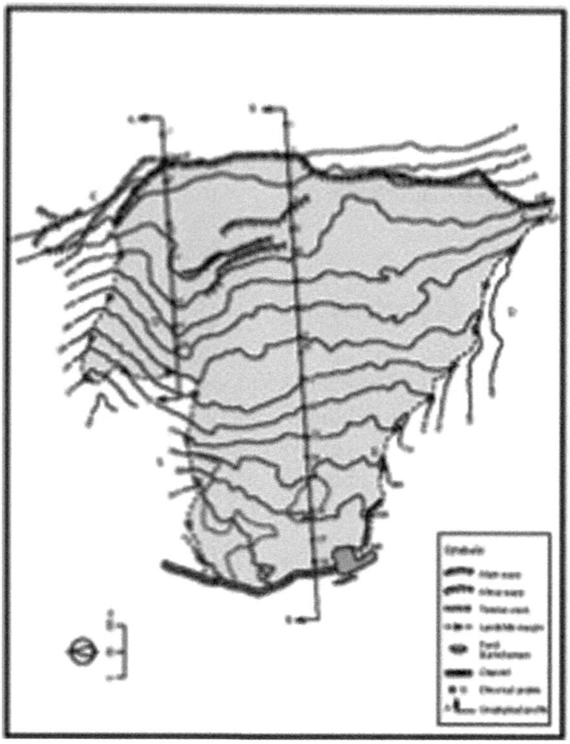

Fig. 3.17 Topographic map of the *Fatalak* landslide after *Manjil* earthquake

rocks underlying slopes. Such a change in the groundwater regime tends to weaken soil deposits or increase pore water pressures, leading to instability of soil masses and land movement which takes place some days after earthquakes The colossal landslide in the village of *Galdian* at the time of the *Manjil* earthquake may be deemed as an example of this type of landslide.

3.11.2 Rockfalls

Rockfalls are still other type of geotechnical hazard brought about by an earthquake. The rockfalls often occur on steep slopes, mountain walls, and road cuts. The amount of rock masses involved is not generally so much, but because of its sudden occurrence during main shaking of an earthquake, it is generally impossible to elude the danger of falling rocks, if it occurs on the spot. Major issue of importance in the rockfall hazard is blockade of roads or highways, paralyzing the traffic which is often fatal for relief and rescue operation. The physical mechanism for occurrence of rockfalls is poorly understood, because of too many factors or conditions involved in individual cases of rock collapse. It is likely that highly fractured rocks or overhung rock masses tend to fall and crumble. If there is heavy rainfall preceding earthquakes, water tends to seep into fissures or cleavages, thereby losing the stability of rock masses.

In the case of the *Manjil* earthquake, rockfalls occurred at numerous places and blocked roadways and passes in the epicentral area. The stoppage of traffic in the trunk line between Rasht and *Manjil* brought about considerable inconvenience for the mass transportation.

3.11.3 Liquefaction

Liquefaction of loose sand and silt soil layers water-saturated due to high level of water table, occurred on an area of 650 square kilometers of the Caspian plain with northeast-oriented strip from *Sangar* on the west almost to the cost of Caspian sea (Fig. 3.18), under the range of ground accelerations of 10–20% g on a distance of 50–80 km from the causative fault of June 20, 1990, *Manjil* Earthquake (Ishihara K

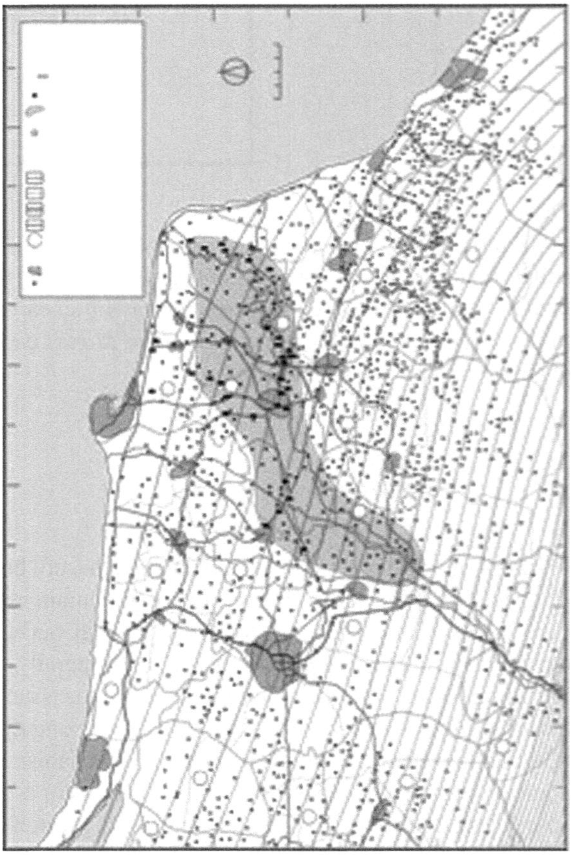

Fig. 3.18 Area of liquefaction in the Gilan plain due to June 20, 1990, Manjil Earthquake and average maximum ground acceleration Distribution (after Haeri 1990, 1998, and after Petrovski et al. 1998)

et al. 1991, Ishihara K 1998; Haeri 1998). Soil liquefaction caused extensive damage to the buildings in the districts of *Astaneh* and *Rudbaneh*, agricultural land and farms in the wider region as well as extensive deformations and damage to the regional and urban infrastructure.

Liquefaction caused extensive damages in a vast area of fluvial deposit of *Sefidrud* and its tributaries. The distance between observed liquefaction in this region and the epicenter is between 50 and 90 km. The most extensive damage occurred in *Astaneh* and *Rudbaneh* some 75 km northeast of epicenter.

Damage in *Astaneh*: Within the city of *Astaneh*, liquefaction caused extensive damages to buildings in two particular parts of the city as shown in Fig. 3.19 Within these two particular parts, the houses collapsed, torn apart and experienced differential settlement. Sand boiled from room floors (Fig. 20), water wells (Fig. 21), and everywhere that could penetrate to release the earthquake-induced pore water pressure.

Damage in *Rudbaneh*. Damage incurred in *Rudbaneh* was almost of the same intensely as that of *Astaneh*. The main road passing through *Rudbaneh* is mainly

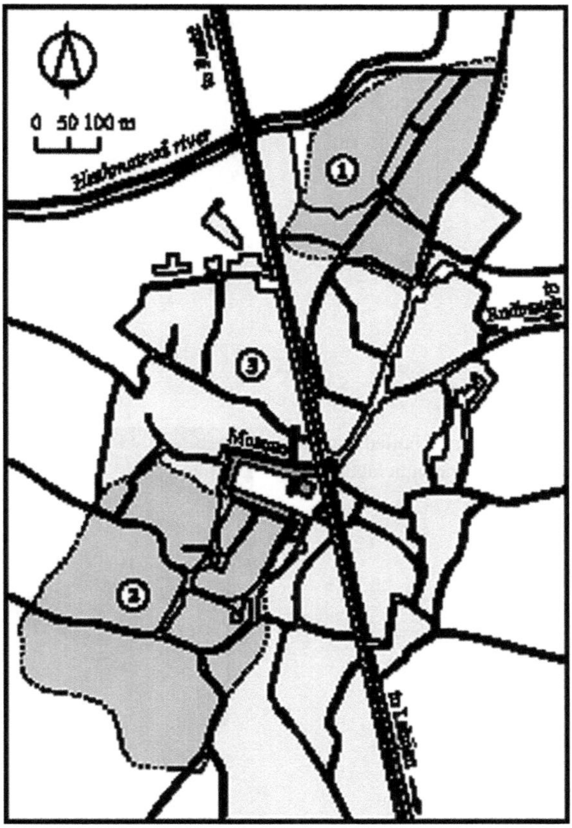

Fig. 3.19 Zones of induced liquefaction in the town of Astaneh by June 20, 1990, Manjil Earthquake

parallel to a river located at the north part of the village. The soil condition in *Rudbaneh* is also similar to the general soil condition of places with the signs of liquefaction in this area; i.e., a top layer of maximum 2 m of clayey silt overlaying loose sand.

Liquefaction of loose sand (Haeri 1998**).**

See Figs. 3.19, 3.20, and 3.21.

From the damage distribution pattern in the city of *Astaneh* and its suburbs, we can conclude that:

1. The conventional wooden rural housing is the best type of the building in such a liquefaction-prone area.
2. The buildings with an integrity in foundation and structure systems behave much safer compared to the buildings with spread footings. Therefore, buildings consisting of spread footings should be avoided at such sites.
3. If the unliquefiable soils are thick enough, the liquefaction may not show at the surface. Therefore, thickening of the unliquefiable soil near the ground surface

Fig. 3.20 Intensive deferential settlement and ground floor slab damage due to liquefaction

Fig. 3.21 Sand boiled water well due to the liquefaction in Astaneh

may be one of the solutions for building construction on a liquefiable soil at this site.

4. There are other countermeasures for liquefaction ground damage; however, any of these measures should be verified for this site with further studies.

Post-earthquake damage evaluations should be organized so that teams may rapidly use a systematic methodology. Basic information from these evaluations should enable local and national governmental authorities to make critical decisions and also to employ economically justified and technically consistent seismic risk reduction measures in a uniform manner for the entire country. If coordinated efforts are made to use the uniform methodology presented in this Part of the Guidelines, more practical and transferable data can be developed that will be of potential use in the seismically active regions of the world.

Post-earthquake damage evaluations should be organized so that teams may rapidly use a systematic methodology. Basic information from these evaluations should enable local and national governmental authorities to make critical decisions and also to employ economically justified and technically consistent seismic risk reduction measures in a uniform manner for the entire country. If coordinated efforts are made to use the uniform methodology presented in this Part of the Guidelines, more practical and transferable data can be developed that will be of potential use in the seismically active regions of the world.

3.12 Uniform Methodology

Principal elements of this uniform methodology and procedure for post-earthquake damage evaluation are presented here. These elements include:

1. Damage and usability classifications for buildings.
2. Procedures for and organization of data collection.
3. Earthquake damage data analysis and data bank organization.
4. Estimation of economic losses, and human fatalities and injuries.
5. Measures for reducing adverse earthquake consequences and for mitigating seismic risk. Connected with these principal elements are earthquake damage evaluation, development of empirical vulnerability, or damage cost functions and analysis of earthquake damage distribution (see Table 3.1).

3.13 Components of Lifeline Systems in the Region Affected by June 20, 1990, *Manjil* Earthquake

The damages to lifeline systems were directly associated with the ground shaking and ground failures. Utilities suffered severe damage due to permanent ground movements and surface faulting. Rescue operations were seriously interrupted by debris

Table 3.1 Empirical performance evaluation and earthquake damage and usability classification of buildings

Damage category and performance level	Damage description	Requirements	Usability category
1. No damage safe occupancy	Without visible damage to the structural elements. Possible fine cracks in the wall and ceiling mortar. Hardly visible non-structural and structural damage	Buildings classified in damage categories 1 and 2 are with decreased earthquake-resistant capacity and do not pose danger to human life. Immediately usable or after removal of local hazard (cracked chimneys, attics or gable walls)	Due. I usable Light non-structural damage, very isolated or negligible structural damage
2. Light damage Operational	Cracks to the wall and ceiling mortar. Failing of large patches of mortar from wall and ceiling surface. Considerable cracks, or partial failure of chimneys, attics, and gable walls. Disturbance partial sliding, sliding, and falling down of roof covering. Cracks in structural members		
3. Moderate damage Life safety	Diagonal or other cracks to structural walls, walls between windows, and similar structural elements. Cracks to reinforced concrete structural members: columns, beams, R.C. walls. Partially failed or failed chimneys, attics or gable walls, disturbance, sliding and falling down of roof covering	Buildings classified into damage category 3 are with significantly decreased earthquake-resistant capacity. Limited entry is permitted, unusable before repair and strengthening. Need for supporting and protection of the building and its surroundings should be considered	Due. II Temporary unusable Considerable structural damage but yet repairable structural system

(continued)

Table 3.1 (continued)

Damage category and performance level	Damage description	Requirements	Usability category
4. Heavy damage Collapse prevention	Larger cracks with or without disattachment of walls with crushing of materials. Larger cracks with crushed material of walls between windows and similar elements of structural walls. Large cracks with small dislocation of R.C. structural elements: columns, beams, and R.C. walls. Slight dislocation of structural elements and the entire building	Building classified into categories 4 and 5 is unsafe with possible sudden collapse. Entry is prohibited. Protection of streets and neighboring buildings or urgent demolition required	Due. III unusable collapsed, partially or totally collapsed structural system
5. Severe damage Partial or complete collapse	Structural members and their connections are extremely damaged and dislocated. A large number of crushed structural elements. Considerable dislocations of the entire building. Partially or completely collapsed buildings		

Guidelines for Earthquake Disaster Management, Volume IV, Part 4 resulted from rockfalls and large landslides. Liquefied soil damaged foundations and resulted in disposition of irrigation canals, broken pipelines, cracked pavements, and filled water wells with boiled sand. After the earthquake, the affected area experienced closures of the roads, suspension of electric power, water and gas supply systems, shut down of telephone lines, and many other failures in the lifeline systems. The impact of disrupted services on the affected communities has been one of the major characteristics of this earthquake.

3.14 Damage to Transportation System

Transportation system of the region comprising roads, bridges, and tunnels suffered severe damage in this earthquake. Landslide rockfalls, and other earthquake-induced ground failures were the primary causes for the interruptions in the highway system. During the earthquake, steel bridges and tunnels performed fairly satisfactory while reinforced concrete bridges sustained some damages. The damaged structures and debris caused interruptions in rescue operations right after the quake, even cleared within several days after the event.

Roads: The extent and types of damages of roads due to the mainshock and aftershocks of *Manjil* Earthquake were primarily different and dependent on the position of the road section. Namely, road sections passing through wide and flat areas mainly damaged due to the ground settlements as a result of liquefaction or other types of seismically induced ground failures. Contrary, the road sections passing through steep hilly and mountainous regions were much severely damaged due to the local and global soil and rock slope instabilities. In total, more than 1200 km rural roads were needed improvement or reconstruction. Modes of failure and damage on roads with examples of slope failure and damage of road pavement and bridge abutments were observed in June 20, 1990, *Manjil* Earthquake.

Bridges: Bridges are usually vulnerable to many earthquake hazards and represent the most vital links in transportation and other lifelines. There were three types of bridges, built more than 20 years ago, in the affected area: steel truss bridges, concrete girder bridges, concrete slab, and old brick masonry arch bridges. Two of the old arch bridges sustained severe damage. Total of 10 bridges, total length each ranged from 30 to 800 m were investigated after the earthquake (Eshghi 1990). In general, steel and reinforced concrete bridges performed satisfactorily. Damage to bridges due to the earthquake resulted mainly from:

1. Severe disruption and settlement of the bridge approach embankment and abutment fills due to the seismic shaking (e.g., Bala-Bala bridge).
2. Liquefaction and lateral spreading of foundation soil (e.g., slab bridges).
3. High acceleration is induced in the bridge structures (e.g., slab bridges).

Damage to superstructures, those to bearing supports, e.g., failures of shoes, breakage of pins, protrusion of rollers, expansion devices and also pounding of longitudinal girders against each other or against abutments, were most prominent.

Some shear cracks occurred at the edges of the bridge seats near the shoes. Differential movements damaged sidewalks, service conduits and guard/handrails (Eshghi 1990).

Tunnels: Along *Ghazvin–Rasht* highway, there were seven tunnels with a total length of about 2 km which effectively resisted damage (Fig. 3.22).

To effectively reduce seismic risk and the vulnerability of the environment to future earthquakes in the world's high-risk zones, there is an urgent need for a comprehensive and cohesive system of planning and design to integrate, facilitate, complement, and strengthen decisions at various levels and different aspects, eventually leading to a safer environment. The catastrophic *Manjil* Earthquake of June 1990 in Northern Iran, in spite of tremendous human and property losses, provided a unique opportunity to study and analyze, among other areas, the role urban planning and design can play in effectively mitigating seismic risk in the urban areas of the seismic-prone regions of the world. In this book, urban planning and design activities to reduce seismic risk will be discussed, and reference will be made to the case of the City of Rasht in Northern Iran.

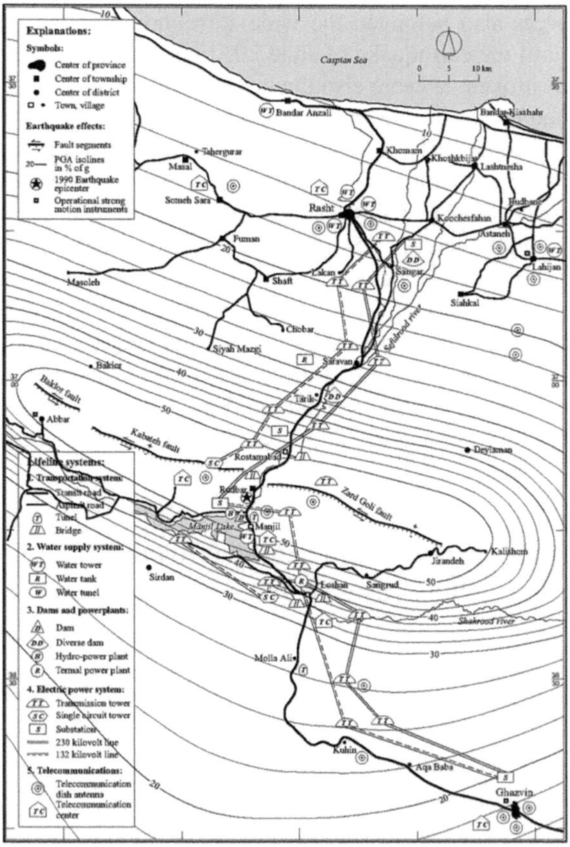

Fig. 3.22 Components of lifeline systems in the region and maximum ground accelerations due to June 20, 1990, earthquake (lifeline components after Eshghi 1992, Ground accelerations after (Petrovski et al. 1998)

Over 500,000 families were left homeless, and 178 village locations were abandoned due to landslides and other ground instabilities. In the epicentral region with ground acceleration larger than 30% g with an area of about 8000 square kilometers, 383 villages and the towns of *Manjil, Rudbar, Loshan, Jirandeh,* and *Abbar*, most of the buildings were destroyed or damaged beyond repair. More than 67 landslides and many rockfalls were triggered in the wider epicentral region, causing destruction and damages to the transportation systems, water supply, electricity, and other lifelines of regional and urban infrastructure.

Most contemporary criteria for earthquake disaster management have been implemented and formulated in the third cycle of development of seismic risk analysis. Regional and urban development planning has been studied in seismic areas as well as the required level of damage control and acceptable level of seismic risk, based on empirical and qualitative performance analysis.

Iran, as very few countries throughout the world, is exposed to strong seismic effects almost along the entire territory of the country. In this part, well-known is a great number of areas that have been affected by occurrences of strong earthquakes, ever since the oldest historic times up until today. However, there are also areas of observed seismic activity but with intensities not of a catastrophic nature. To such areas also belonged the wide surrounding of the Loshan—Manjil—Rudbar area, until the earthquake of June 20, 1990, when in a very short time these towns were destroyed, as were also the numerous village settlements over the territory of Gilan and Zanjan provinces. The earthquake of June 20, 1990, just like the other earthquakes on the territory of Iran, is a classical example of a tectonic earthquake being caused as a result of the contemporary tectonic processes in the earth's crust.

It is the first time that an attempt has been made to prepare an integral series of guidelines for earthquake disaster management, covering the needs of government officials, planners, economists, architects, engineers, and scientists, with the purpose of planning for seismic risk reduction and the practical implementation of methodologies and findings in earthquake-affected regions. This series of guidelines, developed within the project, is therefore expected to be immensely beneficial, not only to the Islamic Republic of Iran, but also to other countries in earthquake-prone regions, particularly in the developing world (Bahrainy 1998).

3.15 Structural Damages Recorded as a Result of June 20, 1990, May Be Classified as Follows

- Foundation failures.
- Cracks in walls.
- Wall separation at corners.
- Cracks around openings.
- Sliding of roof over walls.

- Collapse of walls.
- Collapse of buildings.

3.15.1 Categories of Deficiencies

Based on site visits, visual inspection photographic records, sampling and testing of affected materials and analysis of reported results, the deficiencies are categorized as follows:

Design, detailing, specification deficiencies
Defective materials
Workmanship deficiencies.

3.15.2 Design, Detailing, and Specification Deficiencies

The following deficiencies were noted:

- Wrong positioning of door/window opening.
- Large size openings separated by walls that are too slender.
- Insufficient wall thickness.
- Poor connection details.
- Columns that are not tied.
- Roof that is not anchored to the wall.
- Absence of bracings.
- Absence of tie beams.
- Poor detailing of connections.

3.15.3 Case Histories Which Formed the Basis of the Observations Are Presented Below

Size of wall openings: The collapse of a structure with relatively large door/window openings can be frequently seen. The sum of the two openings usually occupies about two-thirds of the length of the wall. This is not desirable. It should be limited to about one-third of the length of the wall, according to internationally accepted practice. The large openings obviously contribute to the failure.

3.15.3.1 Insufficient Wall Thickness

One of the buildings in *Hassanjani Shalegh* Village consisted of load-bearing brick walls. The height of the wall was about 5 m. The thickness of brickwork was 22 cm.

A separate facing brick wall of 10 cm thick was built without any connection or bond with the structural brickwork. The mortar joints were not of the required quality. The brick bond was incorrect. The failure of the structure has been assessed as due to insufficient wall thickness and other deficiencies.

3.15.3.2 Poor Connection Details

In the case of structural steel framed buildings, detailing of the joints was found to be deficient in many cases, even during reconstruction. Plates 4 and 5 show details of structural steelworks under construction at Rostam Abad for a two-story structure. It may be noted that the detailing of connection is far from satisfactory. Standard details are readily available in reference books.

3.15.3.3 Columns not Tied

In a number of buildings that were inspected, it was found that columns were not properly tied together, leading to weakening.

3.15.3.4 Roof not Anchored to Wall

Some structures in the *Jamshid Abad* area suffered sliding of roof and displacement over walls due to inadequate anchoring to walls. The consequent shear force damages the walls.

3.15.3.5 Absence of Tie Beams

The absence of tie beam or sills at window level is noted. A number of structures at *Jamshid Abad* suffered damage due to the absence of tie beams.

3.15.3.6 Defective Materials

Materials studied at the various sites included the following: adobe, mud mortar, and lime–soil mortar; cement mortar and lime–cement mortar; bricks and concrete blocks; concrete and reinforcement and structural steel. Samples were taken wherever possible and appropriate, and sent to laboratory for testing. In addition, cement, lime, and pozzolana samples were obtained from other sources for tests.

Soon after the most severe earthquake of the century in Iran, in the northern provinces of *Gilan* and *Zanjan*, a quick survey, about the nature of the damages caused to nearly 214,000 buildings, was carried out. It revealed that almost all kinds of buildings had suffered severe damage.

3.15.4 The Main Reasons Could be that

The construction practices were faulty, and the quality of building materials did not conform to standard specifications, or a combination of both the above factors.

Whatever the reason, the fact remained that in the entire area the most common constructions were of wood-framed skeletons and light roofs, and buildings of sun-dried bricks or adobe, and stone and mud with heavy roofs, which hardly followed any modern design principles.

3.16 Seismic Hazard Analysis

The main objective of the seismic hazard analysis is aimed at formulation of seismotectonic characteristics and modeling of seismogenic sources for seismic hazard analysis in the central part of Northern Iran, as an example of the implementation of the methodology of morpho-structural analysis. The geological, tectonic, neotectonic, and seismotectonic characteristics of the central part of Northern Iran, with particular emphasis on the provinces of *Gilan* and *Zanjan*, have been studied on the basis of data and results derived from previous geological, tectonic, neotectonic, and seismological studies in combination with the results and findings obtained from additional field investigations of the *Loshan—Manjil—Rudbar—Lahijan—Rasht* region.

Basic objectives in earthquake protection planning and determination of earthquake parameters for seismic risk analysis on regional level and performance of the analysis for seismic microzonation studies of selected urban areas using recommended records of earthquake time histories are discussed and presented in… with the maps of Planning Scale Earthquake (PSE), Maximum Considered Earthquake (MCE), and Frequent Scale Earthquake (FSE) as well as simulated June 20, 1990, *Manjil* Earthquake as basic earthquake scenarios for analysis of seismic risk in earthquake protection and disaster preparedness planning. Formulation of design criteria for buildings and selection of design earthquakes have been discussed also, considering relevant influence of the ground motions and entire building—foundation—soil system, with presentation of the guidelines for earthquake-resistant design of building structures considering structural systems, structural materials, and non-structural components.

During June 20, 1990, Manjil Earthquake in the affected region and wider, more than 23 strong motion instruments were triggered and relatively good quality records have been obtained on a distance from 8 to about 200 km from causative fault with maximum acceleration amplitudes in the range of 5–55% g. Most significant 45 components of the 15 triggered strong motion instruments have been analyzed in all details and presented in the project technical report as working material. Out of these, only three specific records of *Abbar, Lahijan,* and *Ghazvin* stations, considering particularly their frequency content have been recommended based on the criteria of frequency ranges carrying dominant portions of energy with analysis of three cutoff levels of 30, 50, and 70% on smoothed and normalized Fourier amplitude spectra.

Basic reasons for recommending selected earthquake time histories for determination of earthquake parameters for planning and design are discussed here, considering formulated design criteria of existing and recommended building typology as well as dominant soil conditions of the recording sites, the affected region and analyzed urban areas.

Dominant part of the existing buildings and structures, lifelines and other man-made facilities have been built in the past without implementation of recently developed effective measures for earthquake protection and will be not realistic to expect that upgrading—preventive programs will be implemented in near future if substantial development programs will not be created. Furthermore, recent catastrophic earthquakes in highly developed countries have shown that transportation and other lifeline systems, most essential for the implementation of disaster preparedness and emergency relief operations, can be highly vulnerable and thus not functional even under moderate earthquake shaking. Consequently, it would be most realistic to expect that developed and developing countries will be able to give priority to upgrading—preventive measures only to the essential facilities for disaster preparedness and emergency relief like hospitals and lifeline systems, assuring their continuous functioning and serviceability in post-disaster conditions.

For planning and undertaking measures for earthquake disaster protection within the phases of disaster prevention, preparedness and relief and reconstruction and rehabilitation, an understanding of the potential damage probable earthquakes is of fundamental importance to determining expected levels of environment exposure, and assessment of damage and integral seismic risk. More recently, there have been significant attempts to describe the potential effects of earthquakes and present their impact through scenarios on regional and urban levels in order to provide appropriate basis for the reduction of seismic risk incorporated in development planning and earthquake disaster preparedness organization. Geological, tectonic, seismological, seismotectonic studies, and seismic hazard analyses are used successfully to predict appropriate scenario earthquakes considering expected earthquake effects on the existing and planned elements at risk with formulated level of acceptable seismic risk depending on the development level and economic capacity of the country, region or urban area.

Here the parameters for seismic risk analysis and planning of earthquake protection as well as disaster preparedness planning on the regional level are presented for three earthquake scenarios of Planning Scale Earthquake (PSE), Maximum Considered Earthquake (MCE), Frequent Scale Earthquake (FSE) as developed in Part 2 of this volume, and simulated June 20, 1990, *Manjil* Earthquake.

Preliminary seismic zoning map of Planning Scale Earthquake in the provinces of *Gilan* and *Zanjan* could be considered suitable for the intended use of seismic risk analysis and planning for reduction of seismic risk, i.e., estimation of loss potential of existing and planned building typology as well as regional and urban development planning and design in the region of Gilan and Zanjan provinces.

For planning of development and upgrading of essential facilities important for emergency response such as hospitals, schools, high-rise buildings, fire stations, regional and urban infrastructure including lifelines such as water, electricity, gas and oil supply systems, transportation and communication systems, seismic zoning map of Maximum Considered Earthquake (MCE) for the provinces of *Gilan* and *Zanjan* is elaborated based on seismic hazard analysis. It is very important to recognize that even under such severe earthquake ground motions the essential facilities mentioned above should remain functional in order to fulfill the requirements of emergency response in the region and particularly densely populated urban areas.

3.17 Formulation of Design Criteria for Buildings and Design Earthquakes

The philosophy of earthquake-resistant design for buildings other than essential facilities has been well-established and proposed to prevent non-structural damage in frequent minor earthquake ground shakings, prevent structural damage and minimize non-structural damage in occasional moderate earthquake shakings, and avoid collapse or serious damage in rare major ground shakings. This philosophy is in complete accord with the concept of comprehensive design and damage control design.

To design a building to resist the effect of severe earthquake ground motions necessitates consideration of all possible severe excitations that might occur during a building's service life. Under severe excitation, strength and deformation capacity will depend on the damage (residual state of stress and strain) accumulated from previous medium or severe excitations caused by natural hazards such as earthquakes, windstorms, floods, and other severe events such as blasts and fires. Predictions of a building's response to any environmental hazard require that the static and dynamic characteristics of the resulting excitations be determined (the intensity and its variation with time while acting on a building). In cases involving wind, the intensity and its variation can be defined with acceptable accuracy, given the site and shape of the building; the problem is more complicated with earthquakes since the actual characteristics of the excitations depend on the building's response.

For a given site and known structure, there is no unique design earthquake, and different considerations may be needed to develop design earthquakes for serviceability, damage control, and collapse limit states. The interrelationships between the design earthquake and the requirements of different limit states and other aspects of seismic-resistant design as discussed by have shown the need for a comprehensive approach to the problem of establishing design earthquakes. Specification of design earthquakes must include detail characteristics of the entire building foundation soil system, earthquake risk and related costs, and the extent of damage that can be tolerated.

In a comprehensive approach to the design of structures, it is first necessary to establish the design criteria, i.e., the design requirements at least at three main limit states of the expected behavior of the structure: serviceability, damage control, and safety against collapse. Ultimate limit state design criteria are not only controlled by the energy dissipation capacity of the structural system, but also by damageability. Thus, the selection of design ductility factor without considering structural flexibility, type of structural and non-structural systems, or earthquake type is generally unacceptable. Even for a specific structural system, the acceptable ductility will vary depending on whether non-structural or structural damage controls. If non-structural damage controls, the allowable ductility will decrease with increasing structural flexibility.

The economic impact of designing structures for either seismic resistance coefficients or design ductility ratios higher than those presently assumed should be assessed. Such studies will require examining the uncertainties involved in selecting values of all parameters pertinent to the design process. To do this, the interrelationship of these parameters must be considered. For example, in designing for strength, the computed internal forces determined from critical design excitations should be reconciled with the actual strength of structural elements. Present methods of designing sections, regions and whole structures include several factors that usually lead to over-strength. Thus by taking a conservative approach and looking at each side of the design strength equation independently, one may arrive at an unreasonable, uneconomical design.

In earthquake-resistant construction particular attention should be given to the conceptual design. Conceptual design is defined as the avoidance of minimization of problems created by the effect of seismic excitations by applying the understanding of the behavior rather than using numerical computations. To illustrate this definition, let us consider the case of supply uncertainties, i.e., the lack of reliable equations to predict the supplied strength, stiffness, energy absorption, and energy dissipation capacities, particularly when supplies can be controlled by the presence in the critical regions of high shear forces.

Despite the lack of reliable equations, there have been advanced in understanding of the behavior in such critical regions, particularly in reinforced concrete elements used in plane moment-resisting frames and isolated shear walls subjected to unidirectional loading conditions in the plane. For these elements, both the problems and their sources have been identified. These advances are of paramount importance—because if the designer knows what the problems and their sources are, he has two possibilities. First, he can try to avoid them; but if this cannot be done, then he can try to minimize the problems by proper design of the members.

Considering formulated criteria the basic principles for earthquake-resistant design of building structures can be summarized in the following essential guidelines:

3.18 Essential Guidelines

1. Building and structure should be light (avoid unnecessary masses).
2. Building and structure should be simple, symmetric, and regular in plan and height to prevent significant torsion forces, avoiding large height/width ratio and large plan area.

    - Structure should have sufficient initial lateral stiffness to avoid significant damage under minor and moderate earthquake shaking, and toughness (stability of strength and stiffness) under the repeated reversals of deformations which could be induced by a severe earthquake. The stiffer the structure the less sensitive it will be to the effects of interacting non-structural components, and the tougher it is the less sensitive it will be to the effects of sudden failure of interacting non-structural elements:
    - Structure should have a uniform and continuous distribution of strength, stiffness, and ductility.
    - Structure should have relatively shorter spans than non-seismic-resistant buildings and avoid the use of long cantilevers.
    - Structure should have the largest possible number of structural defense lines; i.e., it should be composed of different ductile structural subsystems which interact or are interconnected by ductile structural elements whose inelastic behavior would permit the structure to find its way from a critical stage of dynamic response.
    - Structures should be detailed so that inelastic deformation can be constrained to develop in desired regions.
    - Structure should be provided with balanced strength and stiffness between members, connections, and support.
    - Strength and stiffness of the entire building should be compatible with the strength and stiffness of the soil foundation.

3.19 Land Instability

Land instability represented by landsliding and liquefaction has been recognized as a major case of distress caused by earthquake in recent times. This aspect was amply demonstrated in the areas ruined by *Manjil* earthquakes of 1990. In fact, damage to private dwellings and lifelines in the city of *Astaneh* was caused by the liquefaction of loose sand deposits existing in some sections of the city. Large-scale landslides took place in the epicentral area and claimed the loss of as many as 500 people as well as destruction of private properties and public facilities. The ground failure as above was a grim reminder of what happened for the umpteenth time in the past and re-emphasized the importance of landsliding and liquefaction as major cause of hazards associated with earthquakes.

Manjil earthquake of June 20, 1990, caused extensive damages and loss of lives throughout the relatively populated epicentral region mainly in the towns of *Manjil* and *Rudbar* and their suburbs. Considerable additional damage occurred further north and west in *Gilan* and *Zanjan* provinces, especially in *Rasht* and *Astaneh*. Liquefaction occurred mostly in *Astaneh* and *Rudbaneh* some 75 km northeast of epicenter. Soil liquefaction caused extensive damages to buildings, farms, and lifelines in a vast area.

Due to the extent of damage occurred in the affected region a number of researchers visited the affected area. In this respect, a geotechnical team visited the site a few days after the earthquake to investigate the geotechnical aspects of this earthquake. Geotechnical aspect of this strong earthquake is a complete set of all possible events associated with any strong ground shaking, i.e., liquefaction, landslide, rockfall, local site effects and soil amplification, and foundation problems. Comprehensive reports in this respect are given elsewhere.

One of the most important geotechnical earthquake engineering considerations of this earthquake was liquefaction of level ground. Liquefaction of level ground has shown to be responsible for many damages incurred to structures and lifelines during moderate to strong earthquakes. Site and laboratory studies on the behavior of loose sands under dynamic loading in recent decades resulted in various theoretical and experimental expressions and design charts to evaluate the potentiality of the liquefaction (Fig. 3.23).

To evaluate the reasons for earthquake-induced liquefaction in *Gilan* and especially in *Astaneh* a careful surface study of the site was performed to map the zones with clear indications of liquefaction. The program of a study of this liquefaction includes subsurface investigation and in situ testing in *Astaneh*. This study has been completed, and the results are presented herein (Haeri 1991a, b) (Fig. 3.24).

In order to estimate the predominant frequency characteristics of surface deposits prevailing in the urban area of the City of R and towns of *Manjil, Loshan,* and *Rudbar*,

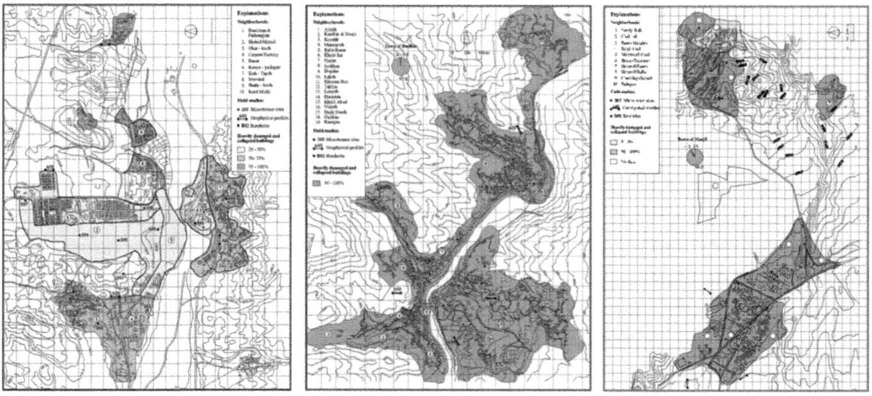

Fig. 3.23 Severe damaged and collapsed residential buildings in the neighborhoods of the towns of *Loshan, Rudbar,* and *Manjil*

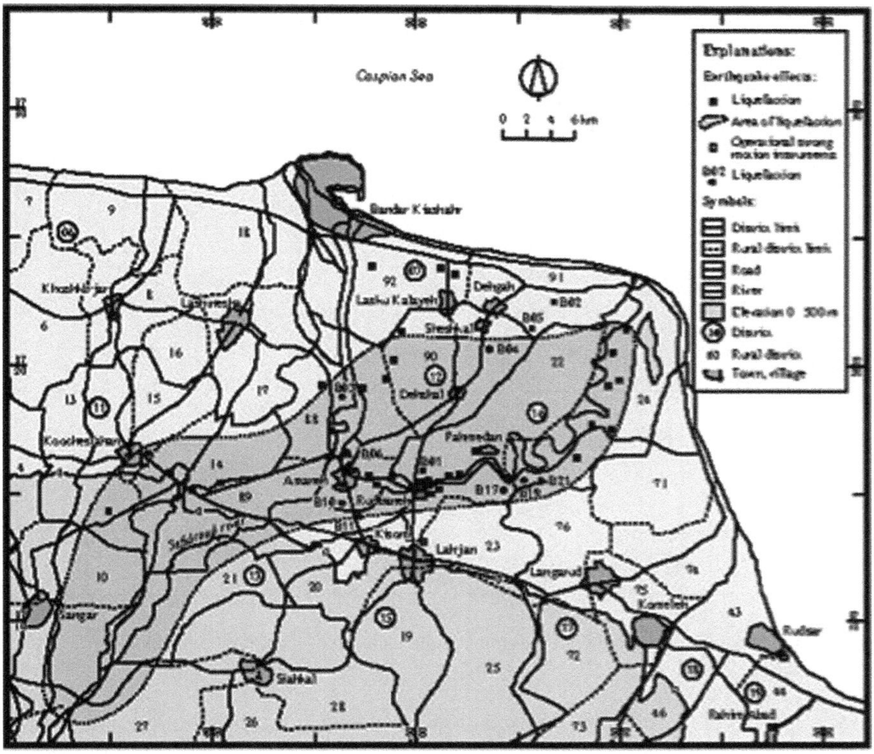

Fig. 3.24 Area of liquefaction in the *Gilan* plain and *Sefidrood* river tributaries due to June 1990 *Manjil* Earthquake with sites of exploration boreholes

in the period September 26 to October 8, 1992, 91 free field two channel microtremor measurements were performed. Considering the size of the City of Rasht, 57 two channel microtremor measurements were performed within the wider urban area of Rasht, whereas 11 in town of *Manjil*, 12 in town of *Loshan,* and 11 in town of *Rudbar* (Fig. 3.25).

Earthquake protection programs are well-understood and implemented with improvements in seismic zoning maps, strong motion instrumentation networks, seismic microzonation studies on urban areas; and site-specific geoseismic studies on important projects, as well as improvements in seismic design and construction codes and regulations. These improvements have mainly been associated with new buildings that are much less voluminous than facilities designed without significant seismic resistance. For the near future, the economic potential of the developing countries in the seismic regions will not likely create conditions for significant seismic risk reductions to buildings, structures, and utilities. Because the seismic hazards are serious and buildings, often old, tend to be of low seismic quality, seismic risk may even increase. Significant economic damage and loss in major earthquakes are expected.

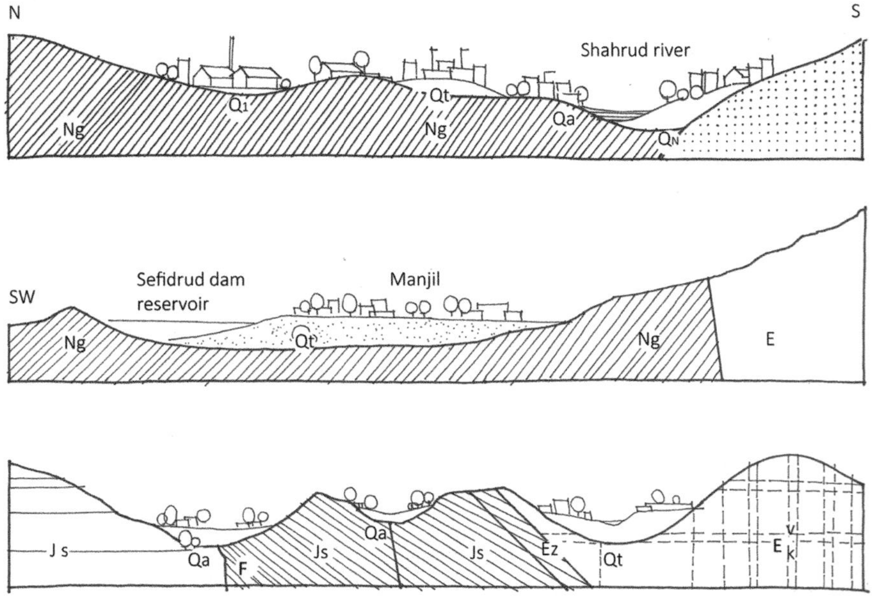

Fig. 3.25 Cross sections of the *Gilan* region and the location of the three towns of *Manjil, Loshan,* and *Rudbar*

Given this high risk and expected damage, the objective of this Part of the Guidelines is to present a uniform procedure for examining and reporting building damage both in urban and in rural areas so that a database on earthquake effects may be established. Also, methods are presented for analyzing earthquake damage and for estimating economic losses. Use of these procedures and methods will yield an adequate volume of data to assist community and national authorities achieve the following earthquake risk reduction program goals.

3.20 Recommended Earthquake Risk Reduction Program Goals

- To reduce deaths and injuries to occupants of buildings that have been weakened or seriously damaged by seismic activity and that with high probability will be subjected to a series of aftershocks within several months after the principal shock.
- To obtain appropriate information on the severity of the disaster in terms of the number of usable, damaged and also dangerous buildings so that people may be immediately protected and housed and so that essential activities may continue in the affected region.

- To develop a database for uniform estimation of economic losses so that an appropriate rehabilitation and assistance program may be devised as the affected region is reconstructed.
- To create a database on earthquake consequences for this and also for other seismic regions.
- To provide data so that for future earthquakes the civil defense system may elaborate rescue operation plans, train staff, and organize supplies.
- To record and classify earthquake damage so that damaged buildings may be repaired and strengthened in an orderly fashion.
- To identify principal elements of earthquake damage and to develop vulnerability relationships for different categories of buildings so that pre-earthquake mitigation programs can incorporate pre-earthquake assessments in planning and implementing short and long-term earthquake risk reduction measures.
- To improve seismic design and construction codes and regulations as well as design and construction practice.
- To improve the scientific basis for physical planning, both urban and general, especially with respect to seismic risk reduction measures in seismically active regions.

Post-earthquake damage evaluations should be organized so that teams may rapidly use a systematic methodology. Basic information from these evaluations should enable local and national governmental authorities to make critical decisions and also to employ economically justified and technically consistent seismic risk reduction measures in a uniform manner for the entire country. If coordinated efforts are made to use the uniform methodology presented in this Part of the Guidelines, more practical and transferable data can be developed that will be of potential use in the seismically active regions of the world.

Principal elements of this uniform methodology and procedure for post-earthquake damage evaluation are presented here.

3.21 Principal Elements Include

- Damage and usability classifications for buildings.
- Procedures for organization of data collection.
- Earthquake damage data analysis and data bank organization.
- Estimation of economic losses, and human fatalities and injuries.
- Measures for reducing adverse earthquake consequences and for mitigating seismic risk.

These guidelines, first of its kind, were intended to help practicing engineers, architects, planners and government officials and decision-makers in Iran, and other counties to design and construct buildings, prepare plans, and make critical decisions which will have a good standard of earthquake consideration. It was, no doubt, for the first time that almost all related disciplines of earthquake disaster management sat together, as a team and brought about a synthesis of knowledge on the subject. The most innovative part of the study was the application of findings in the regional and urban planning and design of the Gilan Province and especially the city of Rasht. The core part of the city, a complex and old mix-use fabric, was studied and redesigned in detail, which could be used as an example of urban design in a seismic-prone area.

Land uses play a significant role in reducing risk and increasing resiliency. Land uses may be divided into sensitive and non-sensitive uses, and on the one hand, they may be divided into vulnerable and damaging uses. Sensitive uses are the ones with significant values, such as education, health, and security centers. Damaging uses are the ones that not only suffer damages themselves, but also inflict damages on other uses too, such as natural gas tanks, gas stations, main gas pipelines, and bridges.

Land uses should be distributed in the city in the mixed use form and on the basis of their dependence on consumers and relative to their spatial structure, i.e., neighborhood, block, etc. Aside from emphasizing on issues such as sensitivity, vulnerability, and damage, other criteria such as compatibility, desirability, capacity, dependency, and meaning may be also used. Obviously, there are also certain other important factors which affect land-use vulnerability, such as topography, slope, geological layers, distance to faults, natural gas pipeline, access network, high voltage power network, etc. Besides, building and human density also affect the number of life losses and amount of damages inflicted by earthquake. A careful design and planning of quality and quantity of land uses and especially their appropriate locations will not only prevent their vulnerability to a large extent, but also prevent the secondary damages and losses, such as fire.

In normal times, open and green spaces are used for recreation, leisure, sports, and air cleaning but when a disaster strikes, they first are used for rescue and relief and immediately after that for emergency and temporary settlement. These are critically important in the old and blighted areas of the city which are extremely dense and usually lack those spaces. Easy, fast, and safe access to these spaces during and after disaster is especially important.

The higher the human density within the land uses, the higher the human losses. This is especially critical in the case of sensitive land uses such as schools, hospitals, nursing homes, as well as residential complexes. Attention should be paid to the fast and safe evacuation standards during disaster. Inappropriate building density may, similar to human density, increase the loss and damages inflicted by earthquake. This will be seen in the high-rise buildings, on the one hand, and the dense and compact areas of the cities, on the other. In either case, providing necessary services to the affected areas during an emergency would be difficult. Unjustified density in the high-risk areas will increase vulnerability.

These types are less vulnerable and more resilient compared to the new types, due to lower density, simple plans, and light roofs. The courtyard immediately next to the building provides the possibility of relief and rescue.

The presence of areas prone to liquefaction in the urban area is a great hazard regardless of which construction works have been pursued in the past, especially at the city's center where major land uses are settled and the texture is old, the thoroughfares narrow, the street sides unsafe, and the density high.

The presence of incompatible land uses within the city's residential quarters, particularly hazardous land uses such as the oxygen factory, the *Gilan* Electrical Factory, the *Pars Khazar* Factory, and the step-up step-down electricity stations create dangers to their surrounding texture.

The settlement of low-income families on the city margins has given rise to a particular housing pattern in these areas which, in disregard of the earthquake hazard and with no observation of required standards pertaining to this factor, increases the vulnerability of the marginal texture; and the lack of appropriate access to health centers and their remoteness exasperates the situation.

Ultimately, the non-observance of building codes, the overhangs in building stories, the narrow thoroughfares, the non-adherence to a hierarchy of access networks, the land-use distribution node, and the inattention toward the vulnerability of urban infrastructures, etc., indicate the generally inconsistent state of city for confronting an earthquake.

Finally, by consideration of the seismic hazard impact and other natural factors in the creation of the city's spatial structure and also by focusing on population issues and the city's present physical state, it is possible to reduce the city's vulnerability to the safe and economically acceptable level in future development planning.

Based on formulated earthquake scenarios of the planning scale earthquake and estimated June 20, 1990, *Manjil* Earthquake the level of the expected regional ground motion parameters are determined for the purpose of seismic risk analysis and regional planning for earthquake protection. In order to proceed with the formulation of the corresponding earthquake scenario for seismic risk analysis and earthquake protection in urban planning and design of the selected urban areas of the city of Rasht and towns of *Manjil, Rudbar,* and *Loshan* as a case studies within the same region of the provinces of *Gilan* and *Zanjan*, seismic microzonation studies are required to be performed for each of the selected urban areas considering modification of regional earthquake ground motions due to influence of topography and local soil conditions (Petrovsky and Bahrainy 1992).

3.22 Assistance in the Implementation of a Post-earthquake Rehabilitation Program

Following the catastrophic *Manjil* earthquake of June 20, 1990, a project was defined as the "Assistance in the Implementation of a Post-earthquake Rehabilitation Program," for which several significant goals were defined including: transfer of knowledge, developing new knowledge. … Volume 7 "Guidelines for reduction of seismic risk in regional and urban planning" is taking on the subject of human settlements and the role they can play in reducing seismic risk.

The project was carried out in the framework of International Decade for Natural Disaster Reduction and was hoped to stimulate interest and activities by countries and governments to implement an integrated approach to disaster mitigation, pre-disaster planning, preparedness, and prevention. The content of the resultant guidelines is as follows:

Volume I: Guidelines on seismic hazard analysis, ground instabilities, and determination of earthquake parameters for planning and design Composed of ten parts: (1) Seismotectonic study and modeling of seismic sources affecting provinces of Gilan and Zanjan, (2) Seismic hazard analysis and seismic zoning of the provinces of Gilan and Zanjan, Appendix: Earthquake catalogue of northwestern Iran, (3) Earthquake parameters for planning and design and selected strong motion records, Appendix: Acceleration time histories and response spectra of the selected strong motion records, (4) General aspects of ground instabilities in the provinces of Gilan and Zanjan, (5) Analysis of liquefaction induced by Manjil Earthquake, (6) Analysis of Galdian landslide triggered by Manjil Earthquake, (7) Analysis of Fatalak landslide triggered by Manjil Earthquake, (8) Seismic microzoning studies of urban areas for seismic risk analysis and earthquake protection planning. Preliminary seismic microzoning of the city of Rasht, (9) Preliminary seismic microzoning of the towns of Rudbar, Manjil, and Loshan, (10) Analysis of microtremors recorded in the city of Rasht and towns of Rudbar, Manjil, and Loshan.

Volume II: Guidelines on the improvement of production and quality control of building materials and construction works Composed of eleven parts: (1) Evaluation of quality of building materials based on earthquake damage analysis, Appendix: Building damage observations and construction material tests, (2) Evaluation of existing production of building materials, (3) Assessment of demand and supply of building materials, (4) Available material resources for dominant building materials, (5) Common types of mortars, utilization of pozzolana in mortar production, (6) Improvement of brick and block masonry, (7) Guidelines on the identification of raw materials, (8) Guidelines on specifications of production and machinery, (9) Guidelines on product quality control methods,

(10) Guidelines on small-scale building site production of building materials, (11) Guidelines on quality control of construction works, maintenance, and repair, Appendix: Recommended literature.

Volume III: Guidelines on earthquake-resistant design and construction of steel and reinforced concrete buildings Composed of four parts:

(1) Earthquake-resistant design and construction of steel buildings, Appendix A: Example of seismic analysis and design of steel frame building, Appendix B Example of earthquake performance analysis of steel frame building, (2) Earthquake-resistant design and construction of reinforced concrete buildings, Appendix A: Example of seismic analysis and design of reinforced concrete frame building, Appendix B: Example of earthquake performance analysis of reinforced concrete frame building, (3) Analysis and design of foundations for earthquake-resistant buildings, (4) Analysis of the influence of soil–structure interaction on dynamic response of buildings, Appendix A: Analysis of the effects of soil–structure interaction on dynamic response

of buildings, Appendix B: Analysis of soil–structure interaction influence on dynamic response of water tower in the city of Rasht.

Volume IV: Guidelines on earthquake damage evaluation, seismic performance analysis and upgrading of earthquake resistance of masonry buildings Composed of six parts: (1) Methodology and procedure for earthquake damage classification and damage distribution analysis, (2) Earthquake damage classification and development of database of damaged buildings after June 20, 1990, Manjil Earthquake (3) Earthquake damage distribution analysis in the region affected by June 20, 1990, Manjil Earthquake, (4) Empirical performance evaluation and development of empirical damage cost functions for rural masonry buildings, (5) Earthquake-resistant construction and upgrading of existing rural buildings, (6) Upgrading of earthquake resistance of engineered masonry buildings, Appendix A: Analysis of observed failure modes and damage of masonry buildings with evaluation of shortcomings in seismic design code implementation, Appendix B: Example of design and analysis of existing and upgraded masonry school buildings, Appendix C: Performance analysis of earthquake-resistant two- and four-story model school building.

Volume V: Guidelines for upgrading of earthquake resistance of steel and reinforced concrete buildings, seismic performance analysis, and development of damage cost functions Composed of four parts: 1. Upgrading of earthquake resistance of engineered steel buildings, Appendix A: Example of seismic analysis and design of upgraded steel frame building, Appendix B: Example of earthquake performance analysis of upgraded steel frame building, 2. Upgrading of earthquake resistance of engineered reinforced concrete buildings, Appendix A: Example of seismic analysis and design of existing and upgraded reinforced concrete frame building, Appendix B: Example of earthquake performance analysis of existing and upgraded reinforced concrete frame thirteen-story hospital building, 3. Development of damage cost functions for earthquake-resistant model public and high-rise office and residential buildings, selected mid-rise and rural masonry dwellings, 4. Earthquake damage cost functions for reinforced concrete, steel, confined and unreinforced masonry public, high-rise office and residential buildings, urban and rural dwellings.

Volume VI: Mapping of elements at risk, seismic hazard, earthquake damage and losses for analysis of seismic risk Composed of four parts: (1) Mapping of elements at risk, earthquake ground motions, damage, and losses in the region affected by June 20, 1990, Manjil Earthquake, (2) Mapping of elements at risk, seismic hazard, and earthquake losses in rural districts for analysis of seismic risk in regional planning, (3) Mapping of elements at risk, seismic hazard, and earthquake losses in the city of Rasht for analysis of seismic risk in urban planning, (4) Mapping of elements at risk, seismic hazard, and earthquake losses in the city Center of Rasht for analysis of seismic risk in urban planning and design, on double executive format.

Volume VII: Guidelines for reduction of seismic risk in regional and urban planning Composed of seven parts: (1) Methodology for analysis of earthquake losses in regional and urban development planning, (2) Earthquake losses and seismic risk analysis in regional planning. Economic and social losses of housing and essential public buildings in rural districts of the region, Appendix: Environmental conditions, elements at risk and their components exposed to June 20, 1990, Manjil Earthquake,

(3) Earthquake losses and seismic risk analysis in urban planning: Seismic risk analysis in the city of Rasht for development before June 20, 1990, Manjil Earthquake, Appendix: Environmental characteristics, elements at risk and their components in the city of Rasht based on census data of 1986, (4) Earthquake losses and seismic risk analysis of the city Center of Rasht, (5) Seismic risk reduction in urban development planning: Planning for seismic risk reduction in the city of Rasht, Appendix: Environmental conditions and development potentials of the city of Rasht, (6) Vulnerability assessment to improve earthquake protection and serviceability of urban infrastructure, Appendix: Failure modes and damage of infrastructure, lifelines and facilities due to June 20, 1990, Manjil Earthquake, (7) Elements for disaster preparedness planning, rescue, and relief operations in urban areas. Case study of the city of Rasht.

3.23 Lessons Learned from the *Manjil* Earthquake of June 20, 1990

- Promote public awareness and capacity building with regard to natural disasters and ways to reduce them.
- Include natural disaster considerations in all planning and design activities at all levels, particularly national development plans, regional plans, and urban comprehensive plans.
- Urban and regional planning and design could be used as effective tools to prevent natural disasters and reduce their risks, results which will be far less costly and more manageable.
- Focusing on prevention and preparedness, rather than post-disaster actions.
- Effective involvement of interest groups, volunteers, and NGOs in the natural disaster reduction activities.
- Transfer of technical knowledge to developers, engineers, architects, planners, and managers.
- Using the lessons learned from past experiences in order to improve the efficiency and adequacy of measures, plans and procedures related to natural disaster reduction.
- Modification of the existing structure of the natural disaster management to eliminate ambiguities, overlaps, and deficiencies.
- Preventing unnecessary expansion and spread of organizations by applying the "flexible organization" idea, so that adapt itself with emerging new situations.
- Increase the preparedness of the national, provincial, and local organizations; and individuals, groups, and NGOs.
- Create cohesion and coordination in the natural disaster management system along with decentralization.
- Pay due attention to the mutual relation between natural disasters, environment, and development.

- Considering the goal of reducing natural disaster risks as an integral part of sustainable development agenda.
- Separating policy-making from implementation.
- Delegating authority to governors in the area of natural disaster management to provide unity and coordination in procedures and decision making and effective control and supervision.
- Delegating disaster management responsibilities to relevant specialized organizations.
- Including the activities of temporary housing in the reconstruction activities and delegating the overall responsibility to a single organization.
- Exploring reliable credit sources to cover the damages of the disaster-inflicted areas.
- To promote the natural disaster insurance culture.
- Rational and just distribution of natural disaster resources among activities, groups, and regions to optimize the results.
- Increasing the efficiency and productivity of assigned resources by preventing repetition and misuse of resources (Bahrainy 1992, Bahrainy et al. 1998).

3.24 Conclusion

The June 20, 1990, *Manjil* Earthquake in northwest Iran, which was the most damaging event to date in the country, was taken as a case study to analyze the impact of the earthquake on built environment and look for ways and means to improve the conditions of the affected area in order to protect it against future earthquakes. This earthquake, with magnitude Ms = 7.4 and estimated epicentral intensity of IX–X degrees of MM Scale, caused 13,911 deaths, 36,693 people treated for injuries and over 8000 people badly injured. It was strongly felt over an area of 600,000 square kilometers, including Tehran, Tabriz, and several other major cities in Iran.

The chapter begins with a review of the spatial–physical conditions of the general area affected, focusing on the spatial structure and regional distribution of settlements. Situation of the towns in the region has been dictated by the natural conditions into a linear form: *Loshan*, *Manjil*, *Rudbar,* and *Rostamabad* towns, and *Rasht* has acquired a central position.

Social characteristics of the area include long history of settlements tribal culture, religious, and cultural variations, high population density in plains and lower density in mountains, wide differences in settlement population in number of plains in comparison with mountains. The main economic characteristics include dominance of agro-economics in plains, importance of industrial economy in industrial belts and centers, dominance of herding economy in mountains, and economic variation of different parts of the region.

Landslides, rockfalls, and liquefaction, are three of the most dramatic events that occurred because of this earthquake. One of the landslides triggered was the *Fatalak*. Liquefaction occurred on an area of 650 square kilometers of the Caspian plain with northeast-oriented strip from *Sangar* on the west almost to the cost of Caspian Sea. It caused extensive damage to the buildings in the districts of *Astaneh* and *Rudbaneh*, agricultural land and farms in the wider region as well as extensive deformations and damage to the regional and urban infrastructure.

Principal elements of uniform methodology and procedure for post-earthquake damage evaluation are presented using four different elements. Also measures for reducing adverse earthquake consequences and for mitigating seismic risk are introduced. Connected with these principal elements are earthquake damage evaluation, development of empirical vulnerability, or damage cost functions and analysis of earthquake damage distribution. A classification of structural damages recorded as a result of June 20, 1990, and the design, detailing, and specification deficiencies, such as insufficient wall thickness, poor connection details, columns not tied, roof not anchored to wall, inadequate bracings, absence of tie beams and defective materials, are shown. Aside from these building deficiencies, major deficiencies were also found in the overall built environment, such as lack of public space, narrow alleys, high building density, lack of hierarchy in the transportation network, and lack of easy and safe evacuation.

On the basis of the lessons learned from this earthquake and also by reviewing the deficiencies found in the construction activities, and an analysis of seismic hazard, a set of design criteria are formulated to be applied in the region. Also, a set of essential guidelines is presented to be followed during the modification of existing areas, reconstruction of the damaged areas or the design of a new settlement.

A brief description of a research project entitled: "Assistance in the Implementation of a Post-earthquake Rehabilitation Program," is given here. The project intended to not only contributes to the reconstruction of the seismic-affected areas, but also to develop a set of guidelines for earthquake risk reduction to be applied in possible seismic-prone areas.

References

Ahari Z et al (1996) Spatial and locational planning and analysis of settlements to mitigate earthquake risk. National report, Volume 5, UNDP—UNCHS (Habitat) Project IRAl90l004, Housing Foundation of I.R., NDPC (Translated in English), Tehran, Spring

Bahrainy H (national director) (1992) Natural Disaster Management. UNDP-UNCH (Habitat) Project IRA 90//004, Assistance in the implementation of Post earthquake Reconstruction Program, 8 vols

Bahrainy H (1998) Urban planning and design in seismic-prone region (the case of Rasht in Northern Iran). American society of civil engineering, urban planning and development division, Dec. 1998, vol 124 no 4

Bahrainy H et al (1998) Proposed structure for disaster management for Iran, Unpublished research report, Ministry of Interior, Iran

Bahrainy H (2003) Natural disaster management in Iran during 1990's—the need for a new structure. ASCE J Urban Plan Dev 129(3):140–160

Berberian M, Qorashi M (1991) The June 20, 1990 Rudbar-Tarom (NW Persia) catastrophic earthquake; a preliminary field reconnaissance report. Volume III, Proceedings of the First International Conference in Seismology and Earthquake Engineering, May 27–29, 1991, IIEES, Tehran

Coburn AW, Petrovski J, Ristic D, Armillas I, Bjering N (1990) Mission Report and Technical Review of the impact of the earthquake of June 20, 1990 in the provinces of Gilan and Zanjan. Earthquake Reconstruction Program Formulation Mission Report, UNDRO, Geneva

Eshghi S (1992) Behavior of lifeline systems during *Manjil*-Iran earthquake of June 20, 1990. In: Proceedings of the tenth world conference on earthquake engineering, Madrid, Spain 19–24 Jul 1992. A A Balkema Publishers, pp 5553–5557

Haeri SM (1990) Liquefaction associated with *Manjil* Earthquake of June 21, 1990, Iran. Civil Engineering Department, K.N. *Toossi* University, Tehran, Iran

Haeri SM (1991a) Liquefaction associated with Manjil earthquake of June 1990, Iran. In: Proceeding of 5th international conference on soil dynamics of earthquake engineering

Haeri SM (1991b) Geotechnical aspects of the *Manjil* Earthquake of June 20, 1990. Int Inst Earthq Eng Seismol Iran

Haeri SM (1998) Analysis of liquefaction induced by June 20, *Manjil* earthquake, guidelines for earthquake disaster management, vol 1, Part 5. UNDP-UNCHS (Habitat) Report

Ishihara K (1998) General aspects of ground instabilities in the provinces of Gilan and Zanjan, Part 4, vol 1. Guidelines for Earthquake Disaster Management, UNDP-UNCHS (Habitat) and Housing Foundation of I.R. Tehran-Skopje, September

Ishihara K, Moinfar A, Haeri M, Towhata I, Tsujino S (1991) Geotechnical aspects of the June 21, 1990 Manjil Earthquake in Iran, Report

Jancevski J, Qorashi M (1998) Seismotectonic study and modeling of seismic sources affecting provinces of Gilan and Zanjan, part 1, vol I. Guidelines for Earthquake Disaster Management, UNDP UNCHS (Habitat) and Housing Foundation of IR, Tehran

Music V (1993) Earthquakes and management of urban environment. Urban planning and design as a part of preventive strategies. International Consultant Report, UNDP - UNCHS (Habitat) Project IRAl90l004, Housing Foundation of I.R., Tehran, October

Petrovski JT (1992) Vulnerability and seismic risk assessment for earthquake disaster management in urban areas. Invited paper, First International Conference on Disaster Prevention in Urban Areas, Tehran, May

Petrovski JT (1998) Earthquake damage cost functions for reinforced concrete, steel, confined and unreinforced masonry public, high rise office and residential buildings, urban and rural dwellings, Part 4, vol. V, Guidelines for Earthquake Disaster Management, UNDP—UNCHS (Habitat) and Housing Foundation of I R, Tehran—Skopje

Petrovsky J, Bahrainy H (1992) Analysis and development of empirical vulnerability functions. Res Rep UNDP-UNCHS Project IRA/90/004

Petrovski JT, Bahrainy H, Music V (eds) (1998) Guidelines for reduction of seismic risk in regional and urban planning. Volume VII of the guidelines for earthquake disaster management, UNDP-UNCHS (Habitat) and Housing Foundation of I R, Tehran—Skopje

Taleb M (1997) Reconstruction of earthquake stricken areas in the provinces of Gilan and Zanjan. National summary report, UNDP—UNCHS (Habitat) Project IRAl90l004, Housing Foundation of IR, NDPC (Translated in English), Tehran, September

Taleb M, Alizadeh M (1997) Economic losses due to June 20, 1990 Manjil Earthquake in Northern Iran. National report, vol. 4, UNDP—UNCHS (Habitat) Project IRAl90l004, Housing Foundation of IR, NDPC (Partially translated in English), Tehran, September

Chapter 4
Urban Design at Macro-Scale: Urban Design in a Seismic-Prone City

Abstract Here, in this chapter, urban design recommendations will be made for a region and city which are prone to seismic risk. Following chapters will deal with urban design solutions for the city center, Bazaar area, access network and open spaces in a city. The city of Rasht will be used as example. Information needed for the regional and city scale study and design in a seismic-prone region includes seismic and geological map and data, hazard assessment, risk assessment, vulnerability assessment, built environment data, earthquake scenarios, urban and regional policies, and an implementation guidelines and mechanisms. In order to integrate seismic risk reduction goals into the long-range development plans, long-term development scenarios are formulated for the city of Rasht and vicinities. Most possible alternatives are centers hierarchy model, radial corridor model, bi-polar metro model and metro-satellite model.

Keyword Urban design · Macro-scale · *Rasht* · General structure · Topography · Urban texture · Open space · Districts · Accessibility · Estimation · Land use · Scenarios

Here, in this chapter, urban design recommendations will be made for a region and city which are prone to seismic risk. Following chapters will deal with urban design solutions for the city center, Bazaar area, access network and open spaces in a city. The city of Rasht will be used as example.

4.1 Information Needed for the Regional and City Scale Study and Design in a Seismic-Prone Region

- Seismicity map.
- Geological condition/soil stability.
- Seismic zonation map.
- Hazard assessment, risk assessment, and vulnerability assessment.
- Land slope/topography.
- Land uses (sensitive, vulnerable, and damaging).

H. Bahrainy and A. Bakhtiar, *Urban Design in Seismic-Prone Regions*,
The Urban Book Series, https://doi.org/10.1007/978-3-031-08321-1_4

- Construction quality.
- Building age and depreciation.
- Construction techniques.
- Building materials.
- Building form.
- Texture (grain, open and green space,…).
- Building density (height) of the buildings.
- Human density, no. of people per hectare.
- Accessibility network (hierarchy, bridges, parking, traffic volume, alternatives).
- Urban utilities (water, sewage, electricity, communication).
- Urban facilities (health centers, fire extinguisher, security facilities).
- Regional policies.
- Location of settlements and infrastructure in the region on the basis of seismic zonation map.
- Mapping the spatial pattern of settlements and infrastructure distribution.
- Mapping the spatial pattern of regional land use distribution.
- Applying the smart growth concept.
- Different earthquake scenarios on the basis of magnitude, location, type, and time of the earthquake.
- Formulating implementation guidelines and mechanisms.

4.1.1 Spatial Structure and Regional Distribution of Settlements

Regional distribution of settlements in the area has historically followed spatial characteristics. From the spatial structural point of view, these characteristics can be divided into environmental, social and economics.

4.1.1.1 A Summary of Environmental Characteristics

- Geographical variations.
- Prevalence of various geographical features such as plains, foothills, and mountains.
- Temperate climatic conditions in plains and foothills and cold climatic conditions in mountains.
- Perennial water resources.
- Access routes and their role in settlement patterns.
- Region's situation with respect to its neighbors.
- Vegetation cover.
- Forest areas, especially in foothills and mountains.
- Agricultural lands in plains.

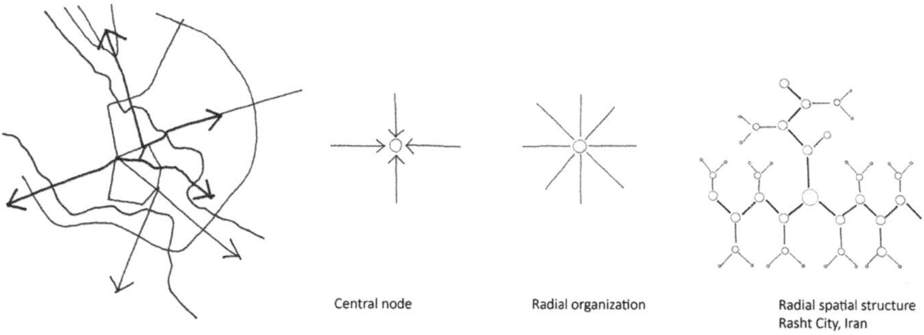

Fig. 4.1 Radical organization and central node as the main factors in the general structure of the city of Rasht

4.1.1.2 A Summary of Social Characteristics

- Long history of settlement.
- Tribal culture.
- Religious and cultural variations.
- High population density in plains and lower density in mountains.
- Wide differences in settlement population in number of plains in comparison to mountains (Kazemi et al. 1997).

4.1.1.3 A Summary of Economic Characteristics

- Dominance of agro-economics in plains.
- Importance of industrial economy in industrial belts and centers.
- Dominance of herding economy in mountains.
- Economic variation of different parts of the region.

4.1.1.4 Central Node and Radial Organization Comprise the General Spatial Organization of the City

See Fig. 4.1.

4.1.2 General Characteristics

4.1.2.1 Settlement Distribution Pattern

Six cities and 855 villages are situated in the area. Village distribution in different rural districts in not even, for instance *Daylaman* rural district with 66 villages, has the maximum number of villages and *Saravan* rural district with 7 villages contains the last number of villages.

The assessment of urbanization and their situation with regard to the geographical position and routes depicts that out of six main towns of the region, four are situated in a linear position, that connects the region to the neighboring region, this line is the life line of the region, and its impact spreads over the region and has a national and presently international importance. Along the *Shahrood* and *Tarom* valleys, there is a major city or large population center.

Over all, out of six towns of the region, only the city of Rasht and town of Sangar are situated at plains and have a mild slope, while *Rudbar, Loshan, Siahkal,* and *Manjil* are situated at foothills and have sharp slopes of higher degrees. *Loshan, Manjil* and *Rudbar*, due to their location at natural passes and national transit system, have strategic importance. The city of *Rasht* also, due to its situation at plains and national transit route, is of great importance.

4.1.2.2 The City of Rasht

Rasht as the center of the *Gilan* Province is situated in a very damp and moderate Caspian climate zone. The city is located 15 km to the north of the *Alborz* range and 20kms to the south of the Caspian Sea. Its geographical coordinates are 49″36′ longitude, 37″19′ latitude, and an aerial distance of 232 km from Tehran. To the north, it connects to *Bandar Anzali* port, to the east to *Lahijan*, to the west to *Fooman*, and to the south to *Rudbar* and *Qazvin*. Its dominant climate features are very damp and moderate climate, moderate temperatures, and high level of humidity and rainfall.

The city's population has been said to amount to some 288,306 and 362,737 for 1986 and 1991, respectively, according to the housing and population census for the same years.

The population is settled on over a 3180 hectares city area. One of the distinguished features of the province and the city is the rate of the existing population per unit area or in other words, the population density that rates second in the country after the capital.

From the cultural point of view, Rasht is significant among the cities of Iran. The literacy ratio is significantly high in this city. From the historical point of view, Rasht is not very old, since the region was sub-merged by the Caspian Sea, and after having dried up, it was uninhabitable for some time. The city has been noted in books since the fourth solar century. One of the oldest writers to describe the city and its hot, damp climate was *Hamdollah Mostowfi* in the 8th solar century. According to him, there were ample sources of cotton and silk prepared in the city for export, the area covered by it was large, and the city is an important urban center in the region. From the cultural point of view, Rasht is significant among the cities of Iran. The literacy ratio is significantly high in this city.

4.1.2.3 Topography

Rasht city is situated 15 km to the north of the *Alborz* range and 20 km to the south of the Caspian Sea. The city has two moderate slopes, one from the south to the north and the other from the southeast to the northwest. The city's slope from the south to the north is 0.26% and from the southeast to the northwest is 0.16%.

The land in Rasht, as implied above, is generally leveled. Its height in the middle of the city to the average surface of the waters of the Persian Gulf is 2 m, which declines to 12 n in the north of the city, and of the whole, the contour line in the city's densely populated areas is 7 n; in the south, southwest, and the east, the land inclined toward the rivers is lower. Thus, it is possible to say that the city is located on an uplifted plain of an east–west direction 4 km long and two kilometers wide. One of the most important effects of landforms is the presence of a slight slope in the middle and northern parts of the city, which nukes surface water collection problematic. On the whole, the situation of the city of Rasht on a plain and a low one at that, and the very moderate slope of this area have, in the past, contributed to the possibility of the city's horizontal expansion. Although such an expansion has, in some cases, lead to the destruction of fanning land, the city's southern lands, owing to their relatively higher level in comparison to other parts, have given rise to an inclination for expansion and this is the outcome of the existing problems originating from the city's relatively level topography. Rasht's topographical features and their impact on the city's form are worth being studied in terms of climate and size of the city with a view to the earthquake factor.

On the whole, owing to the high level of rainfall and the flow of waters on Rasht's southern skirts on a permanent basis, there are ample sources of underground waters in the region and underground water tables in the city are placed very high, such that underground waters in these parts are about 3–7 n below ground depending on the terrain. Underground waters flow from Rasht toward the Caspian Sea, and a great portion of these waters enters the Caspian Sea; however, due to the piezometric pressure of underground waters in Rasht, salt water from the sea cannot penetrate into the city. Also, on the west bank of the *Sefid Rood* river, waters from a 150 km boundary are collected by the *Siyah Rood* and *Gohar Rood* rivers and flow toward Rasht. In reality, the *Siyah Rood* and the Gohar Rood rivers surround Rasht from two sides. The *Siyah Rood* river flows to the east and northeast of the city and the *Gohar Rood* river to its south and west, both emptying into the Anal marshes after having passed Rasht on their way. The waters of the *Siyah Rood* river in Rasht County are almost on the same level as underground waters in the city of Rasht. Of course, without natural and artificial drainage, the fluctuation of river water in different seasons being between one to two meters threatens environmental well-being and consistency. The volume of the waters of both rivers is high, and water flows along them on all seasons.

These rivers bear problems for the city, the most important of which, at present, is the disposal of sewage by different units next to the rivers. This, particularly in seasons when the river's waters are low in volume, besides creating an unpleasant scenery, is likely to give rise to parasitic diseases. Moreover, erratic developments of the river margins in disregard of keeping to the limits are other problems that are worth mentioning. This bears immense disorders at the time of earthquake disaster.

On the other hand, the banks of both rivers have maintained their natural form and remain unchanged which is why they have been turned into sites of garbage disposal.

Despite all the problems associated with the littoral boundaries of both rivers, their passage across the city has also created some potentials. Their elongation and green boundaries can be converted into a very pleasant natural landscape. The conversion of these long and green strips stretching from the south to the north of the city and the beautification of the river margins and the observation of the river boundaries in current and future construction works are essential measures which may be utilized in different ways at the time of earthquake. For example, as far as these measures are concerned, it is possible to allude to such issues as refuge, establishment of sites of emergency settlement, and the settlement of relief teams. It is essential to devote land on the river margins to green space and cultural and leisure activities with a view to the city's shortages in the above contexts. On the other hand, taking into consideration the seasonal flooding of the citys existing rivers, the settlement of the mentioned land uses on the margins of the two rivers and the relocation of residential land uses beyond the river boundary is a must. Furthermore, owing to the unstable nature of the soils around the rivers, any buildings built there would be seriously threatened during an earthquake, and if there is a necessity for erecting buildings within these boundaries, they ought to be constructed in the specific locations and in keeping with earthquake-resistant building codes.

As formerly stated, dedicating these lands to green space and low-density land uses of a cultural leisurely nature is one of the most consistent approaches that could be taken in regard to then, and while responding to situations, it would alleviate the city's shortages in the said contexts.

The inability to absorb sufficient water by the soil layers and the saturation of the surface layers in the city has given rise to numerous lagoons. (1) These lagoons in the city and marginal areas have been seen to cover more or less great expanses. They have given rise to problems, among which it is possible to mention the absence of hygiene in their periphery owing to the accumulation of garbage in these areas.

However, it is obvious that by making the required expenditures in these parts they can be utilized as urban recreational environments; otherwise, it is vital to cover them with soil.

Excluding the rivers, there are significant areas of vacant and fallow land in the city of Rasht. These lands which are privately owned, for the most part, are considered as appropriate sites of the settlement of a portion of the city's population. Although, as formerly pointed out, a part of the lands in the city has been left in the form of lagoons and marshes; it is still possible to allocate them to land uses corresponding to the city's land use shortages provided measures are taken to improve their inadequate state (Fig. 4.2).

Although farming land will not be merely considered as a natural factor, their roles and potentials will be extremely important in promoting agriculture and providing the needs related to this context. These lands cover the areas surrounding the texture for the most part, and it is essential to conserve and protect them and prevent the

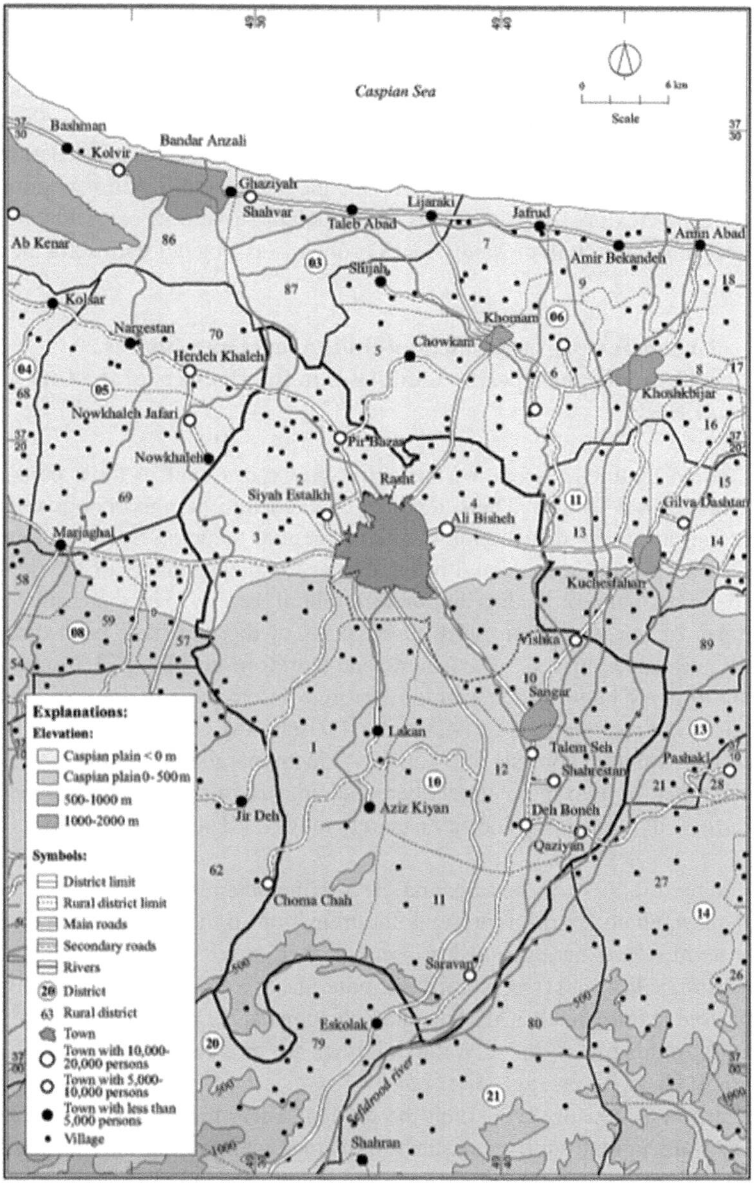

Fig. 4.2 Topography of the city of Rasht and environmental characteristics

erection of buildings upon them. On the whole, the city faces great constraints owing to its being surrounded by fertile and valuable land, orchards, and farming land.

Other portions of farming land consist of those, which are enclosed by the urban texture as a result of the city's horizontal expansion. These types of land presently occupy a somewhat large portion of the urban texture where no construction work is possible and this has created problems for the settlement of the new population inside the urban texture. In other words, the population overflow caused by the city's natural population growth presently faces problems associated with settlement.

4.1.2.4 Urban Texture of Residential Built Area, Open Spaces, and Public and Specific Land Uses in the Districts of the City of Rasht

Rasht's urban texture was one of the radial grid types, and in most parts, it has a dense residential texture. These textures evolved into neighborhoods creating a multiple variety in the distribution of the residential textures. Socioeconomic and cultural disparities are considered to be among the determinants of the density of the residential texture or its distribution throughout the city. The disparities created in the form of the texture in the different parts of the city have been directly or indirectly influenced by the same factors, and therefore, regular grid, irregular grid, radial and organic textures are seen to exist in different parts of the city connected or overlapping one another (Figs. 4.3 and 4.4) (Bahrainy et al. 1998).

The city's structure consists of two rivers flowing parallel of a north–south direction intercepted by two main streets perpendicular to one another of an east–west and a north–south direction. The city's texture may be identified to comprise of three types.

(1) Old texture; (2) new texture; and (3) intermediate texture.

The intermediate texture represents the transition to the new texture. The three types of textures are distinguished in terms of their street design, building heights, construction quality, and type of building materials and their extent of development. The city's old texture that has a native character consists of an organic texture, which is very old. In this part of the city, the buildings are mostly single or double storied and aged 30 years, on average (Fig. 4.5).

Four different housing types may be recognized in the old section, including: (masonry walls in mud mortar, masonry walls in cement mortar and masonry walls in cement mortar, but different plan). These types are less vulnerable and more resilient compared to the new types, due to lower density, simple plans and light roofs. The court-yard immediately next to the building provides the possibility of relief and rescue (Fig. 4.6).

Another feature of this texture is the presence of neighborhoods and neighborhood centers with diverse functions. The neighborhoods that have a special name of their own, in each case, are in fact developed versions of the city's original habitats, which gradually joined to form its original core. The neighborhoods still have some kind of boundary and carry their old names. Green spaces are chiefly distributed in the

Fig. 4.3 Neighborhoods in the city of Rasht, 1986 Census

marginal neighborhoods, and apart from two or three major green spaces, the central neighborhoods have no other areas dedicated to the purpose.

The city center contains the old and valuable texture that is dilapidated and highly vulnerable toward earthquake. Also all major land uses in the city are centered in this part. The criteria for the physical outlook of any city may be sought for in the old parts, in the market and its numerous branches, in the neighborhoods surrounding the market, in the local features, nodes, passages landmarks, and edges.

The passage of the *Gohar* and *Zarjoub* rivers could act as specific physical criteria in the city's spatial structure: And presently, owing to private construction works, and the non-observance of safety limits, these developments stretch up to the edges of the rivers, greatly damaging the city's spatial functional structure. The restoration

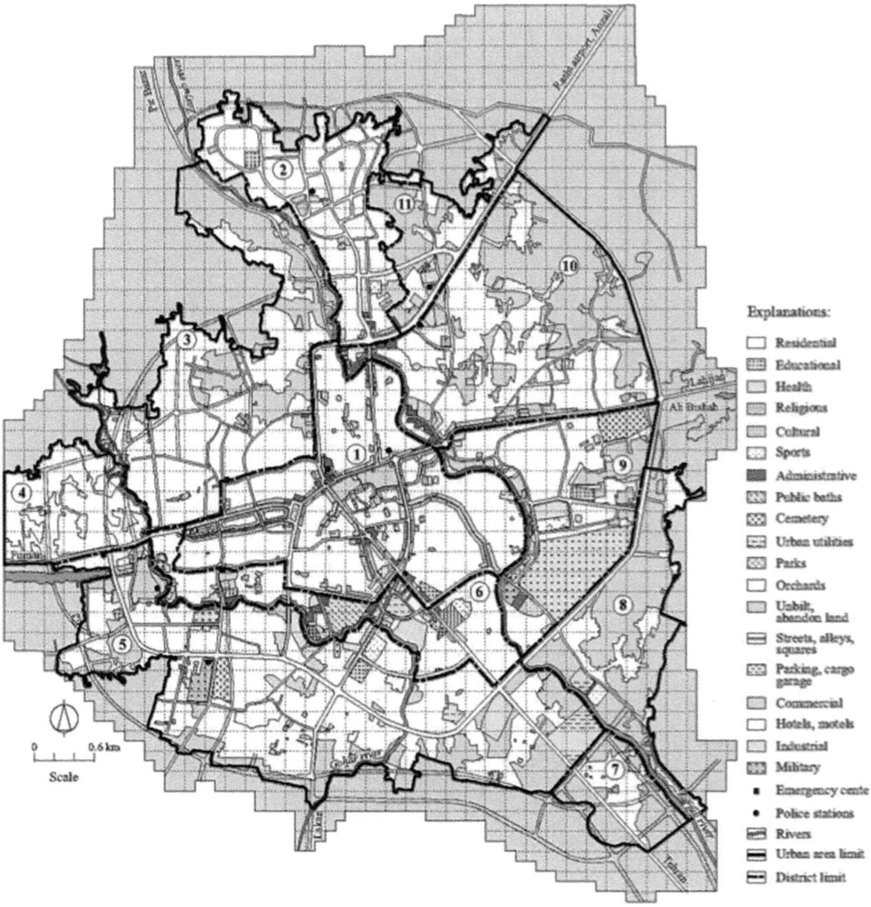

Fig. 4.4 Residential, public, and other specific land uses in the city of Rasht City characteristics Census 1986

and appropriate use of these water routes, which were once used for import and export communications to the northern port (*Bandar Anzali*) of the Caspian coast, are viable (Fig. 4.5).

The presence of the city's central market in this texture is considered as one of the city's distinguished features. The central market where different types of whole sale or retail commercial land uses in the city and its peripheries are centralized is a factor worth considering, and this is the reason for the high density of vehicle and pedestrian traffic in the area at all hours. The market, which is considered as one of the oldest sites in the city, has numerous passages, which stretch from the center of the market on to the residential texture. Some of these passageways are used to act as commercial links between Rasht and other regions, and even now, they still maintain their importance. Each passage has a name, which expresses the type of

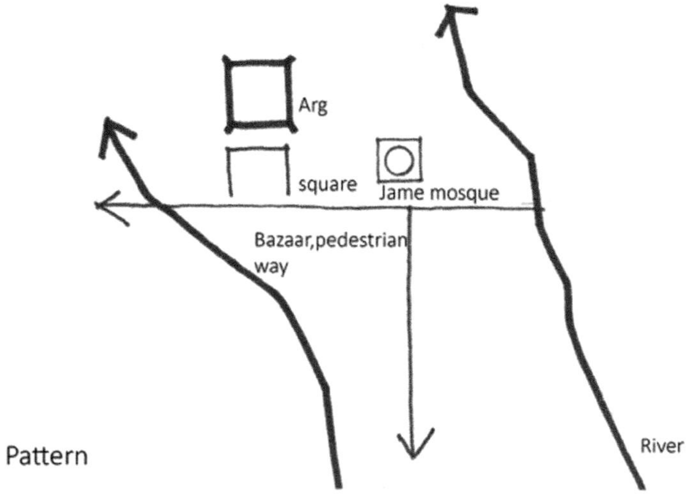

Fig. 4.5 The role of two rivers in shaping the General structure of the city of *Rasht*

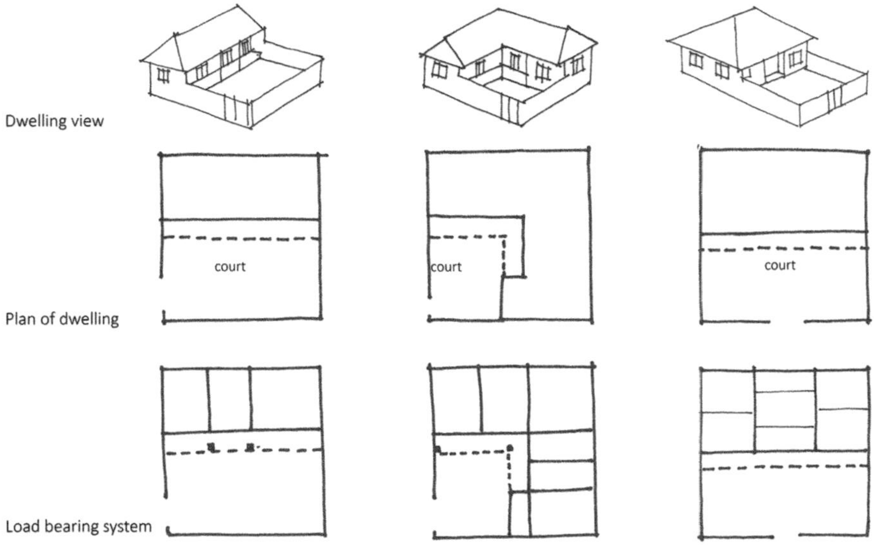

Fig. 4.6 The dominant type of traditional housing in the city of Rasht

activity proceeding within it. The market and its passages face physical constraints at the present time. Changes in types of land uses, transport of merchandise, and daily population nobility are issues, which cannot be appropriately responded to by the existing texture of the market.

The *Sabze* Meidan and *Shahrdari* squares which have been set up quite near to one another are contained within the city's old texture and the Shahrdari square may be considered as the point where the city's two main routes meet. Communication between these two squares is afforded through a main street and several other routes. Both squares particularly the *Shahrdari* square are surrounded by commercial and major administrative land uses, which have intensified the traffic in the area (Fig. 4.7).

Old buildings that are sound and active may be seen around the squares. Most buildings in the area are with two-story height. To be brief, the old texture, despite its existing physical constraints and bottlenecks, remains alive and dynamic, as the heart of the city. This may be attributed to the density of the important land uses settled particularly at the central market and its passages.

Beyond the alleys and the passages belonging to the old texture, changes in the form of buildings, materials and heights are observed. This is expressive of the gradual transition from a historic period to a new one. This part of the urban texture encircles

Fig. 4.7 Urban texture characteristics in the districts of the city of Rasht

the old texture revealing its identity as the intermediate texture. The intermediate texture may be considered as the period of transition from the old to the new.

Communication between the old and the new textures besides being furnished through a number of main streets, is chiefly made through old passages and alleyways, some of which currently function as minor vehicular networks. New streets surround the intermediate and new textures and separate them. The southern sector of the old and intermediate textures *Manzariyeh* is known as the first area of the new texture containing relatively high buildings constructed according to standard and fashionable in appearance.

The second sector of the new texture may be found to the north of the city. The *Golsar* area is one of the neighborhoods where the affluent citizens of the city have settled. Here, most buildings are in the form of detached villas and its grid texture has given rise to an expansive area. On the whole, the features of the aforementioned districts may be summarized as below:

- Presence of relatively wide streets.
- Grid nature of the texture.
- Presence of stylish villas.
- Absence of zoning.
- Newness of texture.

Although the area surrounding the *Golsar* and *Manzariyeh* neighborhoods is considered to be parts of the new texture, yet the construction quality, the population groups settled there, the streets, and other factors chiefly point to the presence of a marginal texture.

In the northern sectors of the city, horizontal expansion has continued to some of the surrounding rural areas. In other words, despite their situation in the urban area, these districts are still considered as parts of the rural texture owing to their construction quality, ways of life, and ways of earning a livelihood.

In the southern part, the same has happened only in a different way. The trend of migration from different locations and towns to Rasht has lead to the formation of an immigrant texture, which despite the presence of many relatively wide streets lacks suitable facilities and adequate building quality

One of the other features of Rasht's texture is the settlement of minor and major industries inside the main texture and around it. There are a great many factories and industries that are settled along the margins of the city's main street, which leads to the point of departure to Tehran. Public land uses are also settled there. These land uses have settled for up to several kilometers outside the city.

As a general conclusion, city of Rasht texture may be identified as follows:

- The city has an irregular radial grid network.
- The city's structure consists of two main streets (north–south and east–west) intersecting each other perpendicularly and two rivers that flow in a south–north direction, and also a circular freeway from the east to the west.
- Rasht's situation in a very damp and moderate Caspian climate has given rise to a particular building form with pitched roofs.

- For this reason and the need to create air currents, and also due to the high level of underground waters, buildings in the city are generally low, rising to a maximum height of 8 stories.
- Three old, intermediate, and new textures may be identified and differentiated.
- Among the features of the old textures, it is possible to name the central market and the numerous old neighborhoods.
- The new texture is distinguished by its grid streets which are relatively wide, and its low buildings.
- The crossing of the two rivers through the city along the south–north axis has practically divided the city into three sectors connected by bridges lying across the rivers at different points.
- The settlement of industries and manufacturing workshops along main routes especially the city's southern route is conspicuous.

The central sector mainly comprises districts 1 and 6 (see Fig. 4.8). The presence of the old market and the centralization of urban services and land uses resulting in attraction of people from other parts of the city have given rise to problems of vehicular traffic in these parts. Moreover, the increase in population density the compact and minuscule nature of the texture and the smallness of the differentiated components aggravate the situation. The texture in these areas is chiefly organic and dense with unsuitable access to the texture, except for some parts which have regular and dense textures with unsuitable accesses.

Fig. 4.8 Ten districts of the city of *Rasht*

On the whole, this part is not seismic resistant, owing to its age. The presence of the market and also the centralization of other urban land uses, and on the one hand, the shortages in the capacity of the transportation met works, have given rise to increased traffic density ultimately leading to an increased vulnerability, and on the other, owing to the inadequate capacity of transportation routes and narrow thoroughfares within the residential blocks which are easily obstructed as result of the collapse of old buildings forming their sides, relief operations will face nanny problems. Moreover, owing to the denseness and the minuscule nature of the texture and the shortage of the ratio on built-up space, the narrowness of the thoroughfares causes problems in association to the escape and refuge seeking of the inhabitants during a disaster.

The intermediary sector comprises the main parts of districts 2, 3, 5, 9, and 10. These have better conditions in terms of traffic and land use centralization and are able to receive the services of the central sectors or that of other parts.

This part consists of different types mainly comprising of dense organic textures with one-way access, with the difference that its density is not as high as that of the central texture and its relative density is more moderate. The presence of non-built open spaces among the texture is noteworthy. In view of these explanations, the intermediary texture is altogether less vulnerable both because of its lower age as compared to the texture constituting the central sector of the city and its larger share of open spaces.

In parts where the textures are in grid form, relief operations during a crisis and after it would be more efficiently fulfilled, and apart from zone 3, which is unable to easily communicate through a ring road with the area where health centers are located, the other areas are completely accessible owing to their particular situation and access to the city's internal met work and the presence of the external ring road. Narrow and convoluted thoroughfares within the blocks may also be seen in some parts of this area, and there is the problem of access to the blocks to some extent; nevertheless, owing to the presence of significant areas of open space in the texture, there would almost be no shortage of open spaces with potentials to serve as sites of evacuation at tines of earthquake disaster.

The marginal sector with traces of built-up sites randomly dispersed in open non-built space mainly comprises districts 4, 7, and 8. This part, owing to the shortage of urban, local, and district services, has great problems receiving services from the central sector. Some of the characteristics of this texture are low population building densities and accessible open spaces.

The recent nature of construction works in parts of the marginal sector may be considered as a positive factor in terms of identity; however, the settlement of the low-income classes on the city's margin would lead to the creation of a certain pattern of non-standard construction works which would in turn definitely result in increased vulnerability.

The presence of an open texture (of both types) and open spaces makes evacuation seeking possible; however, owing to the presence of low-speed and low safety routes, the dispatch of relief units to remote areas would give rise to basic problems in terms of relief during and after a disaster.

From the point of view of reconstruction, there are not many difficulties, except in some parts, where the streets are too narrow to allow for the passage of special vehicles.

4.1.2.5 Open Spaces

Map (Fig. 4.9) shows the distribution of open and green space in the city of Rasht. While, in consideration of the extensive area in green space in the city and its suburbs, the current per head may meet the city's demands, and most likely, from the urban planning viewpoint, it may be adequate up to a point, however, in view of the earthquake issue and its consequences, and design based on reducing the city's vulnerability, and all the criteria and standards ought to be reviewed in terms of the earthquake hazard impact and expected earthquake losses.

Experience gained from the June 20, 1990 *Manjil*-Earthquake clearly signifies that fatality outside the houses secured a high rate due to the narrowness of the streets, the frailness of the walls, and the absence of open spaces near residences, to be used as space for evacuation.

Fig. 4.9 Existing open and green spaces in the city of Rasht

Thus, the adequate proportion of the required open space with the population density the type of texture, and the age of buildings and their uniform distribution in the city are seen to be vital and logical.

The squares are geometrically classified in two groups:

1. Squares that have been preconceived and have a specific geometric type.
2. Squares that lack a specific geometric shape located in the old texture and the neighborhood centers.

The squares that have a specific geometric shape have been built in the old texture as proscribed by the comprehensive plan, on the way of the city's main transportation met works, acting as solvents for the problems of vehicular traffic. Following the origination of these squares, in most cases, other land uses have also been established, whereas the squares and plazas in the old texture have come into existence as the outcomes of their surrounding land uses, and they, in fact, serve as resting places where the locals gather and also as places where services offered by the surrounding land uses are rendered.

To name a few of the geometrically shaped squares, it is possible to point out *Meidan Shardari, Sabze Maidan, Meidan Imam Khomeini, Meidan Zarjoub, Meidan Enghelab, Meidan Shahid Baiban Gard, Meidan Shahid Ansari, Meidan Pasdaran,* and also the squares in the ring road and the entrances to and exits from the city.

These plazas and crossways have the potential to be converted into neighborhood centers owing to their accommodating a series of special land uses, such as commercial, educational, and religious land uses (e.g., mosques and shrines), and even when open space is required, it is possible to site select suitable areas in their periphery, which may both be used as sites to which public forces may be directed and also as spaces for refuge at time of disaster.

The statistics for open space in each district signifies its inadequate distribution around the city. The general open space per head in Rasht is 7.83 m^2 per person; however, the high per head rate in some areas is not a sign of the presence of open space, which may be used by the public in these districts. As an instance, the high open space per head rate in district 9 is due to the private open spaces belonging to military bases and factories.

In district 6, the City Park and the open space in the factory grounds and the abandoned land have given rise to the high rate of the open space.

The appropriate distribution of space in the city demands a strong transportation network covering all parts of the city. Only through the presence of a standard access network communication between open spaces with major open spaces, open spaces and main thoroughfares, open spaces with the whole city, and open spaces with one another is achieved. One of the most important issues is to furnish communication with open spaces after earthquakes, which emphasizes the need for establishing a resistant and safe spatial network, while the space itself should be made safe so as to furnish locations where evacuation may be performed.

By considering the per head and the coverage of proportion to density texture, etc., it is possible to discover the weaknesses and strengths of each space and its needs, and it is also possible to achieve an adequate level of spatial distribution by providing a series of principles and standards and generalizing them to the whole city.

4.1.3 Transportation Network

The structure of Rasht's main urban network is formed by two streets perpendicular to one another (north–south and east–west) and a freeway in the form of a ring road stretching from the western front and ending in the northeast, connecting the two perpendicular streets (Figs. 4.10, 4.11 and 4.12).

The two vertical and horizontal routes meet at the market and the *Shahrdari* square, which is the most important part of the city. The city's vertical route, Imam Khomeini avenue, may be considered as the city's major street, since this route is,

Fig. 4.10 Main transportation network in the city of Rasht

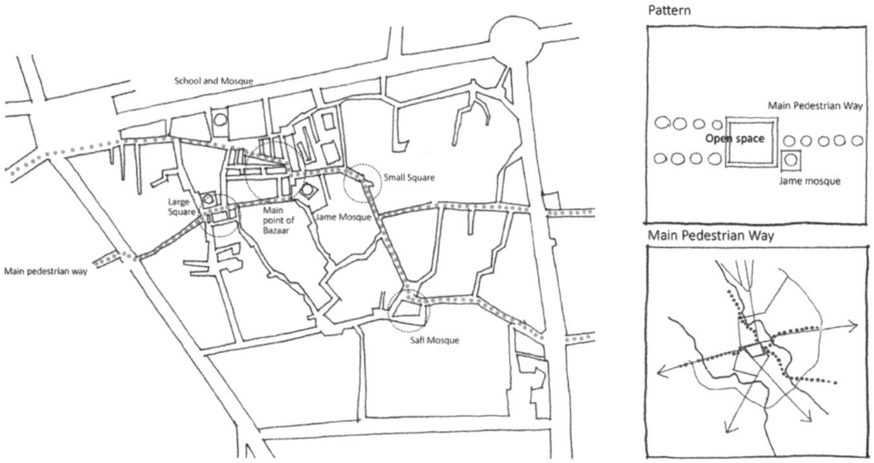

Fig. 4.11 General pattern and pedestrian ways as determining factors in shaping the main structure of the city

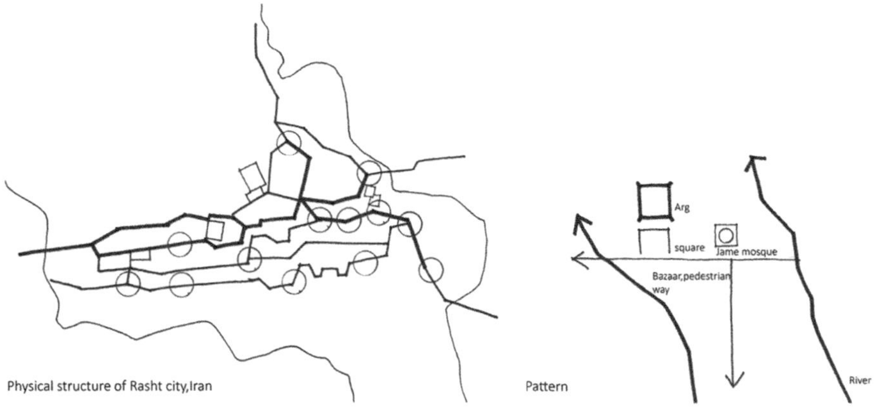

Fig. 4.12 Physical structure of the city of Rasht and role of two rivers in dividing its texture into three parts

on the one hand, the only urban transportation route to the towns of *Rudbar* and *Qazvin* and the city of Tehran, and on the other hand, it is covered on both sides from the outer bounds of the city to the *Shahrdari* Square, by important and public land uses. By a brief survey of the city's nap, the density of public and important land uses surrounding this street and within the old texture may be generally ascertained. Obviously, the transportation requirement of the mentioned land uses with other parts has also given rise to the creation of a number of new routes. As the city expanded and the central texture was left with no more room, public land uses settled around it.

It is not possible to consider any particular order on which to base the distribution nude of existing land uses, but the impact of each on the network is worth deliberation and study. The division of the transportation network can also, in relation to the justification of land use settlement, be used as a basis, though the discovery of a target-oriented system in association to the expansion of the network seems to be somewhat difficult in recent years.

Rasht's transportation network is comprised of different routes classified as below:

Passageways: These are the same old routes in the city, a number of which have been formed radially around the market and others as a horizontal series of lines. Although the passageways are not rich in terms of their width, however, they are important owing to their communicational role and the land uses that surround then. Commercial land uses are present form on the main sides of these passageways on ground level. Commercial land uses diminish as one moves away from the market toward the residential texture. In the central texture of the city educational land uses (primary and secondary schools) and administrative land uses are seen to exist along its main passageways.

Passageways may be generally classified into three groups, based on the type of surrounding land use, as follows:

1. Passageways that are considered as continuations of the market corridors and have a commercial function for the most part. Land uses bordering these passageways are usually commercial and service related. As mentioned earlier, these passageways branch off the market, and as they proceed toward the residential texture they change their scale from wholesale to retail trade such as to be able to supply the daily needs of the residential part of this texture. These types of passageways usually encounter spatial openings on their way. These spatial openings are known as centers of neighborhoods. On the sides of these spaces, public land uses of a local scale such as those of a mosque, a baker's shop, and a grocer's shop have been established. These passageways, despite having vehicular communication with other parts and their current requirements, face bottlenecks at various points and in spite of their limited widths appear as a major network.
2. Passages that have lost their functional importance (*Sahgarisazan*) and at present act as communication links between the main streets in the city and the different neighborhoods within the central texture. These passageways chiefly act as side streets in neighborhoods, and the land uses bordering this type are often residential and occasionally educational or administrative. They are currently being used both as vehicular and pedestrian routes while they are unable to answer for the density of traffic that needs to pass through then.
3. The third type is passageways that are, at present, only being used as pedestrian routes linking different points of the urban texture, but lack public and significant land uses.

The three types of streets have similar width (2–12 ms), building age, and the height of their sides. City's main streets are also divided into three groups according to their land uses:

1. Main streets that have a functional role are considered as urban nudes and edges.

2. Main streets that have urban transportation roles and are not important in terms of their surrounding land uses but from the point of view of affecting communication between important land uses and different urban points.
3. Streets that have both the characteristics mentioned above. The most important of these are *Imam Khomeini* and *Taleghani* avenues, which are both important in terms of their transportation role and their surrounding land uses.

4.2 Urban Design to Mitigate Seismic Risk

4.2.1 Vulnerability of City's Different Textures (Petrovsky 1992)

The settlement of low-income families on the city margins has given rise to a particular housing pattern in these areas which, in disregard of the earthquake hazard and with no observation of required standards pertaining to this factor, increases the vulnerability of the marginal texture, and the lack of appropriate access to health centers and their remoteness exasperates the situation.

Ultimately, the non-observance of building codes, the overhangs in building stories, the narrow thoroughfares, the non-adherence to a hierarchy of access networks, the land use distribution node, the inattention toward the vulnerability of urban infrastructures, etc., indicate the generally inconsistent state of city for confronting an earthquake.

The north–south extension of the *Zarjoub* and *Gohar Rood* rivers has resulted in the segmentation of *Rasht's* urban texture into three sectors (Fig. 4.5). This is not such an important matter in a normal situation but assuming that an earthquake was to occur and connection between the three sectors be severed (through the vulnerability of the connecting bridges), undesirable consequences could result. For instance, the differing vulnerability levels of each sector and the absence of the required facilities in each would give rise to intensified vulnerability. In other words, each sector would act in isolation and relief operation centers would not be able to service other parts of the city. One of the existing problems is the distribution of public land uses, particularly that of relief-oriented land uses across the city. The city is not balanced, and as may be seen from the map of the city, (Fig. 4.13) a great share of public land uses in different contexts such as educational and medical land uses is centralized in a small area. Although this could, on the one hand, be considered as an advantage, yet in view of Rasht's texture and its segmentation by the rivers and the unbalanced distribution of public land uses in a specific hierarchy, in an earthquake situation, the consequences of disaster are doubled in each of the three sectors. Another point, which is worth consideration, is the city's size. The greater the size of the city, the more difficult would servicing and coordination between the units across the city be, considering that public land uses are centralized in specific locations in disregard of adherence to any hierarchy between the units affiliated to each.

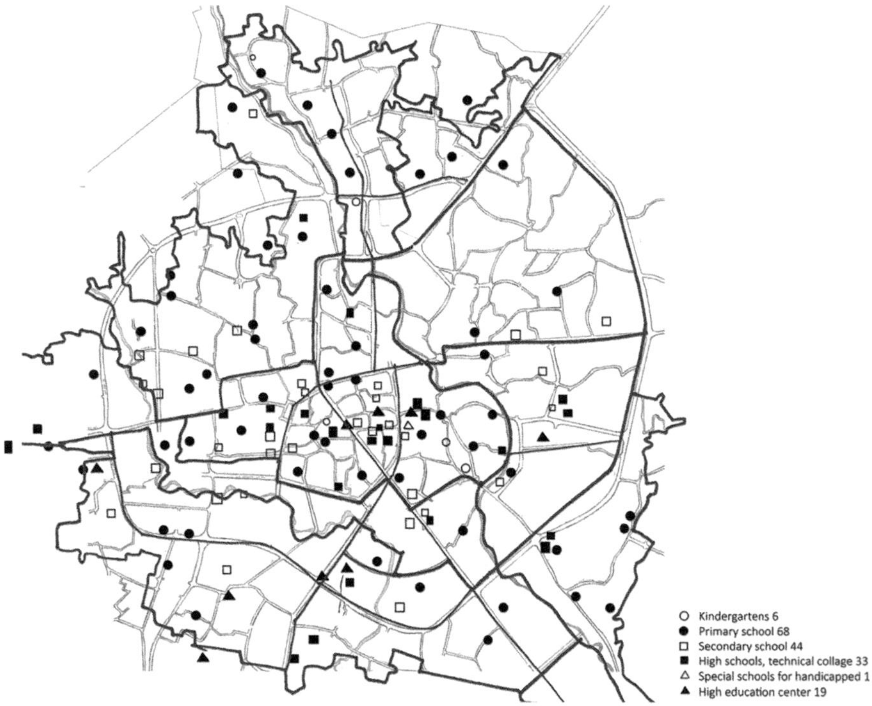

Fig. 4.13 Health and social facilities in the city of Rasht

Therefore, it is necessary for each of the three sectors to act independently when confronted with a hazard and their need for relationships with the other sectors be minimized. Such a relationship would only justifiable for the central units of every organization. This could also be worth noting in regard to the administrative and managerial structure of the city. If a major share of the city's public land uses located in certain parts and their affiliated units were, in view of population and seismotectonic features, etc., to be distributed in each of the three sectors of the texture, during an earthquake probably causing a breakdown of communications between different parts, it would be possible to achieve minimum coverage throughout the entire city.

Creation of multifunction centers in the city's three sectors, particularly with a range of important activities during a crisis in observation of a hierarchy is considered as one of the ways to deal with the above problem. This could be realized by distributing land uses that are deemed as important during a crisis such as those of a medical, relief, and open space nature in each of the three sectors (Petrovsky 1992).

4.2.2 General Characteristics and Typology of Rasht *Urban Texture Based on Accessibility Criteria*

The following types of textures may be recognized:

- District one: Regular texture, open, adequate accessibility.
- District two: Regular texture, open, inadequate accessibility.
- District three: Regular compact texture, adequate accessibility.
- District four: Regular compact texture, inadequate accessibility.
- District five: Organic open texture, adequate accessibility.
- District six: Organic open texture, inadequate accessibility.
- District seven: Organic compact texture, adequate accessibility.
- District eight: Organic compact texture, inadequate accessibility.

Types 2, 3, 5, 6, and 7 can be seen in Rasht. The textures with no adequate access, which are also compact, are the most vulnerable textures.

Urban texture may be divided in three sections according to various factors, such as the distribution and area of each type, location, service distribution and special uses, building density, building age, and population density.

The central area includes zones 1 and 6. The existence of old Bazaar and the concentration of uses and services have led to draw residents from other parts of the city and therefore causing difficulties for the pedestrians as well as motor vehicles to use the area. Higher population density, compact and fine grain texture and small plot sizes have intensified the problems. The dominant texture in this area is generally organic, and compact, with inadequate access into the area. Deteriorated buildings due to the old age, high density, and inadequate accessibility will all make this area the most vulnerable for all stages of the disaster.

Middle area includes the main part of zones 2, 5, 3, 9, and 10. From accessibility point of view and concentration of land uses this area is in better situation and can receive necessary services from the center and other areas. This area includes different texture types, but mostly regular compact and organic compact, with one-way access. Its compactness, however, is not the same as the central area, and its density is lower too. Open and un-built spaces within the area are considerable. So this area is generally less vulnerable.

In the case of grid system, rescue and relief activities can take place more efficiently, because of good location and access to the internal city network and the external ring road is completely accessible.

The fringe area, with scattered built-up spots in the un-built open space includes zones 4, 7, and 8. Poor urban services in this area make it extremely dependent on the central area for its daily needs. The new construction, with adequate open spaces and low population density in this area, lead to lower vulnerability. However, low-income residents in the area will result in special pattern of construction that is generally under-standard and therefore quite vulnerable (Petrovsky et al. 1998).

Commercial uses include the following:

- Central *Bazaar*, which is located in the central core of the city and aside from commercial activities, plays other social, religious, and recreational activities as well. The organic, compact, and normally under-standard construction of this area makes it the most vulnerable area of the city against earthquake and fire.
- District *Bazaar*, which includes small and large commercial units, outside of the central *Bazaar*.
- Neighborhood *Bazaar*, which is combined with housing.
- Passages, which are the new form of *sara.*
- Temporary *Bazaar*, daily, weekly, or seasonal.

4.2.3 Methods and Factors for Estimating the Damage to a Building and to a Group of Buildings

The method used is based on the amount of economic damage, or the cost has to be paid to rebuild similar building. The factors involved include the following (see Table 4.1):

(a) The factors which affect the building, but are not part of the building:

1. Land slope (L_1).
2. Soil type (L_2).
3. Building height (L_3).
4. Building plan (L_4).
5. Construction quality (L_5).

(b) The factors which are part of the building and their damage means damage to the building:

1. Building system (L_6).
2. Floor structure (L_7).
3. Façade materials (L_8).

Following the quantification of qualities, the damage ratio is calculated by using this formula (see Table 4.2):

$$(\text{LR}) = \left(L_1 \times L_2 \times L_3 \times L_4 \times L_5 \times \frac{1}{4}[(0.645 \times L_6) + (0.323 \times L_7) + (0.032 \times L_8)]\right).$$

The result is known as the damage ratio, and it ranges between zero and one. The ratio higher than one is considered as one. In the following table, the qualitative interpretation of the damage ratio is given.

Table 4.1 Converting quality factors into quantity

Factor number	Affecting factors		Damage weight
	Factors name	Details	
1	Land slope	0–15°	1
		16–30°	1.1
		30<	1.2
2	Soil type	A.F 2>	1.2
		A.F 2–4	1.3
		A.F 4<	2
		A.F = Amplification factor	
3	Building height	One story	1
		Two story	1.2
4	Plan	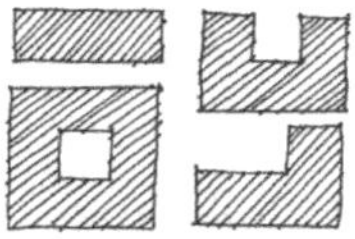	1
			1
			1.1
			1.1
5	Construction quality	Good	0.6
		Average	0.8
		poor	1
6	Construction system	Steel frame with brace	1
		Concrete frame with brace	2
		Brick load bearing wall	4
		Cement load bearing wall	3.5
7	Floor construction system	Steel beam with arched roof	3
		Wood beam with clay	4
		Reinforced concrete beam with hallow block	3
		Scaffold with metal sheet	1
		Dale reinforced concrete	1
8	Façade materials	Fixed bricks–stone	0
		Non-fixed bricks–stone	1
		Mud	0.5

Table 4.2 Interpretation of damage ratio (LR)

Damage ratio (LR)	Interpretation
$LR \leq 0.25$	Low damage, usable but requires repairs, repair is possible at the same time as evacuation
$0.26 \leq LR \leq 0.5$	Medium damage, repair is possible after evacuation
$0.51 \leq LR \leq 0.75$	Severe damage, evacuation is compulsory, and reconstruction is required

4.2.4 Texture Vulnerability Methods

Combining various factors

- External factors:
  - Land slop (L_1)
  - Soil type (L_2).
  - Dominant height of buildings (L_3).
  - Plan (L_4).
  - Dominant construction quality (L_5).
  - Population density in the district (L_8).
  - Texture complexity in the district (L_9).
- Internal factors:
  - Construction system of buildings (L_6).
  - Floor structure (L_7).

Qualitative data is then converted into quantity and by using the following formula, damage ratio is calculated (see Table 4.3):

$$(\mathrm{LR}) = \left(L_1 \times L_2 \times L_4 \times L_5 \times L_8 \times L_9 \times \frac{1}{4}[(0.666 \times L_6) + (0.334 \times L_7)]\right).$$

4.2.5 Transportation Network in the City to Mitigate Seismic Risk (Detail Information on the Subject is Provided in Chap. 6)

Transportation network in the city of Rasht consists of **passages**, which of which are the old alleys around *Bazaar* in the radial form, and others are in parallel linear form. In spite of their limited width, they play significant role in city's transportation. **Highways** were proposed and built following the ratification of the comprehensive plan in 1972, in order to reduce traffic volume in the central area and increase the speed to connect different parts of the city. **Main streets** were built from 1925 on and in fact are the arterial network of the city. **Secondary streets** connect neighborhoods with the main streets. Their widths range from 4 to 12 m (Fig. 4.14).

One of the important factors that disrupt rescue and relief after a disaster is population congestion in the affected area. In these situations, having the area under control is a priority. The most significant element that has to be put under control is transportation network. The more secondary roads and passages in the area, the more difficult will be control. As the number of streets and alleys in the affected area are

Table 4.3 Converting qualitative attributes of districts into quantitative

No. of factor	Affecting factors		Damage ratio
	Name of the factor	Description	
1	Land slope	0–15°	1
		16–30°	1.1
		30° ≤	1.2
2	Soil type	A.F 2>	1.2
		A.F 2–4	1.3
		A.F 4<	2
		A.F = Amplification Factor	
3	Building height	One story	1
		Two story	1.3
4	Plan	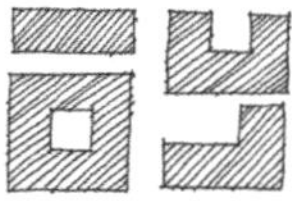	1 1.1
5	Construction quality	Good	0.6
		Average	0.8
		poor	1
6	Construction system	Steel frame with brace	1
		Reinforced concrete frame, with brace	2
		Brick load bearing wall	4
		Cement block load bearing wall	3.5
7	Floor construction system	Steel beam with arched ceiling	
		Wood beam and clay	3
		Reinforced concrete	4
		Beam with hollow block	3
		Scaffold with metal sheet	1
		Dale reinforced concrete	1
8	District population density	Less than 70 in hectares	1
		7–150 in hectares	1.2
		150–250 in hectares	1.4
		250 and more	1.7
9	Texture complexity	- Open and regular with appropriate access	0.6
		- Open and regular with inappropriate access, organic open with appropriate access	0.8
		- Open, organic with inappropriate access, regular compact with appropriate access	1
		- Organic compact with appropriate access	1.2
		- Regular compact with inappropriate access	1.6
		- Organic compact with inappropriate access	1.8

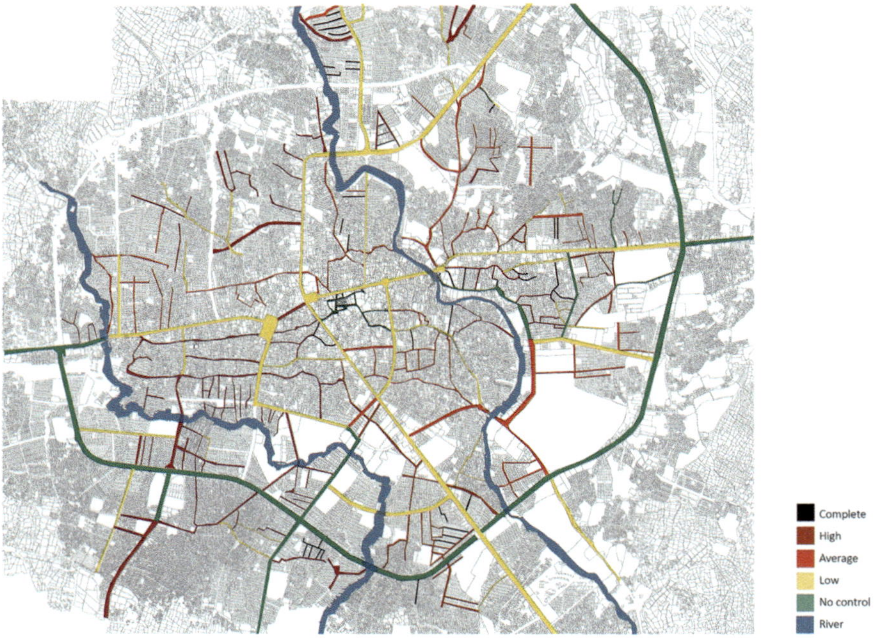

Fig. 4.14 Controllable passages in case of disaster in the city

limited or could be controlled, rescue and relief activities will function better, and as a result, the vulnerability will be reduced.

One of the factors which increases vulnerability and causes damages and life loss is the height of passages' walls. The higher the walls and shorter the width of the passage vulnerability will increase. The collapse of walls in the passages will block access which will inhibit rescue and relief activities with unfortunate consequences. Buildings/walls ages, pavement material, lighting system, water distribution system, sewage disposal system, and the details of street elevations are also factors with various impacts on vulnerability of passages (Bahrainy et al. 1977b).

4.2.5.1 Some General Principles to be Applied in Access Network Design to Reduce Seismic Vulnerability

- Shorter distances between different land uses.
- Network hierarchy.
- Shorter distances between critical land uses during disaster.
- Fast access to important land uses.
- More possible control.
- More possibility of escape and evacuation through the network.
- Adequate speed, shorter length, and more safety of the network.

- Less traffic in the network (%).
- Lower traffic (users) intensity.
- Higher ratio of open space over built-up areas.
- More direct connection between important uses.
- More connection between different parts of the network.
- Less the number of cul-du-sacs in the network.
- More vehicular access to the network.
- Less the number of intersections.
- More independent network for each district.
- The lower the ratio between the length and width of passages.
- The more the ratio between the width to half of the walls' height of the passages.
- The more seismic codes are applied in the construction of utilities and facilities.

One of the design objectives is to make modifications in the texture and make optimum use of the existing condition. Geometrical modification of roads, particularly the main highways in the city has the priority. The only highway in the city is frequently interrupted by intersections, which causes traffic jam and blockage. Elimination of intersections, by building overpass where necessary, is recommended. Three alternatives are proposed here (Fig. 4.15).

In the following alternative (alt. 2) attention is paid to the relation between land uses and movement pattern before earthquake and after, on its basis which a new

Fig. 4.15 Improvement of existing network and proposing a new one (alt. I)

Fig. 4.16 Proposed road network in Rasht (alt. 2)

network is proposed. In the proposed network the old passages in the city's central district are used as local access which are connected to collectors to provide accessibility to internal areas. The new belt around the central area will provide easy and speedy outside access to the central area and will also provide access to different locations of the city without the need to go through the central area (Fig. 4.16).

Alternative 3 (Fig. 4.17) is in fact the combination of two previous alternatives. Effort is made here to, besides making optimum use of the existing situation, the proposed modifications being designed in a way not lead to major structural changes and respond to the present needs of the city, considering earthquake disaster and providing connection between critical and sensitive points during crises. On this basis, four vehicular paths are provided on both sides of the two rivers from south to north, connecting neighborhoods with different parts of the city, and horizontal parallel paths for outside to outside and inside to inside all create the main structure of proposed network, based on the existing pattern.

4.2.5.2 Detail Road Modification in the City of Rasht to Mitigate Seismic Risk (More Detail Proposals on Open Spaces and Road Network are Provided in Chap. 6)

The following case is given here as an example (see Fig. 4.18).

Fig. 4.17 Proposed road network in Rasht (alt. 3)

Fig. 4.18 Proposed modifications in a typical road network in the city

4.2.6 Topography

The problems stemming from the city's topography have already been alluded to, of which it is possible to point out the most important which is surface water collection. The importance of this issue could be doubled when we consider the city's high volume of annual rainfall. Among the other problems related to this theme, one could point to the emergence of numerous stagnant waters (lagoons) throughout the city, a part of which relates to the type of topography. Furthermore, another factor influencing the city's horizontal expansion could be its topography.

Farming land may be divided into two types:

1. Land inside the urban texture.
2. Land covering the city's periphery.

One of the effective applications of these types of land, particularly that of the first category, is during an earthquake. In view of the conditions arising during an earthquake situation, there is need of open space with varying dimensions and scales. The utilization of open spaces for emergency and temporary settlement the settlement of relief teams, etc., was formerly discussed in brief. Farming land dispersed in the texture provides suitable open space in an earthquake situation owing to its size and distribution; nevertheless, the issue of these lands and their being put at the public's disposal at times of emergency should be clarified and defined through the enactment of specific regulations and rules (Fig. 4.2).

The land surrounding the city is forever a restraining factor toward its horizontal expansion. Although this may, on the one hand, be deemed as a problem, however, from the point of view of the earthquake factor, and in view of the fact that the larger the city, the greater the area prone to seismicity and the greater the probability of damage, the presence of farming land impeding the city's growth is considered as a positive factor. Hence, this point should be taken advantage of and the grounds for the physical expansion of the city sought elsewhere.

The only existing forest near *Rasht* is 280 hectares of land to the south of the city. This area is considered as an advantage which could contribute to the creation of a recreational/cultural environment for the region in normal situations. Moreover, during an earthquake, it could act as a suitable location in which camps and temporary settlements could be set up.

Topography and natural feature are determinant factor in risk level and eventually the vulnerability of different areas in a region. Any construction on the elevated areas with steep slopes is threatened, and conditions, such as soft and watery soil permits landslide and liquefaction will occur.

Presence of areas prone to liquefaction in the urban area is a great hazard regardless of which construction works have been pursued in the past, especially at the city's center where major land uses are settled and the texture is old, the thoroughfares narrow, the street sides unsafe and the density high.

The presence of incompatible land uses within the city's residential quarters, particularly hazardous land uses such as the oxygen factory, the *Gilan* Electrical

Factory, the Pars *Khazar* Factory, and the step-up and step-down electricity stations, creates dangers to their surrounding texture.

The settlement of low-income families on the city margins has given rise to a particular housing pattern in these areas which, in disregard of the earthquake hazard and with no observation of required standards pertaining to this factor, increases the vulnerability of the marginal texture and the lack of appropriate access to health centers and their remoteness exasperates the situation.

Ultimately, the non-observance of building codes, the overhangs in building stories, the narrow thoroughfares, the non-adherence to a hierarchy of access networks, the land use distribution node, the inattention toward the vulnerability of urban infrastructures, etc., indicate the generally inconsistent state of city for confronting an earthquake.

Finally, by consideration of the seismic hazard impact and other natural factors in the creation of the city's spatial structure and also by focusing on population issues and the city's present physical state, it is possible to reduce the city's vulnerability to the safe and economically acceptable level in future development planning.

Different types of urban texture in the city of Rasht are categorized as follows:

1. Regular open texture with suitable access.
2. Regular open texture with unsuitable access.
3. Regular dense texture with suitable access.
4. Regular dense texture with unsuitable access.
5. Open organic texture with suitable access.
6. Open organic texture with unsuitable access.
7. Dense organic texture with suitable access.
8. Dense organic texture with unsuitable access.

Types 2, 3, 5, 6, and 7 are chiefly observed in the urban texture of Rasht.

It may conclude that the regular open texture with suitable access and the open organic texture with suitable access will not create any particular problem; however, dense textures with suitable access are only problematic in terms of earthquake since their share of open space is minimal, but textures with unsuitable access which are also considerably dense are the most vulnerable of textures.

4.2.7 The Role of Land Use in Seismic Risk Reduction

Land uses play a significant role in reducing risk and increasing resiliency. Land uses may be divided into sensitive and non-sensitive uses, on the one hand, and vulnerable and damaging uses, on the other hand. Sensitive uses are the ones with significant values, such as education, health and security centers. Damaging uses are the ones that not only suffer damages themselves, but also inflict damages on other uses too, such as natural gas tanks, gas stations, main gas pipelines, and bridges.

Land uses should be distributed in the city in the mixed use form and on the basis of their dependence on consumers and relative to their spatial structure, i.e., neighborhood, block, etc. Aside from emphasizing on issues such as sensitivity, vulnerability and damage, other criteria such as compatibility, desirability, capacity, dependency, and meaning may be also used. Obviously, there are also certain other important factors which affect land use vulnerability, such as topography, slope, geological layers, distance to faults, natural gas pipeline, access network, and high voltage power network. Besides, building and human density also affect the number of life losses and amount of damages inflicted by earthquake. A careful design and planning of quality and quantity of land uses and specially their appropriate locations will not only prevent their vulnerability to a large extent, but also prevent the secondary damages and losses, such as fire (Bahrainy et al. 1997a).

4.2.7.1 The Role of the Public Services

Activities at key public facilities during disaster emergencies are categorized along a time axis as emergency response, restoration, and reconstruction. Emergency responses activated immediately following the outbreak of a disaster are of vital importance; in that, their implementation greatly effects the total amount of damage sustained. In particular, rescue, evacuation and refuge, and emergency medical care are especially important, and in order for them to be effectively implemented, it is essential that quick and efficient measures must be taken by administrative authorities, fire departments, and medical organizations. In addition, enough space must be provided for evacuation and refuge at public facilities such as schools. For this reason, core public facilities such as administrative buildings, hospitals, and schools in large heavily populated urban regions should be utilized as the "bases for disaster operations." This will require the incorporation of higher levels of disaster protection in these critical facilities compared to the general standards.

In the case of the city of Rasht, the shortage of public services and their inappropriate distribution based on population distribution has resulted in a number of problems. The concentration of the city's major land uses at the dense center of the city where the streets are incapable of responding to the passage of people coming to the area from the other parts, while creating day-to-day urban mobility problems, causes many deaths and impedes relief operations in a crisis situation owing to the excessive concentration of population at one point.

The presence of major health centers in one part of the city, the issue of relief dispensed to all parts, and the important land uses cited the need for safe and high speed routes. At present, the city has been divided into three sectors by the rivers. If there is a crisis and communication through the bridges breaks down, access to health and relief centers from all parts of the city would face serious difficulty.

Presence of areas prone to liquefaction in the urban area is a great hazard regardless of which construction works have been pursued in the past, especially at the city's center where major land uses are settled and the texture is old, the thoroughfares narrow, the street sides unsafe and the density high.

The presence of incompatible land uses within the city's residential quarters, particularly hazardous land uses such as the oxygen factory, the *Gilan* Electrical

Factory, the *Pars Khazar* Factory, and the step-up and step-down electricity stations, creates dangers to their surrounding texture.

4.2.8 Building and Human Density

The higher the human density within the land uses, the higher the human losses. This is especially critical in the case of sensitive land uses such as schools, hospitals, nursing homes, and residential complexes. Attention should be paid to the fast and safe evacuation standards during disaster. Population density has risen from 98 persons per hectare in 1986 to 123 persons per hectare in 1991 Available statistics indicate the highest rate of migration to Rasht in the province from the surrounding regions. This is the outcome of the settlement of light industries or agriculture-related industries and the spread of service activities. Doubtless, the increasing population density would, in view of the constraints toward the citys horizontal development, create numerous problems (Kazemi et al.).

Inappropriate building density may, similar to human density, increase the loss and damages inflicted by earthquake. This will be seen in the high-rise buildings, on the one hand, and the dense and compact areas of the cities, on the other. In either

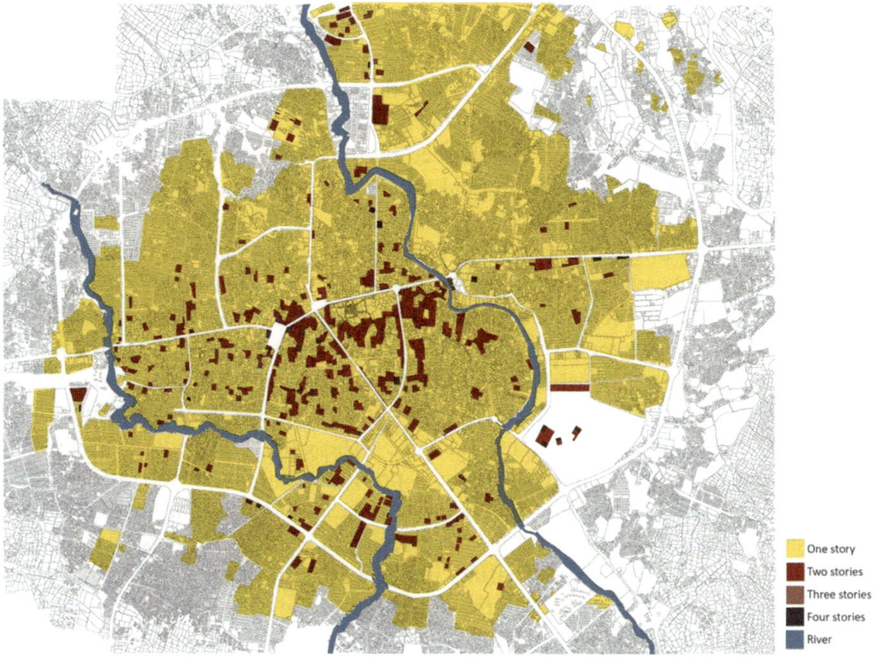

Fig. 4.19 Existing building density in the city of Rasht

case, providing necessary services to the affected areas during an emergency would be difficult. Unjustified density in the high-risk areas will increase vulnerability.

Increasing human density within land uses may lead to higher number of life losses. This is particularly more important in the case of sensitive uses such as schools, nursing homes, and residential complexes. Fast and safe evacuation standards during an earthquake need to be taken into account (Fig. 4.19).

4.2.9 Long-Term Development Scenarios for the City of Rasht

The contemporary urban form of the city of Rasht is in part derived from its historical antecedents. Historically, Rasht developed from a small village to a successful market town, located at the crossroads as major east–west and north–south routes. The east–west route was older and claimed to be a part of the original "silk route" to China. It connected *Foman* and *Lahijan*, known for their feudal rivalry. Then, almost four hundred years later , Rasht was only a small village. With the establishment of a port of *Pirbazaar* to the north, and the arrival of Russian traders, the north–south route became dominant.

The urban pattern of the central business district and the older neighborhoods near the city center still reflect the urban form of the indigenous market town with its crossroads as the two major axes. These axes have become the major north–south and east–west highways today. A three-quarter ring road intercepts these through traffic and helps it to bypass the core city. Thus, a three-quarter ring road superimposed on the original axes essentially captures the basic circulation pattern. Until the turn of the century, the built-up area was defined by the two rivers. Today, the city has grown beyond these rivers and additional radials and spurs have been added to the basic pattern.

The urban form of the city has three basic patterns: the older indigenous form, the contemporary developments of the last two or three decades, and a transitional mixed pattern reflecting renewal and transformation of fringes from the traditional to contemporary development patterns. These differences are reflected in the streets layout, building heights, construction standards, and intensity of development. The older, indigenous pattern includes most of the central business districts and the older neighborhoods near the center and the main axes. These neighborhoods and the older business area are characterized by organic, pre-industrial street network, densely packed housing stock and neighborhoods, or *mahallas* organized around a mosque and cluster of shops. From the seismic hazards point of view, these areas of residential and business districts of a pre-industrial vintage are the most vulnerable.

The more recent development in the periphery of the city—along the ring road and along the radial highways leading out of the city—is still inchoate in form. Most of the land use can be characterized as highway commercial or light industrial, although scattered residential developments can also be found along these arterial corridors. Most significant and planned increment of new residential of new residential development has occurred to the north along a new spur off the *Bandar Anzali* route. These are the districts of *Busa* and *Shahrak Golsar* which includes multi-story residential blocks and large homes surrounded by ample open space and new street layouts. The candidate for a future secondary center seems to lie to the south, along the highway to *Rudbar*, *Manjil,* and *Loshan*. Here, along the highway, a university campus is already located, and a major site for an industrial park is currently under development (Banerjee 1993, 7).

There are several immediate issues that should be considered in developing urban from and urban design proposals for the city. First and foremost is the question of sustainability of urban development from the seismic point of view. We can borrow the concept of "carrying capacity" from the field of ecology to define the natural limits of possible built form based on seismic consideration. Obviously, the seismic microzonation for the city and its surrounding area showing liquefaction and high vibration intensities should dictate the pattern and intensity of built form possible. Based on preliminary seismic microzonation, it is determined that the northwest (NW) quadrant has serious liquefaction problem as well as entire southern part of the urban area. Only in the northeast quadrant toward the road to the airport and *Bander Anzali,* preliminary seismic microzonation determines firm non-liquefaction soil conditions.

From the seismic safety point of view, there are several other issues about the existing built form of *Rasht*.

1. Rasht is a mono-centric city, and a vast majority of the business, retail, administrative, and institutional activities are concentrated in the center. This makes the administration and economy of the city particularly vulnerable, if the center is badly damaged.
2. This possibility is likely since the center of the city is also the oldest district, with pre-industrial, pre-modern urban form, characterized by older, and (some dilapidated) building stock, and narrow streets and alleys.
3. It is also this center where the pedestrian presence is the highest, especially in the evening hours when people congregate to shop and socialize. The public spaces are crowded, especially in the Bazaar area. If a major earthquake were to occur at that time of the day, it could wreak havoc in terms of death and injury.
4. There are several older neighborhoods with a built form of singular vintage. These neighborhoods face many of the same problems discussed previously.
5. Some of these neighborhoods are not well served by arterial streets and are not particularly accessible. The existing network of narrow alleys and passageways is likely to be blocked by fallen debris or collapsed structures, thus causing death and injury or trapping panicked residents. Also, they are not wide enough to allow emergency vehicles to get through.
6. Although the city has some formal parks and gardens, all neighborhoods are not equitably served by these existing open spaces. Indeed, there may be an overall deficit of open spaces strictly from the perspective of recreational and public health needs of the resident population alone. Moreover, when one considers

the benefits of open spaces from the standpoint of accommodating emergency evacuation during major earthquake and aftershocks, the deficit in open space becomes a poignant issue.

Obviously in developing appropriate scenarios for urban form alternatives for Rasht, we must consider demographic and economic variables as well. The following basic question could be raised:

1. What are, for example, the future growth scenarios for the city?
2. To what extent regional and national development plans and spatial strategies will affect Rasht?
3. Will Rasht continue to remain a third-tier city in the Iranian urban hierarchy or emerge as a more important urban center as a result of the new economic and political order of the former Soviet republics along the Caspian coast?
4. What are the likely corridors of development in the regional context?
5. How large and how fast is Rasht likely to grow?
6. What would be the population size of the Rasht area in 2021?
7. How many cars will be owned then?
8. What will be the nature of economy and industrial base?
9. What should be the future of agricultural land that surrounds Rasht?
10. What should be the future of older neighborhoods? Conservation or clearance, and renewal?

Some of these are questions of values. Others are of judgment and projection.

4.2.10 Long-Term Development Scenario

The major seismic concerns in urban form and urban design for the city of Rasht fall in two basic categories.

Centralized conceptual scheme for prevention of large-scale disasters and rational use of natural resources.

1. Retrofitting existing urban form to minimize physical damage, injuries, and loss of life.
2. Developing urban form and urban design guidelines to minimize seismic risk in the future development of the city.

First, let us consider existing infrastructure—i.e., the lifeline utilities, the network of highways and boulevards, system of open spaces, and the like. It should be possible to identify, either analytically or based on expert judgment, its strengths and weaknesses in serving effectively in case of emergencies. For example:

1. Does the existing network of highways offer a means for quick ingress of regress for all population?
2. Which districts and neighborhoods are poorly served?
3. Are the hospitals, clinics, and the like adequately served by the existing network?

4. What are the vulnerable links in the network?
5. What strategic improvements (street widening, improved intersection design, strengthening bridges or overpasses, construction of new but critical arterial road links, etc.) can secure and augment the existing network?

A second strategic action is to take some immediate remedial actions in the other neighborhoods, which are most vulnerable to major disaster. Non-engineered structures, narrow alleys and passageways, lack of public open spaces, etc., make these districts both vulnerable to earthquake damage and vulnerable for post-disaster emergency actions. Therefore, strategic improvements such as selective street widening, creating neighborhood open spaces, structural improvements of narrow alleys to make them safe escape routes, and the like are needed. Examples of such improvements are discussed in the following section.

One of our operating assumption here is that given the nature of building construction, in most cases people are better off outside in an open space than remaining inside the building. This makes availability of open space (large enough to be unaffected by falling debris) in the immediate vicinity of one's home a desirable feature of the urban form. If we examine the urban form of Rasht, this objective seems unattainable for the most part as long as existing open spaces are defined strictly in terms of parks, playgrounds, gardens, and the like. But for emergency planning, the concept of open space needs to be broadly defined, so that emergency open space needs can be met by such spaces a: unused or underused rights-of-ways and the like. The difference in the figure-ground relationships between solid and void in urban form can be quite significant, depending on the definition of open space. However, from the urban design point of view, we need to make sure these are accessible and available during such periods. There is no question that the supply of public open spaces needs to be expanded significantly, both at the neighborhood and at the city scale. All public open spaces should be equipped with emergency toilet and sanitation facilities, space for emergency shelter, and the like. Furthermore, position of additional open spaces will contribute further to enhance the visual appearance and the quality of life in the city (Banerjee 1993,7).

The first alternative considered for long-term development is clearly the recently completed comprehensive plan, which is based upon certain assumptions and objectives. It presents a set of defined policies. In Fig. 4.20, functional urban region of the city of *Rasht* for development scenario within existing 10 districts is presented.

The year of 2021 is used for long-term development planning scenario, and the year of 2006 is used for disaster preparedness scenario planning.

Generally, in view of the existing constraints, the horizontal expansion of *Rasht* would be difficult, and apart from an available area of vacant land within the urban fabric, it would be almost impossible to attempt further development.

Development constraints in this part are the following:

1. Presence of extremely fertile land to the north, west, and east of the city.
2. Area of land dedicated to livestock farming and agriculture and the *Sefid Rud* Livestock Company in the south of the city.

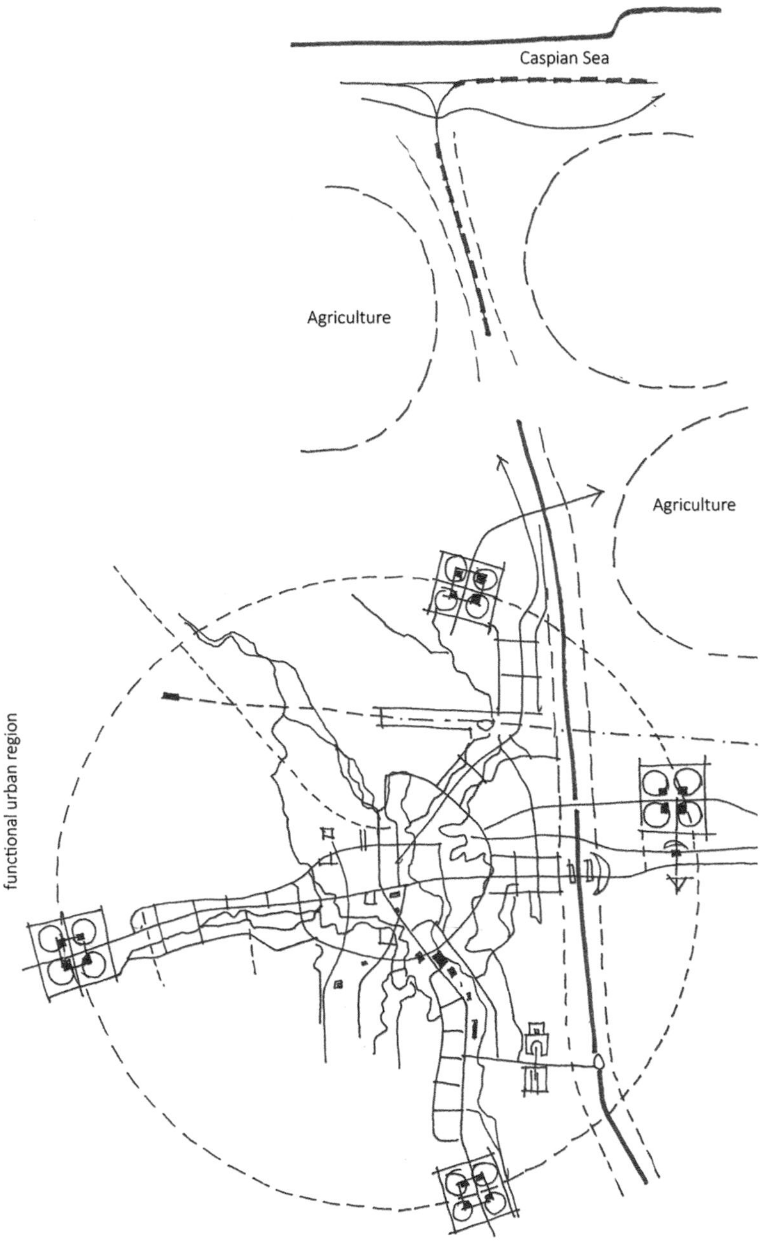

Fig. 4.20 Rasht functional urban region of metro-satellite development scenario (after Music 1993)

3. Presence of the airport and its development plan in the north, as well as farming land.
4. Passage of the highway on the east, besides the presence of farming land.
5. Presence of major and minor industries in the south.

The general development policy of the city is, in view of the above constraints and the increasing population trend, based on two directions:

1. Vacant and abandoned land inside the texture composed of agricultural or orchard land uses to be given first priority in view of the final plan.
2. The city's vertical expansion remains as the only alternative for population settlement, considering the latter's growth rate. Although there is potential for development in a part of the lands located on the south of the city, nevertheless, this has to be attempted in regard to the high liquefaction potential and required costly foundation of buildings and structures.

Here, it is necessary to point out the fact that the city's jurisdiction determined by the comprehensive plan approved in 1990 ought to be reviewed. The above mentioned areas have been determined as described below.

"The areas of jurisdiction and services are, in view of Rasht' nature, identical and the area that was approved in the plan starts in the north of the airport, and from the west to the existing ring road and the land inside the urban boundary and also along the road to Tehran up to the boundary of proposed land uses, and from the western part of the proposed land use along the T ring road the medical land use, and from the east up to the border of the proposed land uses."

4.2.10.1 Other Alternatives of Long-Term Development

Here, we will consider three different long-term and strategic form options for development of the city of Rasht, based on the assumption that decentralization, dispersion, and de-concentration along with ample and easily accessible open spaces are most likely to mitigate negative consequences of a major earthquake. These options, however, do not preclude other policy options like seismic microzoning and building codes, density standards, etc., which could further augment seismic safety objectives of future urban growth and development (Banerjee 1993, 7; Bahrainy et al. 1998).

Centers Hierarchy Model

Centers hierarchy is a model (Fig. 4.21), which envisages a transition from a mono-center to a poly-center city. The original core of the city could still remain the dominant center, but the future growth of the city new centers can be planned to accommodate new businesses, shopping, and other facilities. Businesses, which need additional space for expansion, could eventually relocate to these centers.

There may even be two different versions of the Centers hierarchy model. In one model, all new centers can be seen as sub-regional and unspecialized, all located in

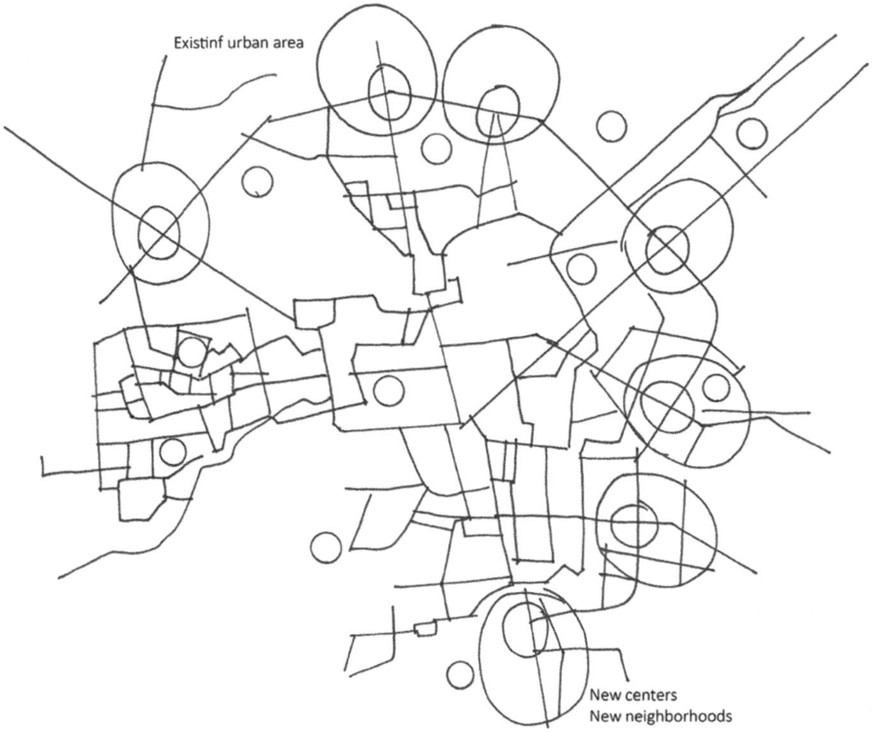

Fig. 4.21 Centers hierarchy model

the periphery, simply offsetting the centripetal tendencies of a mono-centric city. In a variant of this, the Centers hierarchy model could include both second- and third-order centers, namely district and neighborhood centers complementing the main center. This latter variant could embellish the existing structure of neighborhood centers in the older parts of the city.

Radial Corridor Model

Radial corridor model (Fig. 4.22) simply acknowledges the existing trends of strip development along the major arterial routes radiating from the center of the city.

Since development follows access opportunities, this type of corridor development is quite common, especially with increasing car ownership. If the corridors are intensively developed, they can also become transit corridors. The nemesis of corridor development is the highway strip, which tends to be visually unattractive and can cause pedestrian and traffic safety hazards. Yet, if properly planned and designed, radial corridor development model can have several advantages from the

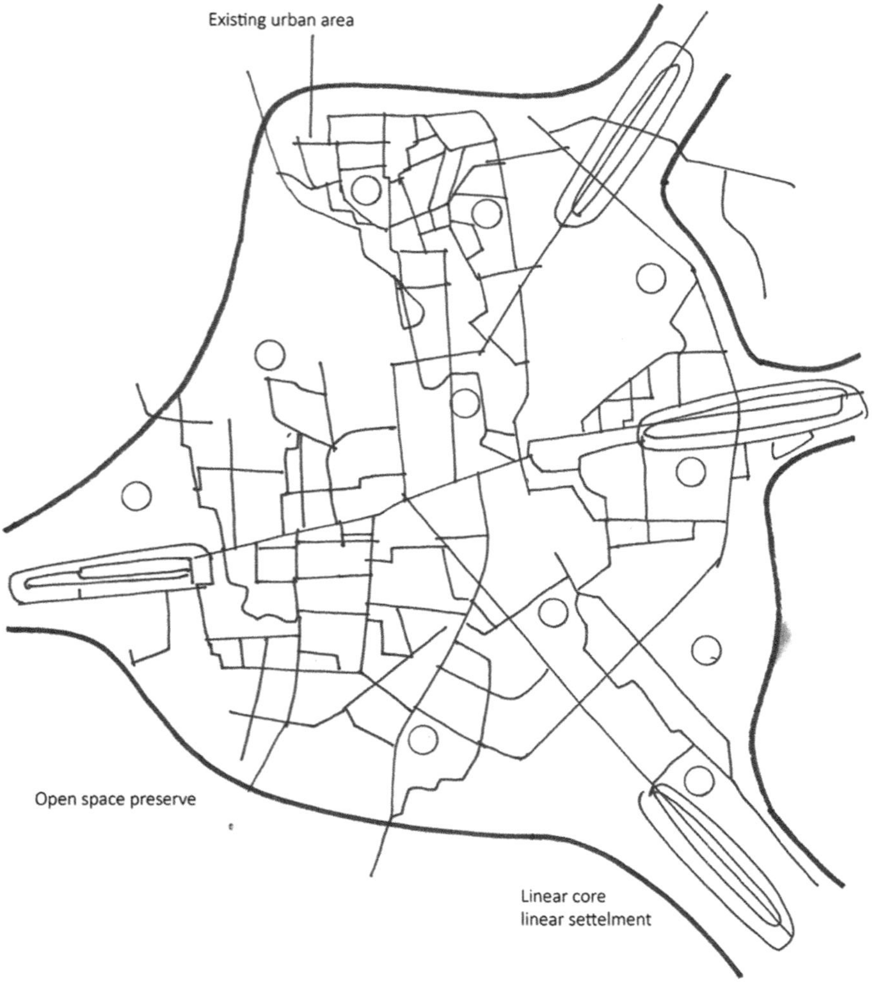

Fig. 4.22 Radial corridor model

seismic point of view. For instance, radial corridors if limited to a certain width can be separated by wedge-shaped open space preserves consisting of existing agricultural land.

The corridors thus can serve to centralize business, shopping, housing, and public amenities, on the one hand, and make open space accessible to all, on the other hand.

Bi-Polar Metro Model

Bi-polar metro model (Fig. 4.23) represents a somewhat longer term development option for the city of Rasht area. It proposes development of several sectors of

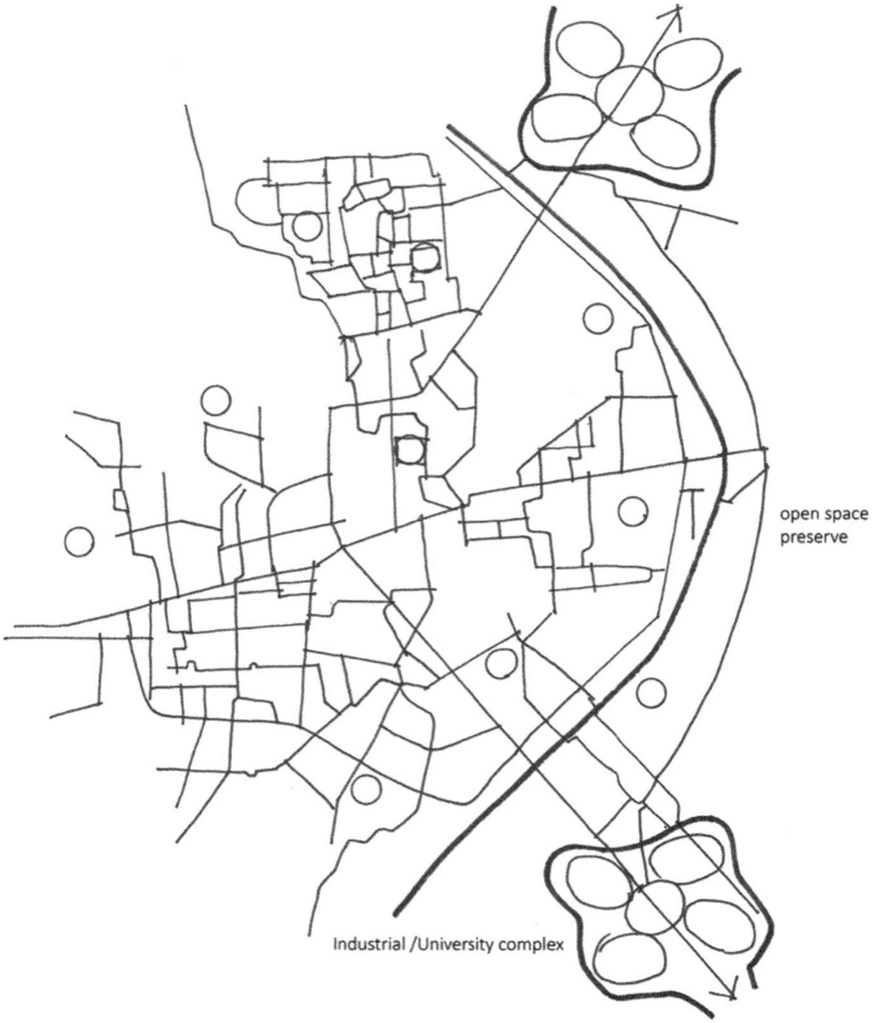

Fig. 4.23 Bi-polar metro-sector model

the metropolitan area. It assumes that the north–south axis is likely to be the most dominant growth axis because of the future importance of access to the Caspian coast from Tehran.

It is possible that the metro-sectors will be served by two major poles of growth, the airport complex to the north and the university-industrial park complex to the south. A north–south bypass might be necessary to redirect through traffic, at the same time serving the two poles. This highway will become the major axis of future development and growth. The metro-sectors will be organized in clusters of neighborhoods surrounding the major business districts (Banerjee 1993, 7).

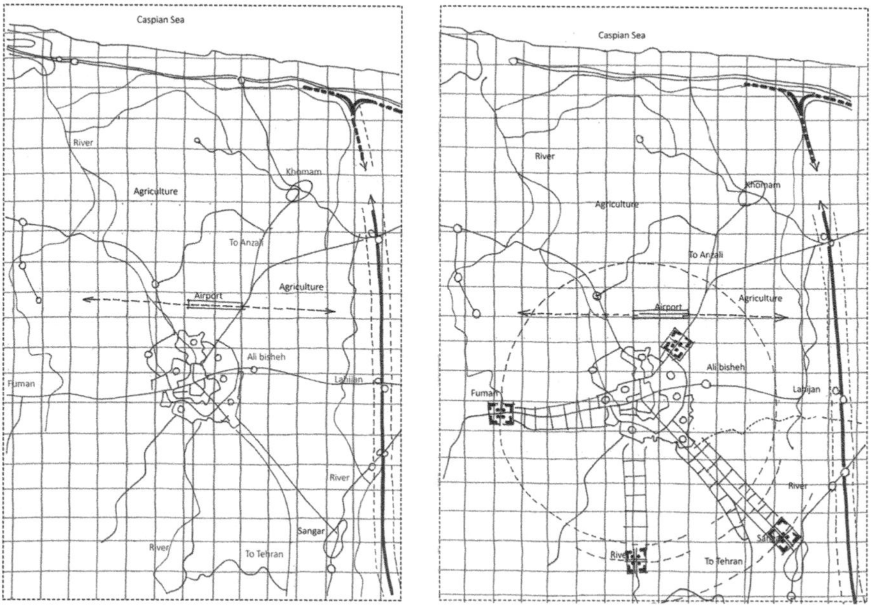

Fig. 4.24 Functional urban region of the city of *Rasht* for development scenarios in 2021

Metro-Satellite Model

Metro-satellite model (Fig. 4.24) essentially represents a much long-range future where the growth will be directed to four planned satellite new towns located at the intersection of the axial highways and peripheral highway (Music 1993, 1998). The new towns will be separated by green belt, open space, and agricultural land. A north–south bypass is proposed in this option also. The scenario encompasses a much larger metropolitan frame, almost of a sub-regional scale. This scheme would require considerable investment in new infrastructure, while preserving natural resources, prime agricultural land, and other resources needed for emergency use.

For the purpose of reduction of earthquake consequences and mitigation of seismic risk, short- and long-term studies and actions should be organized by responsible government authorities and professionals. The basic steps of these studies and actions are summarized for short-term and for long-term needs, respectively (Bahrainy et al. 1998).

4.2.11 Short-Term Studies and Actions for Reduction of Earthquake Consequences

1. Classification of buildings, structures, and local and regional infrastructures according to the usability and level of damage based on the uniform methodology developed here for damage classification.
2. Planning temporary housing, organizing of medical centers, supplies, schools, and other public activities.
3. Studies of earthquake effects and damage distribution.
4. Seismological studies based on existing and temporary installed, seismic stations.
5. Immediate installation of strong motion accelerographs and seismoscopes for recording stronger aftershocks.
6. Seismic records data collection and analysis for elaboration of seismic design criteria for repair and strengthening of damaged buildings and structures.
7. Elaboration of requirements and instructions for repair and strengthening of damaged buildings and structures.
8. Reconsideration of physical and urban plans with mapping of the spatial distribution of earthquake effects.
9. Estimation of earthquake damage losses and planning of financial and legal actions for reduction of earthquake consequences.
10. Urban planning for construction of new settlements for housing, medical centers, schools, and public utilities based on existing immediate needs, existing usable buildings, and anticipated future urban development.
11. Actual repair and strengthening of damaged buildings and demolition of heavily damaged buildings, with associated site investigations and designs for repair and strengthening.

4.2.12 Suggested Long-Term Studies

For physical and urban planning, design, and construction in highly seismic regions, the following recommendations may be made:

1. Improvement of the network of seismological stations with telemetered and computerized systems for rapid collection and analysis of earthquake data.
2. Statistical studies on instrumental and historical seismological data and preparation in a format for rapid computer analysis used for scientific and applied research.
3. Development of a neotectonic map with dynamic evaluation of the neotectonic processes within the seismic regions of the country.
4. Development of a seismotectonic map of the country.
5. Development of seismic hazard map of the country for different levels of seismic risk for planning, design, and construction of different types of structures.

6. Development and installation of a strong motion network.
7. Physical planning of seismic regions based on damage evaluation and vulnerability studies.
8. Evaluation of expected vulnerability and acceptable seismic risk levels along with requirements for seismic protective measures.
9. Elaboration of codes, instructions, and manuals for aseismic design and construction of different types of structures, retrofitting of existing structures, and other specific needs.
10. Development of seismic microzoning maps for significant urban areas.
11. Planning, design, and construction studies for structures of vital importance.
12. Elaboration of laws and regulations for counter measures against large-scale earthquakes (Petrovski et al. 1998).

4.3 Conclusion

In this chapter, urban design recommendations are made for a region and city which are prone to seismic risk. Following chapters will deal with urban design solutions for the city center, Bazaar area, access network, and open spaces in city. The city of *Rasht* will be used as example. Information needed for the regional and city scale study and design in a seismic-prone region includes seismic and geological map and data, hazard assessment, risk assessment, vulnerability assessment, built environment data, earthquake scenarios, urban and regional policies, and an implementation guidelines and mechanisms.

General characteristics of the area consists of settlement distribution pattern (six cities and 855 villages are situated in the area), radical organization, and central node as the main factors in the general structure of the city of Rasht and urban texture of residential built area, open spaces, and public and specific land uses of the city of Rasht.

The city's structure consists of two rivers flowing parallel of a north–south direction intercepted by two main streets perpendicular to one another of an east–west and a north–south direction. The city's texture may be identified to comprise of three types.

1. Old texture; 2. new texture; and 3. intermediate texture.

Texture of the city of has the following characteristics:

- The city has an irregular radial grid network.
- The city's structure consists of two main streets (north–south and east–west) intersecting each other perpendicularly and two rivers that flow in a south-north direction, and also a circular freeway from the east to the west.
- Rasht's situation in a very damp and moderate Caspian climate has given rise to a particular building form with pitched roofs.
- For this reason and the need to create air currents, and also due to the high level of underground waters, buildings in the city are generally low, rising to a maximum height of 8 stories.

- Three old, intermediate, and new textures may be identified and differentiated.
- Among the features of the old textures, it is possible to name the central market and the numerous old neighborhoods.
- The new texture is distinguished by its grid streets, which are relatively wide, and its low buildings.
- The crossing of the two rivers through the city along the south–north axis have practically divided the city into three sectors connected by bridges lying across the rivers at different points.
- The settlement of industries and manufacturing workshops along main routes especially the city's southern route is conspicuous.

Using accessibility as the criteria for urban texture in the city. Following types may be recognized: regular texture, open, with adequate or inadequate accessibility, regular compact texture, adequate or inadequate accessibility, and organic open texture, with adequate or inadequate accessibility and organic compact texture, with adequate or inadequate accessibility.

Urban design to mitigate seismic risk begins with analyzing the vulnerability of city's different textures: The settlement of low-income families on the city margins has given rise to a particular housing pattern in these areas which, in disregard of the earthquake hazard and with no observation of required standards pertaining to this factor, increases the vulnerability of the marginal texture and the lack of appropriate access to health centers and their remoteness exasperates the situation. Open space and transportation network require special attention due to their critical role in risk reduction during and after disaster.

A very important question always raised following a disaster is how to estimate the damage to the buildings and group of buildings. So methods and factors for estimating the damage to buildings and to a group of buildings (texture) are presented in this chapter.

Some general principles to be applied in access network design to reduce seismic vulnerability include the following:

Shorter distances between different land uses.

Network hierarchy.

Shorter distances between critical land uses during disaster.

Fast access to important land uses.

More possible control.

More possibility of escape and evacuation through the network.

Adequate speed, shorter length and more safety of the network.

Less traffic in the network.

Lower traffic (users) intensity.

Higher ratio of open space over built-up areas.

More direct connection between important uses.

More connection between different parts of the network.

Less the number of cul-du-sacs in the network.

More vehicular access to the network.

Less the number of intersections.

More independent network for each district.

The lower the ratio between the length and width of passages.

The more the ratio between the widths to half of the walls' height of the passages.

The more seismic codes are applied in the construction of utilities and facilities.

Balanced distribution of land uses, particularly public services, and also appropriate building and human density play significant role in seismic risk reduction.

In order to integrate seismic risk reduction goals into the long-range development plans, long-term development scenarios are formulated for the city of Rasht and vicinities. Most possible alternatives are Centers hierarchy model, radial corridor model, bi-polar metro model, and metro-satellite model.

References

Bahrainy H et al (1997a) Planning and design of specific land uses in the city of Rasht for reducing of earthquake vulnerability. National Report, Volume 52, UNDP—UNCHS (Habitat) Project IRA1901004, Housing Foundation of I.R., NDPC (Translated in English), Tehran, November

Bahrainy H et al (1997b) Planning and design of transportation network in the city of Rasht. National Report, UNDP—UNCHS (Habitat) Project IRA1901004, Housing Foundation of IR, NDPC (Translated in English), Tehran, August

Bahrainy H et al (1998) Synthesis on redesign of the city of Rasht. National Report, UNDP—UNCHS (Habitat) Project IRA1901004, Housing Foundation of IR, NDPC (Translated in English), Tehran, December

Banerjee T (1993) Urban form and seismic vulnerability. Earthquake protection through urban design and planning. International Consultant Report, UNDP—UNCHS (Habitat) Project IRA1901004, Housing Foundation of IR, Tehran, October

Kazemi M et al (1997) Socio-economic structure of the city of Rasht. National Report, vol 34, UNDP—UNCHS (Habitat) Project IRA1901004, Housing Foundation of IR, NDPC (Translated in English), Tehran, January

Lynch K (1960) The image of the city. MIT Press

Music V (1993) Earthquakes and management of urban environment. Urban planning and design as a part of preventive strategies. International Consultant Report, UNDP—UNCHS (Habitat) Project IRA1901004, Housing Foundation of IR, Tehran, October

Music V (1998) Alternative development scenarios of the city of Rasht and its functional urban region. International Consultant Report, UNDP—UNCHS (Habitat) Project IRA1901004, Housing Foundation of IR, Tehran, December

Petrovski JT (1992) Vulnerability and seismic risk assessment for earthquake disaster management in urban areas. Invited paper, First International Conference on Disaster Prevention in Urban Areas, Tehran, May

Petrovski JT, Bahrainy H, Music V (eds) (1998) Guidelines for reduction of seismic risk in regional and urban planning. Volume VII of the Guidelines for Earthquake Disaster Management, UNDP-UNCHS (Habitat) and Housing Foundation of IR, Tehran—Skopje

Chapter 5
Urban Design Solutions to Mitigate Seismic Risk at the Small Scale: *Bazaar* Area, Rasht

Abstract The existing conditions are a result and a continuation of development planning processes which seldom account directly or indirectly for natural hazards, a fact that is striking, since many of the objects of development planning are precisely the same as those of disaster prevention planning. The favorable achievements of otherwise good planning can be, and often have been, wiped out in minutes because of failure to account for the disaster potential of natural hazards. Tendency to view such misfortunes as unavoidable tragedies, although to substantial degree in fact they are, is a testimony to poor planning that ignores that natural disasters are an integral feature of the colonized environment and that the extent of damages and losses caused are to a significant degree a function of decisions made (or not made) during the process of human settlement development. Small scale is in fact the most effective and commonly used scale in which urban design, as an activity that manipulates physical environment to achieve certain goals, and particularly reducing seismic risks may be applied. "Manipulating the physical environment," encompasses a broad set of activities in a wide range of physical settings. "Manipulation" may include direct physical design and construction, regulatory measures to guide physical changes over time, economic and community development efforts, regional growth strategies, and institutional measures such as funding programs that ultimately result in physical actions. Disasters cause formal and functional disorder in the built environment. Formal implies urban form and space, and functional implies urban activities. Urban design, through its rules and regulations, intends to prevent disorder in case of a disaster. As discussed in chapter one of this book, one of the determinant factors of disorder or un-sustainability of human settlements is basic needs, in which safety and security play significant role.

Keywords Urban design · Micro scale · Medium scale · *Bazaar* · *Rasht* · Existing conditions · Typology · Architecture · Vulnerability · Design ideas · Square

H. Bahrainy and A. Bakhtiar, *Urban Design in Seismic-Prone Regions*,
The Urban Book Series, https://doi.org/10.1007/978-3-031-08321-1_5

5.1 Introduction

Natural disasters can strike settlements of all sizes, but increasing concern is being felt for the possible effects on large agglomerations which are becoming a characteristic feature of developing countries. Disasters of all kinds can cause great losses of human life. Of course, they cause substantial damage to property and can have serious impacts on economically productive activities.

The existing conditions are a result and a continuation of development planning processes which seldom account directly or indirectly for natural hazards, a fact that is striking, since many of the objects of development planning are precisely the same as those of disaster prevention planning. The favorable achievements of otherwise good planning can be, and often have been, wiped out in minutes because

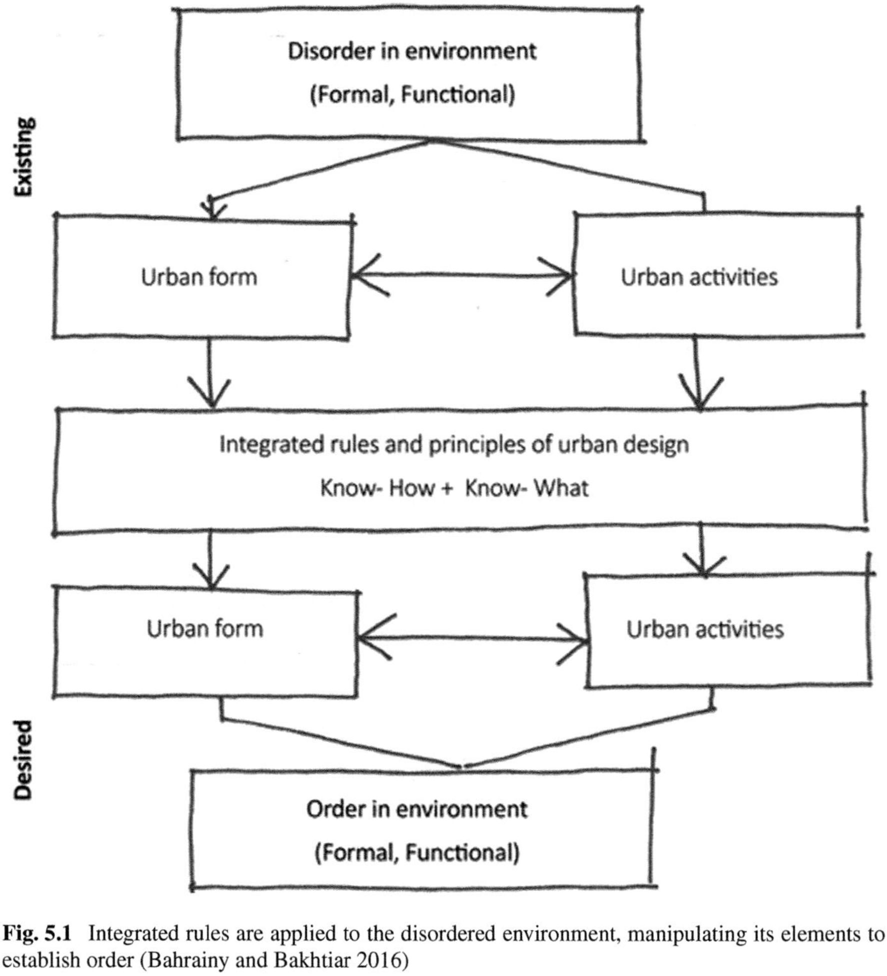

Fig. 5.1 Integrated rules are applied to the disordered environment, manipulating its elements to establish order (Bahrainy and Bakhtiar 2016)

of failure to account for the disaster potential of natural hazards. Tendency to view such misfortunes as unavoidable tragedies, although to substantial degree in fact they are, is a testimony to poor planning that ignores that natural disasters are an integral feature of the colonized environment and that the extent of damages and losses caused are to a significant degree a function of decisions made (or not made) during the process of human settlement development.

Small scale is in fact the most effective and commonly used scale in which urban design, as an activity that manipulates physical environment to achieve certain goals, and particularly reducing seismic risks may be applied. "Manipulating the physical environment," encompasses a broad set of activities in a wide range of physical settings. "Manipulation" may include direct physical design and construction, regulatory measures to guide physical changes over time, economic and community development efforts, regional growth strategies, and institutional measures such as funding programs that ultimately result in physical actions (Owen 2019).

Disasters cause formal and functional disorder in the built environment. Formal implies urban form and space, and functional implies urban activities. Urban design, through its rules and regulations, intends to prevent disorder in case of a disaster (Bahrainy and Bakhtiar 2016). As discussed in chapter one of this book, one of the determinant factors of disorder or un-sustainability of human settlements is basic needs, in which safety and security play significant role (Fig. 5.1).

To show how this could technically being achieved, a real case, which was studied during and after the June 20, 1990, *Manjil* earthquake will be presented here in this chapter. Following a review of the existing conditions of the area, the vulnerability of selected elements will be analyzed and on their basis some recommendations will be made to reduce the seismic risk in the area.

5.2 Existing Conditions

To determine the vulnerability of the area and propose design ideas to reduce seismic risk, an analysis of the existing conditions is necessary, and this includes location of the area in the city, typology of the city structure (large scale), typology of the *Bazaar* structure (macro, medium, and micro scales), area characteristics, and general patterns of the physical elements. Following an analysis and evaluation of the existing conditions of the area, appropriate design recommendations will be made to reduce the vulnerability of the area against earthquake.

5.2.1 **Bazaar** *Location in the City*

The central area of Rasht is located in the zone 1 of the city (see Fig. 5.2). The existence of the old Bazaar and the concentration of uses and services in this area have led to draw residents from other parts of the city and therefore causing difficulties for the pedestrians as well as motor vehicles to use the area. Higher population density, compact and fine grain texture, and small plot sizes have intensified the problems.

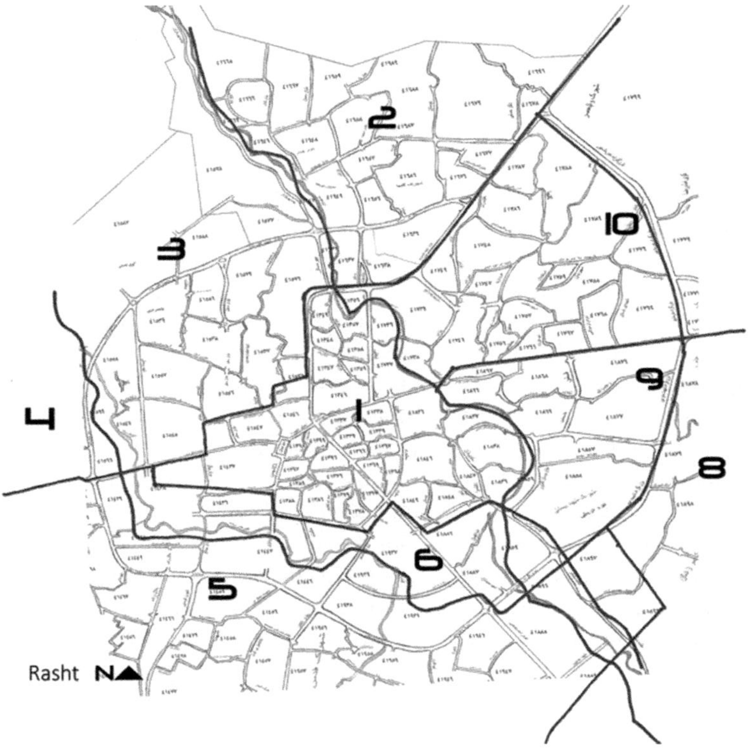

Fig. 5.2 The ten districts in the city of *Rasht*, and the location of district one, where *Bazaar* is located

The dominant texture in the area is generally organic, and compact, with inadequate access into the area. Deteriorated buildings due to the old age, high density, and inadequate accessibility all have made the area the most vulnerable for all stages of the disaster management (Kazemi et al. 1997).

5.2.2 Typology of the City Structure, Macro (Large) Scale

The main city corridors which are the same as the main *caravanserai* roads and roads for transporting goods and also canals for discharging surface water into the city rivers and pedestrians connection between neighborhoods and linear configuration of commercial spaces make the main structure of the city. These corridors go through the center of neighborhoods, where local mosques, public bathes, and educational centers are located. In the city center, the corridors will end in Bazaar, where they break into two main axes. One leads toward the grand square and then connects to

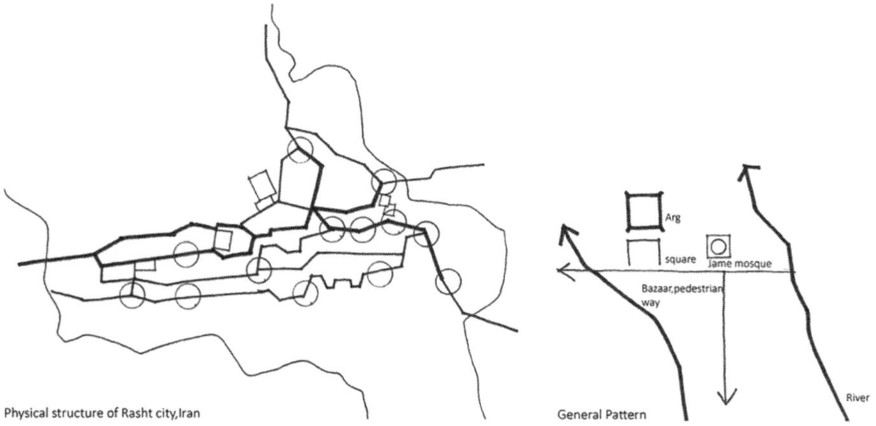

Fig. 5.3 The physical structure and general pattern of development in the city of *Rasht*

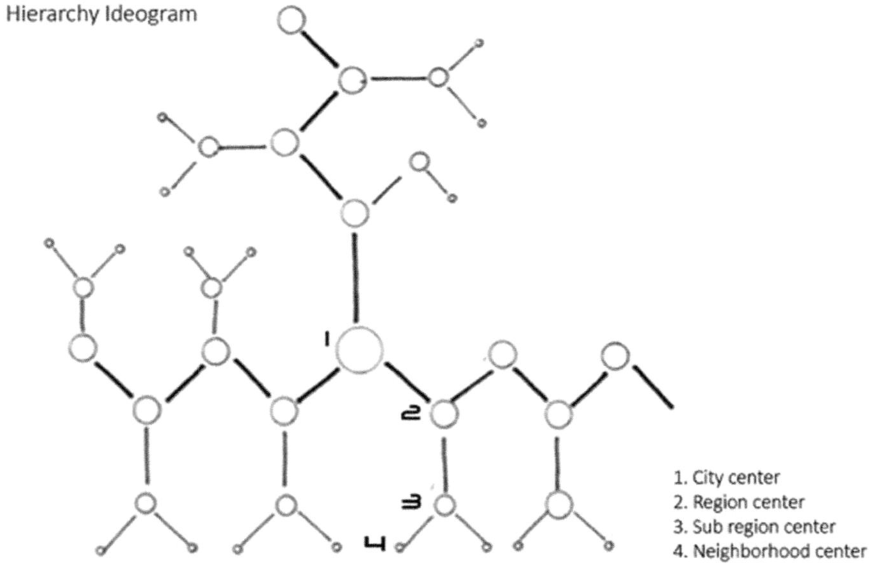

Fig. 5.4 Hierarchy of settlement patterns in the region

the western axes. The other continues through *Bazaar* and connects with western axes (Figs. 5.3, 5.4, 5.5, and 5.6).

Responding to the above three requirements define a circular layout with three main turning points for Bazaar: Municipality square, *Kaseforoshan* Jame mosque, and the grand square. This layout will provide the Bazaar area with a unique identity.

Main eastern axes joined the Bazaar fabric from the east side through two passages. These passages are connected to the small square through *Mesgaran* passage. Main *caravanserais* are located in this square in the radial form. The passage is divided

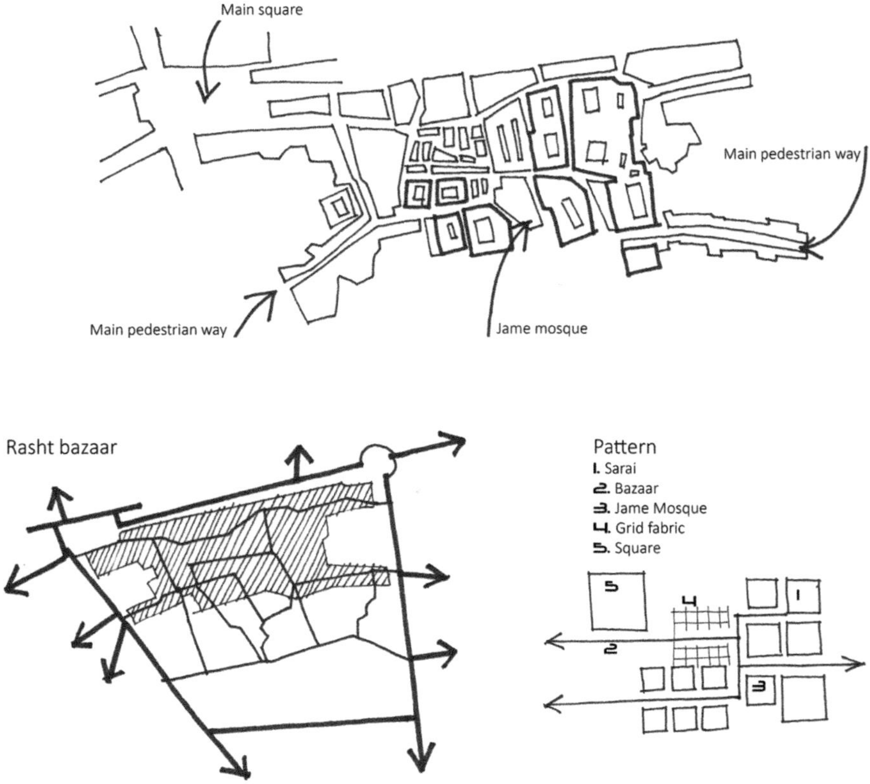

Fig. 5.5 *Rasht Bazaar* and its main components

into two main passages at the location of *Kasehforoshan Jame Mosque.* One branch called *Bazzazan* passage goes through the grid texture and ends at the main city square—municipality square. The other branch called *mikhforoshan*, going through different *caravanserais,* connects to *mahiforoshan* passage and will eventually exit the *Bazaar* area at the Western end. The significance of these passages is due to the existence of religious and cultural centers, such as mosques, religious schools, and tombs, on the one hand, and the concentration of commercial activities, on the other hand. This special and unique pattern includes turning points, dynamic and static nature, and unity of the whole *Bazaar* (Kazemi et al. 1997).

The city center contains the old and valuable texture that is dilapidated and highly vulnerable toward earthquake. Also, all major land uses in the city are centered in this part. The criteria for the physical outlook of any city may be sought for in the old parts, in the market and its numerous branches, in the neighborhoods surrounding the market, and in the local features, nodes, passages landmarks, and edges (Majidi et al. 1997).

The area is the main dominant element of city center with three unique characteristics and requirements:

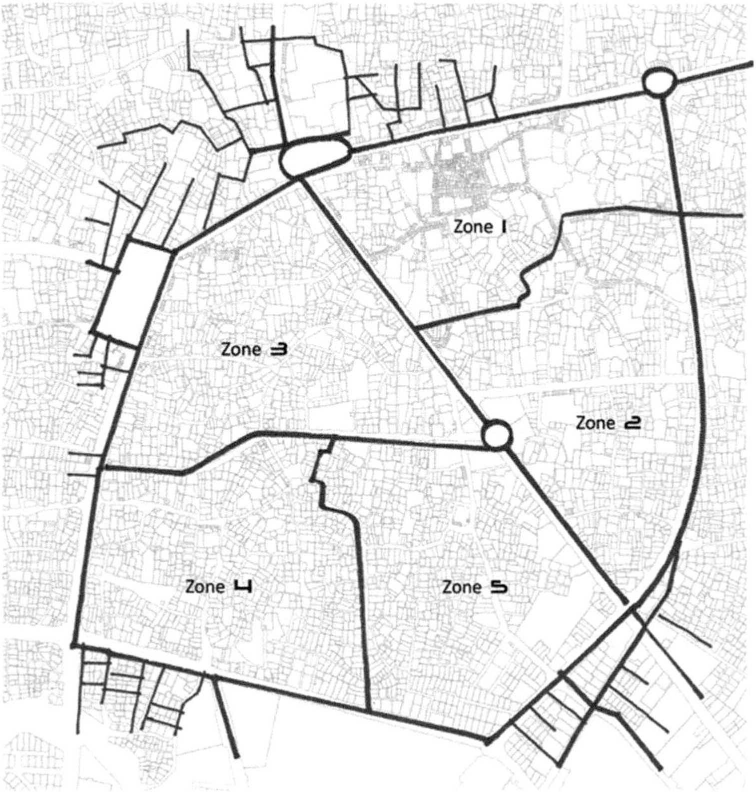

Fig. 5.6 Five zones in the city center of Rasht. Zoning are based on integrated seismic risk: high risk in zone 1 and low risk in zone 5

- Connection with the city's main square and coordination with city center.
- Connection with city's main axes.
- Its cohesion as a single and unique complex has to be preserved.

The shortage of public services and their inappropriate distribution based on population distribution has resulted in a number of problems. The concentration of the city's major land uses at the dense center of the city where the streets are incapable of responding to the passage of people coming to the area from the other parts, while creating day-to-day urban mobility problems, causes many deaths and impedes relief operations in a crisis situation owing to the excessive concentration of population at one point (Bahrainy et al. 1997).

The presence of areas prone to liquefaction in the urban area is a great hazard regardless of which construction works have been pursued in the past, especially at the city's center where major land uses are settled and the texture is old, the thoroughfares narrow, the street sides unsafe, and the density high. Ultimately, the non-observance of building codes, the overhangs in building stories, the narrow thoroughfares, the non-adherence to a hierarchy of access networks, the land-use distribution mode, and the inattention toward the vulnerability of urban infrastructures, etc., indicate the generally inconsistent state of codes. From the point of view of existing land uses, district 1 is one of the most important districts in the city. The district covering an area of 446 ha has a population density of 173.9 persons per hectare (1991 statistics). By a glance at its history, it is possible to conclude that this district is in fact considered as the primary nucleus of the city.

Although *Rasht's* market has not played as significant a role as markets in the towns of central Iran (desert) in forming the city, its development mode, and rate of expansion; yet it is known as the most important center of commerce; it functions on a regional scale, and even when considering the city's tourist function, its scale would rise to one of a national character.

Although the market passages have no roofs as in the markets of desert regions, still, they have the same complex and dense fabric which in accordance with their commercial character overflow with crowds during the daytime and are almost deserted at night. The age of the buildings, Rasht's inadequate climate conditions (too much rainfall, very high humidity, high level of underground waters), and the erratic utilization of the buildings in the central part coupled with dense and non-preconceived buildings that have also been constructed recently in this part have made district 1 one of the oldest of existing textures in the city. Therefore, the reduction of building density and consequently human density in district 1 is brought to attention as a hypothesis and the relocation of the important land uses in these parts if possible that create a high density to other districts as well as restraining vehicle traffic (in quantity and quality) would also be a great contribution to the prevention of further damage to buildings in this district (Kazemi et al. 1997).

The *Enqelab* Square is yet another open space in the city situated in district 1. This square may not, as the formerly mentioned squares, have any connections with the historic city however, it is noteworthy from other aspects owing to its close bond with the city's main transportation network.

On the whole, district 1 faces a shortage of open space and such a land use should be further provided for it. The liquefaction potential in a major part of this district and on the other hand, the potential capability for constructing buildings of a medium height in another part, has given rise to the hypothesis of transferring the building density from the liquefaction prone area to another area (by increasing the number of stories in buildings) which would eventually lead to reduced density.

As stated in the previous section, owing to the presence of important land uses next to open spaces and thoroughfares in district 1, these routes and spaces are overflowing with crowds during the daytime up to dusk. Therefore, if an intense earthquake were to strike at these hours, casualties would be great. Hence, in order to reduce probable casualties caused by such an earthquake, attempts should be made to both reduce human density in the district (by transferring many of the land uses) so that less people are subjected to danger and minimize the vulnerability of the buildings that exist there.

Transportation network: The major transportation network in district 1 consists, for the most part, of streets whose traffic reaches full capacity in daytime (such as *Imam Khomeini*, *Saadi*, *Motahhari*, Dr. *Shariati*, *Alamolhoda* streets). Therefore, in designing the urban network this should be borne in mind; and by designing appropriate connective rings, the traffic load should be reduced.

The district one's minor transportation network of alleys, passages, and market passageways, which are confronted with many problems. Since this network has developed organically, it has complexities of its own that create danger particularly at times of crisis. Along a great length of the network, vehicular traffic is problematic, still, in recent years; these movements of traffic have been made by different means. The incompatibility of thoroughfares with vehicular traffic has both resulted in the aging of buildings lying next to thoroughfares and made vehicular movement difficult. For this reason, district 1 is the only district in the city in which relief centers are unable to render their services in a satisfactory manner (Bahrainy et al. 1997a, b, c, d).

Since most buildings standing beside minor thoroughfares are old, and therefore, in not too good condition, they are highly likely to collapse and block the thoroughfare when earthquake strikes. The other aspect of the problem in minor thoroughfares of district 1 is the presence of historically valued buildings situated along their margins which despite all the mentioned shortcomings makes any change in the texture problematic.

The market, as the most important land use situated in this district, consists of a number of parts. The main market—regional scalechiefly extends around the *Shahrdari* Square and the market functioning on an urban scale manifests itself in the *Sabzeh* Meydan Square, the *Alamolhoda* Street, and the *Shahrdari* Square. The two sections mentioned above have created the main structures of the market. As is observed, there are branches leading off the main market which may rather be identified as district markets. The *Saghari Azan* and the *Khahar* Imam markets are two such examples. Both are very old (the oldest textures in the city) and have a high population and building density. Still, the architectural value of these two parts is of great importance, and obviously, they would impose particular conditions on any policies adopted for disaster confrontation. The passages in these two markets are used for vehicular traffic despite their inadequacy for such a function. This has given rise to a number of problems both for the texture and for those who use it. To preserve the texture of the *Saqarisazan* and the *Khahar* Imam Passages, which are almost unique in Rasht, an adequate transportation network ought to be designed in consideration to the earthquake factor; and it is believed that, to this end, it would be best to eliminate vehicular traffic from the market passages altogether.

Important office buildings are another land use that are chiefly centralized in district 1. Such high referral administrations as, the Disciplinary Forces Central Headquarters, the Central Municipality, the Telecommunications Head Office, the Post Center, the main branches of banks, the Education Head Office, etc., have many times multiplied the density of this district. These administrations are, for the most part, located next to the *Shahrdari*, *Sabzeh* Meydan, and *Enqelab* Squares.

Cultural buildings such as the Museum, the National Library, and especially the cinemas (almost all the cinemas in the city are located in this area) are also centralized

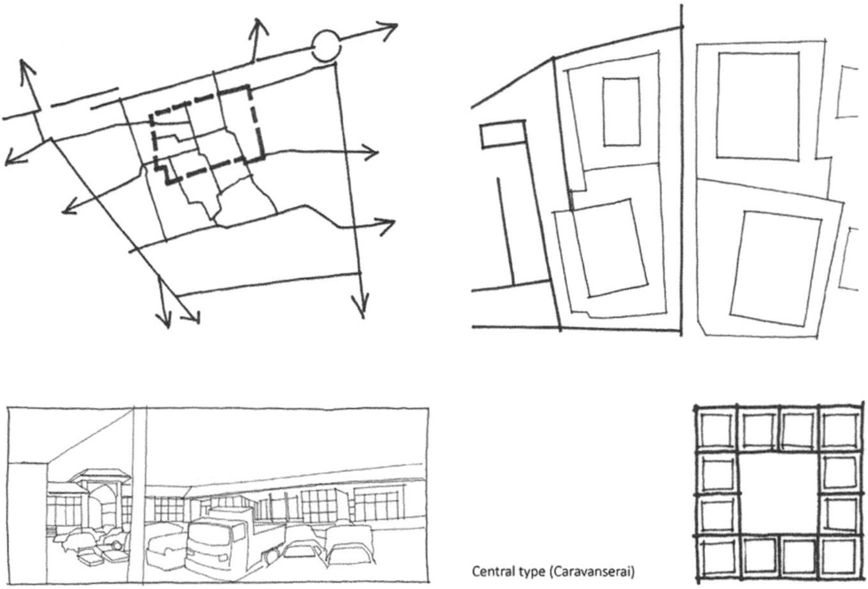

Fig. 5.7 Central type (*caravanserai*)

in district 1, in their turn adding to the human density. The inadequate condition of the cinemas which are high-density land uses could create a variety of incidents—from the point of view of safety during an earthquake (Bahrainy et al. 1997a, b, c, d).

5.2.3 *Typology of the Bazaar Structure (Medium Scale)*

The Bazaar texture has been the scene for the occurrence of various activities at different scales and variety of forms through its history. Bazaar activities are reflected in diverse physical types. These types may be differentiated through diversity of aggregation patterns and juxtaposition of essential components and smaller scale of the texture and the type of commercial activities.

Here, the dominant types will be discussed (Majidi et al. 1997).

Central type (*caravanserai*), grid type, linear type, and dense and integrated blocks

Central type (*caravanserai*): The main element of this type or the essential unit is "central court." The retail units in this type are located along the "central court yard." All units have direct access to the open space in the court. The main functions of caravanserai are often wholesale and warehousing. Lack of adequate parking space in the area has forced the customers to use the courtyard as parking space (Fig. 5.7).

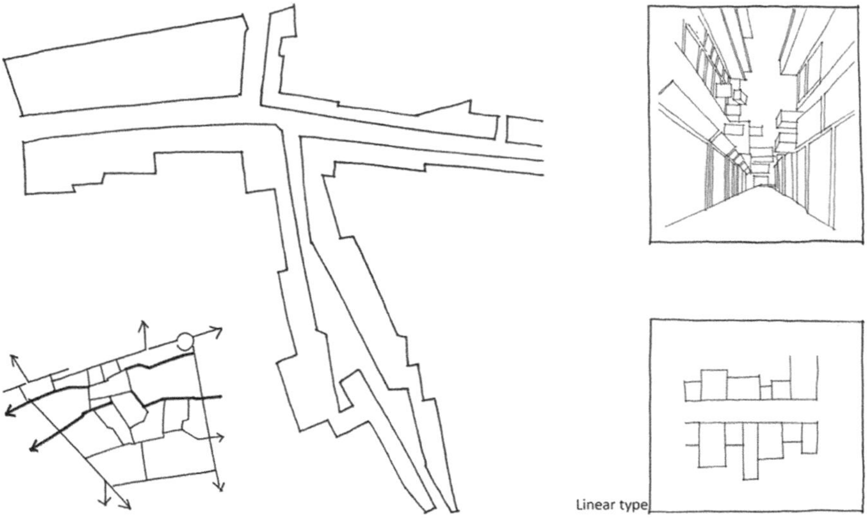

Fig. 5.8 Linear type

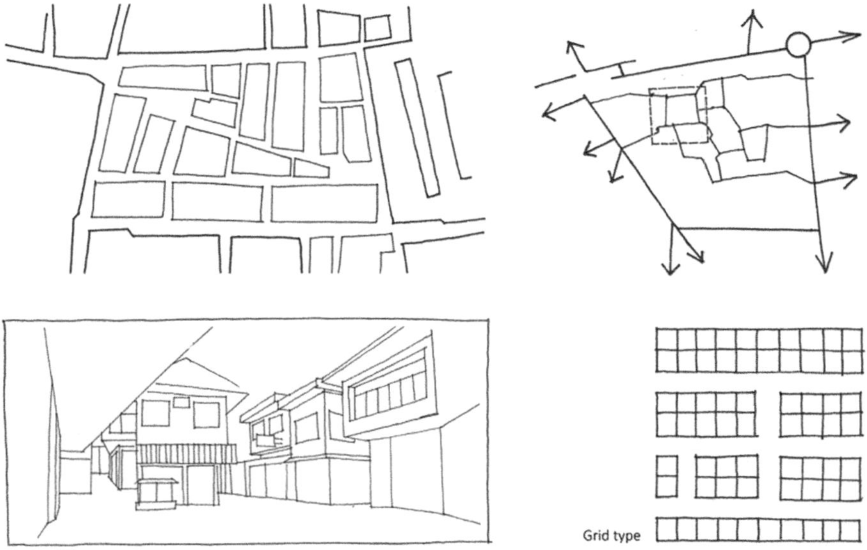

Fig. 5.9 Grid type

Grid type with low depth: This type includes dense bocks of adjacent units which are arranged on the basis of a grid pattern. In this type, the depth of enclosed spaces is very limited. Access to the enclosed spaces is possible directly from the passages. In some cases, a narrow stairway connects the ground floor to the upper units. Due to the limits of the blocks dimensions, height increase is also limited. However, the ratio between height and width of the passages is relatively high, which is due to the low width of the passage. The main activities of this type retail clothing, gold and jewelry, and stationary (Figs. 5.8 and 5.9).

Linear type

This type includes various branches of *Bazaar* in which smaller units with different depth are located on both sides of a passage on the basis of a linear pattern. Access to the enclosed spaces often takes place directly, but in the case of the ones with higher depth it is done indirectly. The passage in this type also lacks open space. The extended overhangs to some extent protect open spaces from rain and sunshine. This type includes the secondary branches and cul-de-sacs. The dominant function of these branches is retail and in some cases wholesale. Different sections belong to a specific group of activities which are called under that group (Fig. 5.8).

Dense and integrated block type (passages)

This type includes enclosed and integrated complexes in which there are smaller units stationed on the basis of diverse patterns of grid, linear, and rarely central. Duplication of the plans on the upper flours creates higher density. Access to the internal units is possible indirectly through circular, central, or linear open or semi-open route. There is no direct access from the complex to outside. The dominant activities in this type is retail in the lower floor and light workshop on the upper ones (Fig. 5.10).

5.2.4 Typology of Bazaar Texture (Micro Scale)

Essential units comprise most area of Bazaar.

Type 1. Business activities are separated in a deep enclosed space. This type includes grocery and food stores, bakeries, and some clothing stores.

Type 2. Unseparated business activities in a deep and completely enclosed spaces. General stores, clothing, and furniture are of this type (Figs. 5.11, 5.12, 5.13, 5.14, 5.15, and 5.16).

Type 4. Joint business stores in low depth and enclosed spaces. Boutiques, clothing, garment, and shoes are located in these spaces.

Type 6. Separated business activities in the entrance of semi-enclosed spaces. This type includes fruits and vegetable and fish stores.

Type 8. In this type, shops are located in two stories. The shops on the ground level are one meter below the ground level. The main functions of these spaces have been kitchen and storage in the past, which continues up until now. These spaces have a direct access to the courtyard. The shops on the second floor are usually larger in size and consist of two sections: storage at the end and the main section in the front. These shops are connected to the courtyard through *eivan*. The main functions of these shops are storage and often wholesale offices.

No doubt the principal activity of Bazaar is commercial, but the diversity of views and townscape, enclosures, non-commercial activities such as public baths, religious places, beautiful window shops, and attractive pavements, and also the possibility for pedestrians' walking and standing makes it quite attractive for all groups (Fig. 5.17).

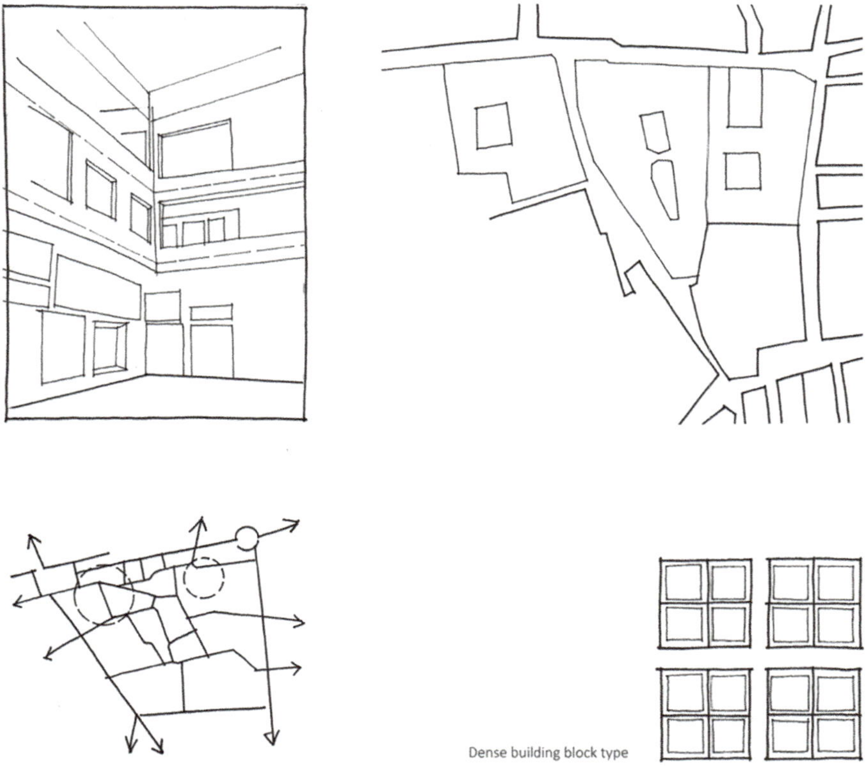

Fig. 5.10 Dense building block type

Fig. 5.11 Existing land use of the *Bazaar* area

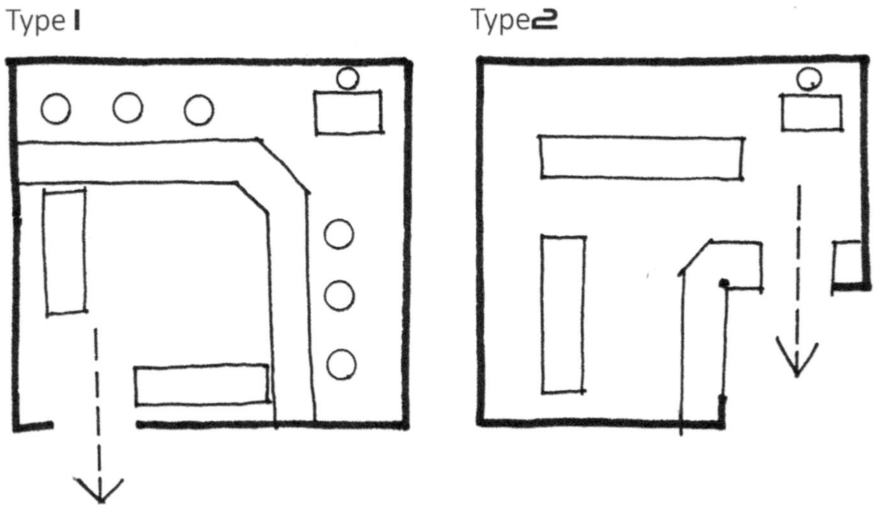

Fig. 5.12 Type 3. Separated business stores, with low depth and enclosed space. These include gold and jewelry, stationary, and clothing

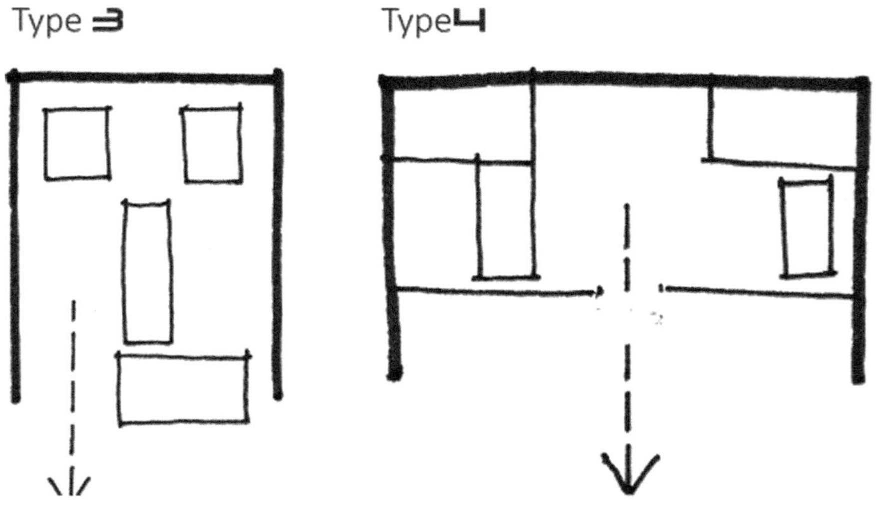

Fig. 5.13 Type 5. Joint business activities in semi-enclosed spaces. Crystals, tools, rugs, furniture, and plastics are located in these spaces

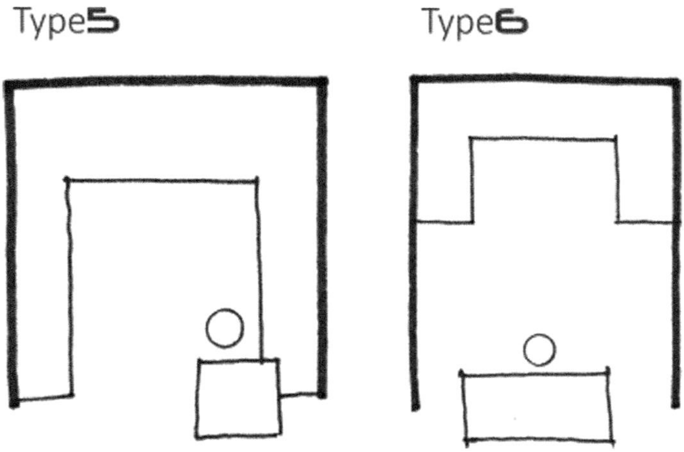

Fig. 5.14 Type 7. Separated business activities in open space which includes fruits and vegetable kiosks

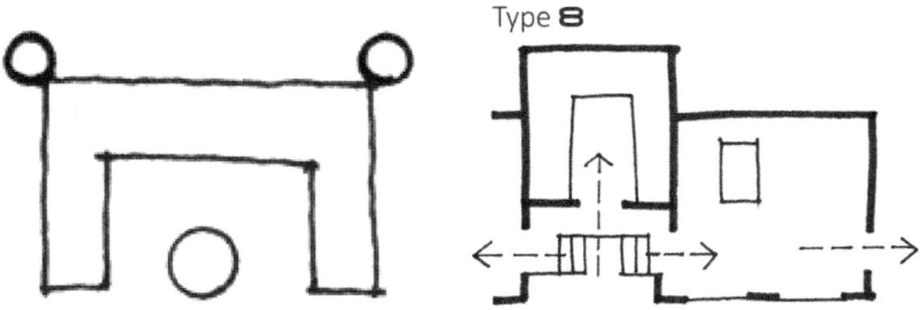

Fig. 5.15 Type 9. Retail and wholesale stores. These stores are usually full of semi-wholesale and retail goods which cover most of the store's space and even beyond that. Type 10. Retail stores in the "passages." Dense and integrated blocks (passages)

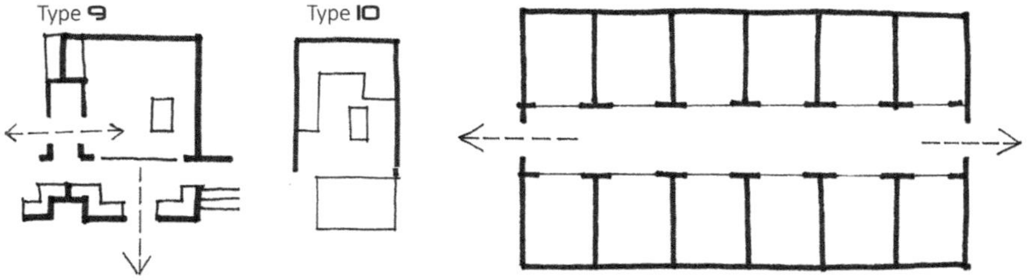

Fig. 5.16 Corridors as horizontal access, stairs as vertical access

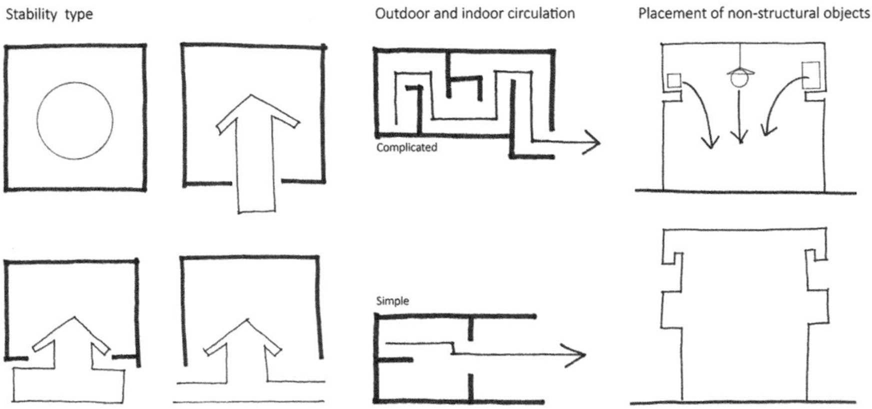

Fig. 5.17 Closed-open space relation with no or limited access between the two and location of non-structural elements in the building

5.2.5 Area Characteristics and Its Architecture

Rasht Bazaar which is the physical reflection of the commercial identity of the city of Rasht and plays a significant role at the regional scale, and it also plays its role as the only active and vibrant center of various commercial, religious, and to some extent administrative activities. Aside from securing the needs of residents of various parts of the city, it is also a place for exploring, choosing, and purchasing goods, and destination for many daily trips for work, visit, walking, and leisure (Music 1993, 1998).

Two main types of architecture may be recognized in the city: rural type and urban type. Characteristics of rural type are as follows: outward looking dwellings, long terraces (*eivan*), paying attention to the surrounding and making efforts to establish relation with it, building is surrounded by open space, simplicity, purity and lightness, and a consoled *eivan*, mass in the lower section and space in the upper level, low dept of closed space, simple proportions based on basic modular square, compatibility with climate, and open space between buildings for air flow (Figs. 5.18, 5.19, 5.20, 5.21, 5.22, and 5.23).

Fig. 5.18 Examples of rural architecture in the city of *Rasht*

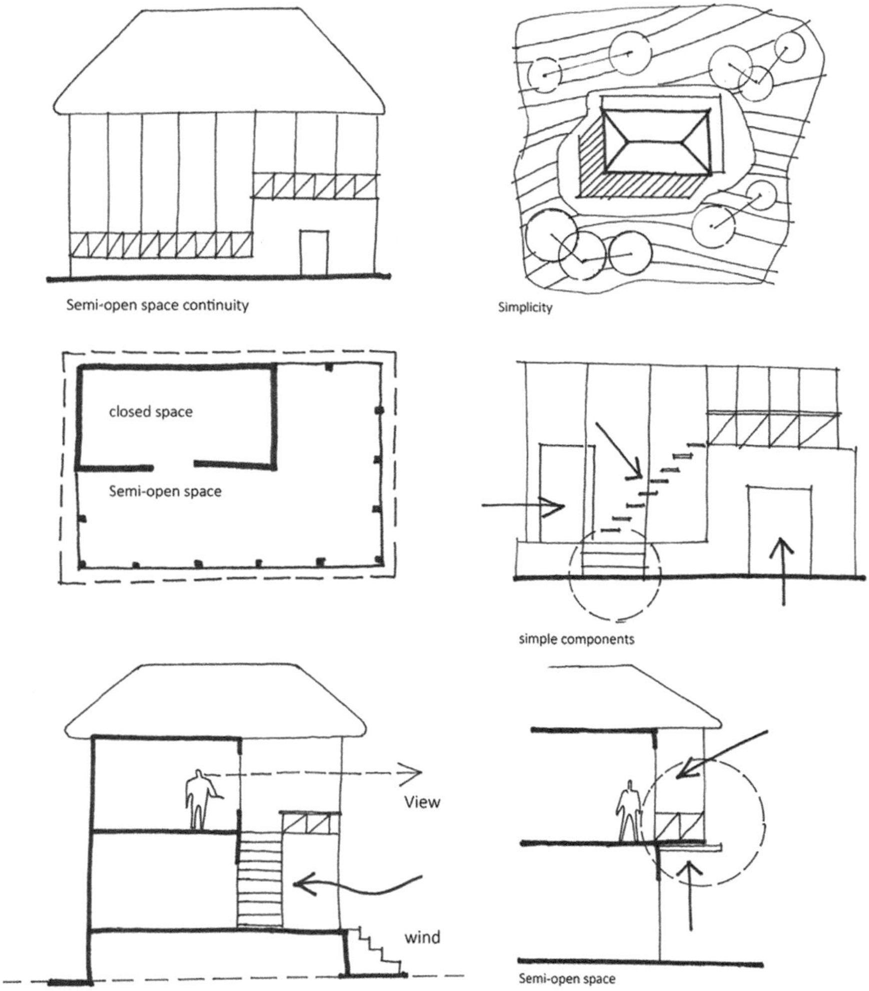

Fig. 5.19 Devoting a large portion of the house to semi-open space. Full lower section and void upper section. Light building material. Easy visual and physical access between house and surrounding. Non-load-bearing posts for terraces and console terraces

Urban housing in the city of *Rasht*. Three types of housing may be recognized: Qajar period, which is basically indigenous, First Pahlavi period, which is a combination of German, Russian and indigenous architecture, contemporary architecture, with no identity, no compatible with local climate and cultural values (Fig. 5.24).

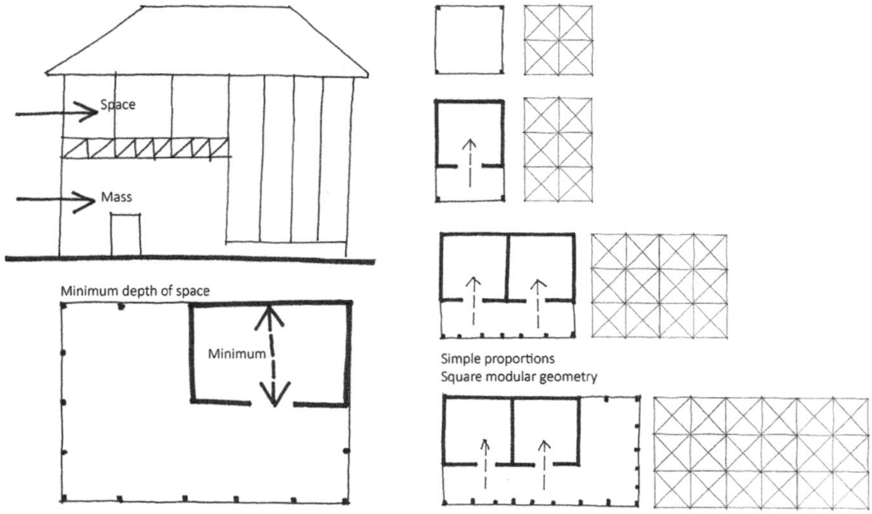

Fig. 5.20 Low depth and enclosed spaces

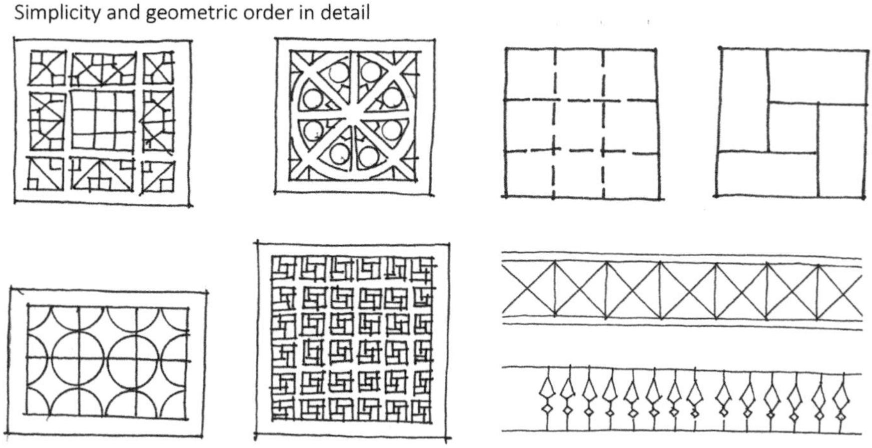

Fig. 5.21 Simple proportions, square as basic module

5.2.5.1 Building Density

Inappropriate building density may, similar to human density, increase the loss and damages inflicted by earthquake. This will be seen in the high-rise buildings, on the one hand, and the dense and compact areas of the cities, on the other hand. In either case, providing necessary services to the affected areas during an emergency would be difficult. Unjustified density in the high-risk areas will increase vulnerability.

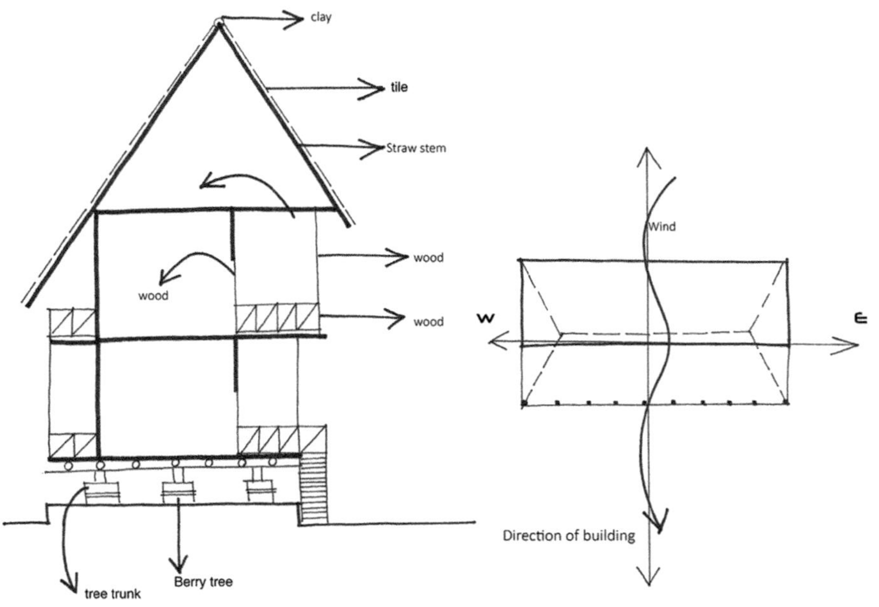

Fig. 5.22 Lightness, simple light, and exposed details such as stairways and doors

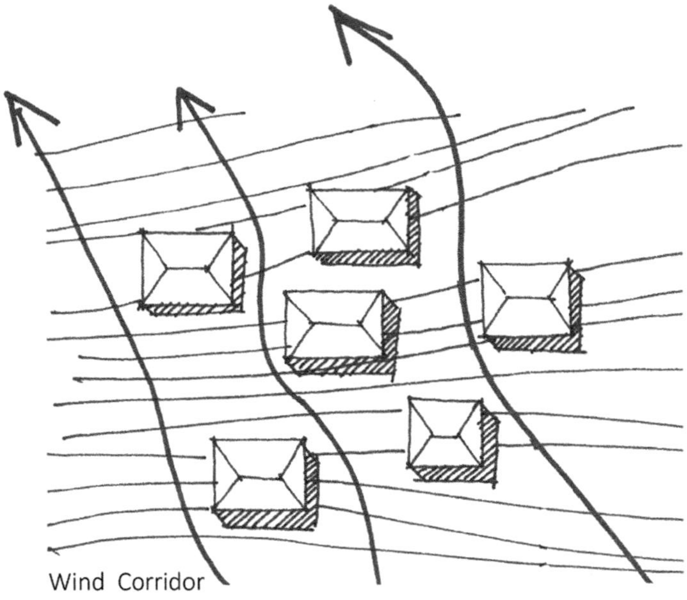

Fig. 5.23 Building is surrounded by open space

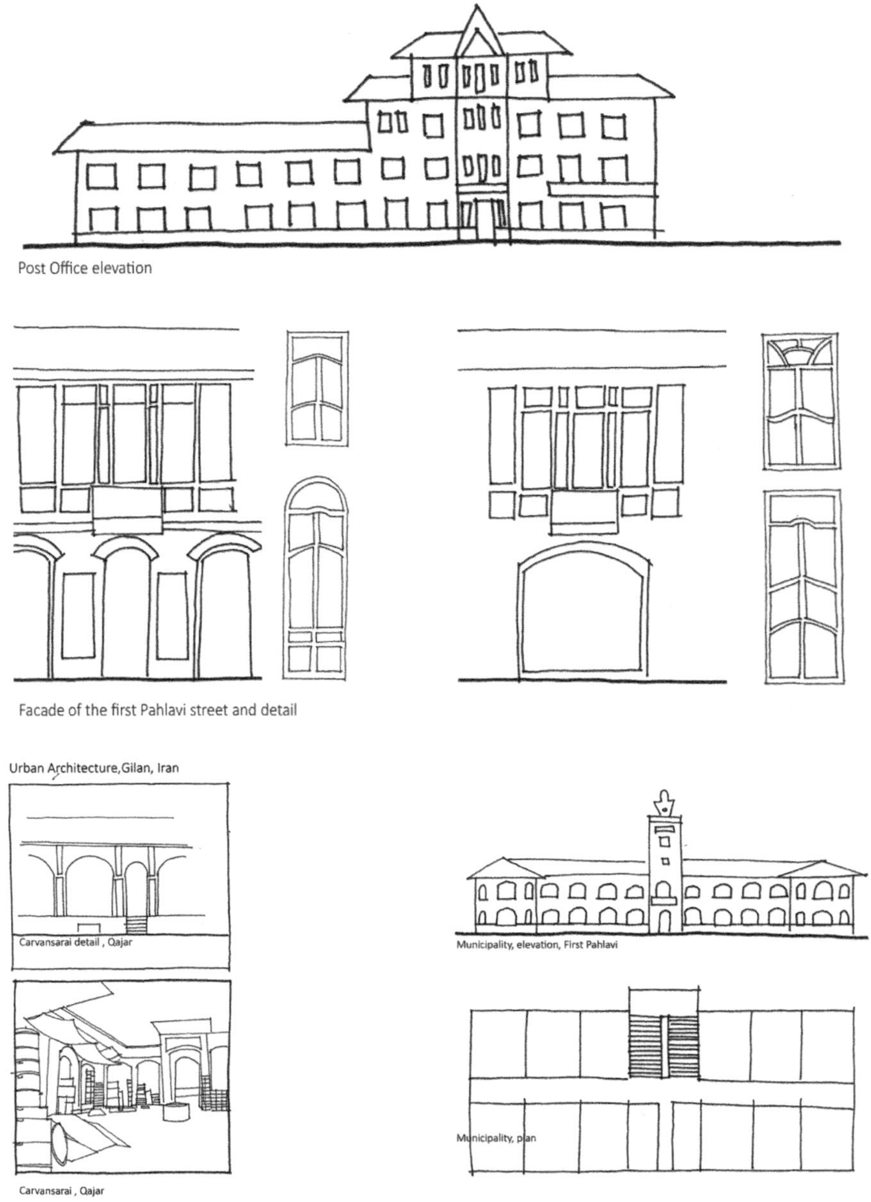

Fig. 5.24 Urban architecture: Qajar and First Pahlavi

Fig. 5.25 Building density in the Bazaar area

Building density in the area varies from low to high, according to the location, age, accessibility, land use, and lot size. But as map shows the dominant building density is one and two stories. Higher density buildings are located next to the major roads (Fig. 5.25).

One of the most important problems in this area is motor vehicle access into the area, when takes place causes serious problems for pedestrians, due to lack of safety. High population density, especially in the open spaces, local roads, and alleys, is another difficulty.

5.2.5.2 The General Pattern of the Physical Elements

Bazaar area is the focal point and the convergence of city passages, with the following characteristics:

- The dynamic nature of the main city passages.
- The static and stable nature of the converging point for city passages (Figs. 5.26, 5.27, and 5.28).

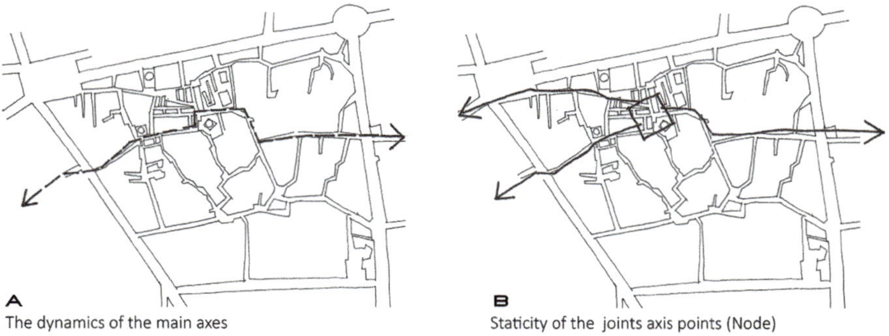

Fig. 5.26 The dynamic and static elements in the *Bazaar* area

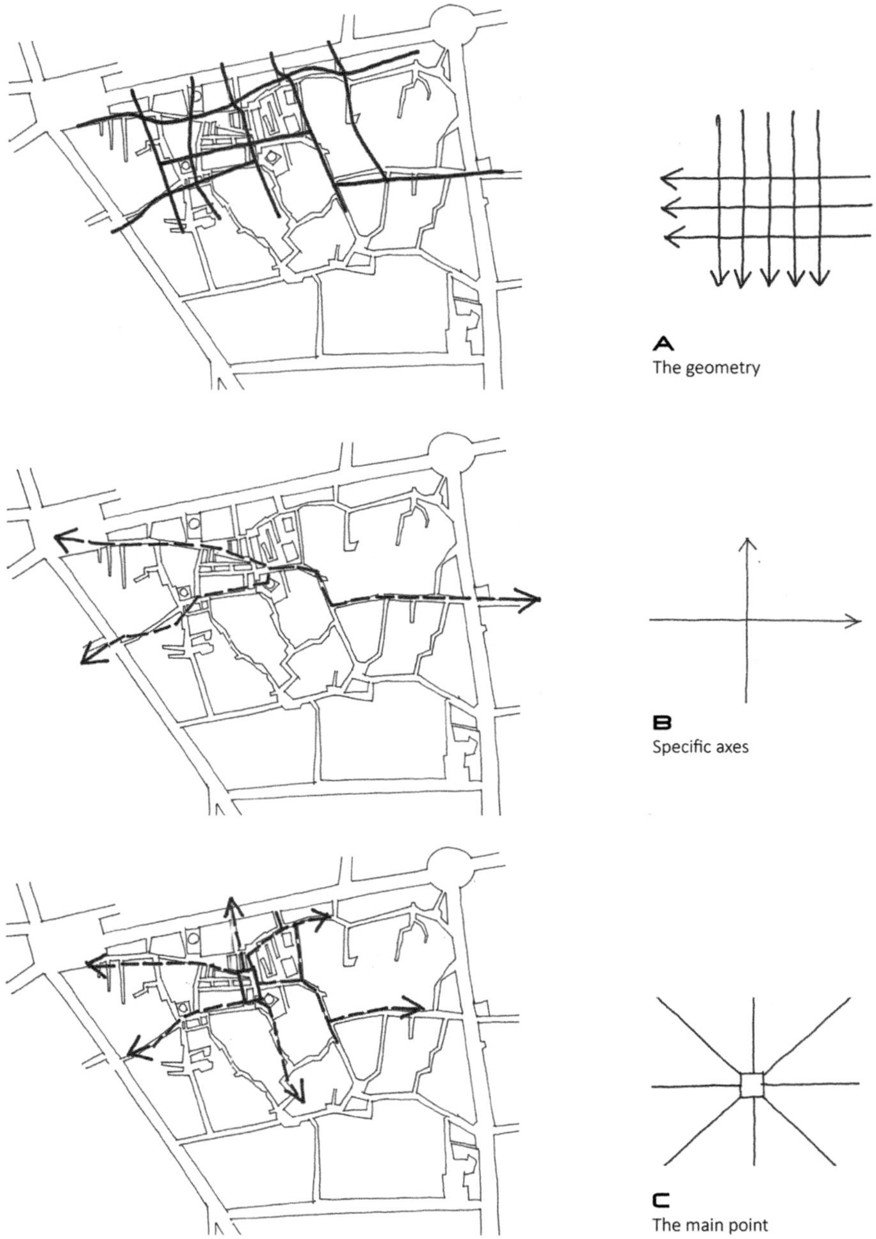

Fig. 5.27 The irregular grid system, two main axes, and the focal point of the Bazaar area structure. Main components of the *Bazaar* area: pedestrian way, mosques, large square, school, main square, and Bazaar itself

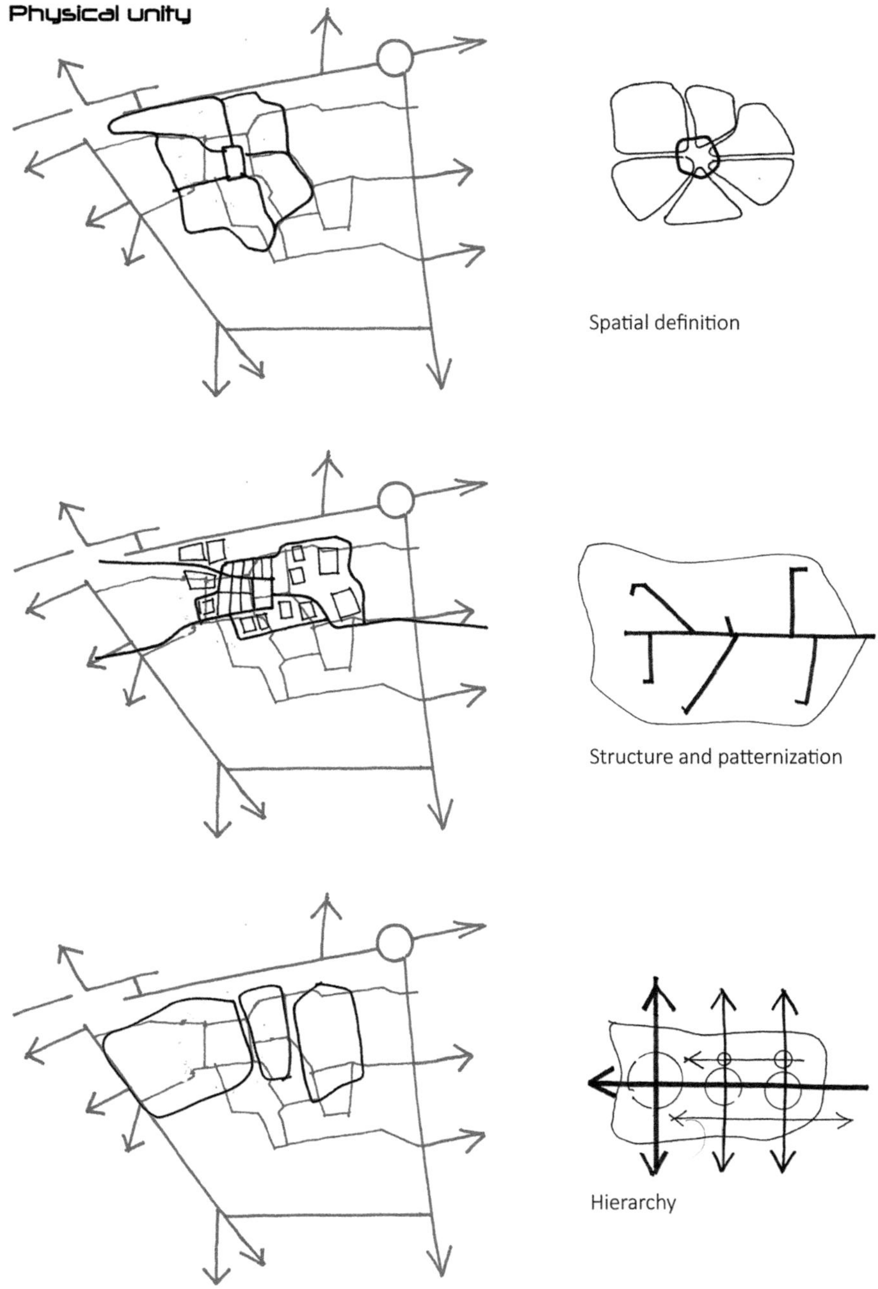

Fig. 5.28 Spatial definition and hierarchy of the *Bazaar* area

It seems that the physical structure of the city of Rasht follows central–radial pattern. The main components of this radial pattern include the central element and the main roads that make the connection to different directions.

The central element implies stability and firmness, while the linear elements (roads) indicate movement, dynamism, and continuity. It seems that due to the existing conditions and the hierarchy of urban structure, on the one hand, and the need for the unity of the whole urban fabric, on the other hand, his pattern has been repeated in the smaller scale along the main roads (Figs. 5.18, 5.19, and 5.20).

One of the other characteristics of the city of *Rasht* is that secondary roads which are parallel and perpendicular to the main road make the framework for city development. The described pattern along with the following elements makes the physical structure of the city (Fig. 5.29).

Bazaar complex, *Jame* Mosque, the main administrative centers of the city such as municipality, police stations, banks, and also the city square, which plays the role of central element for the city and is connected to the city through main pedestrian-oriented commercial corridors and has followed the radial–hierarchical order are the essential elements of the city. The secondary centers along the passages make the

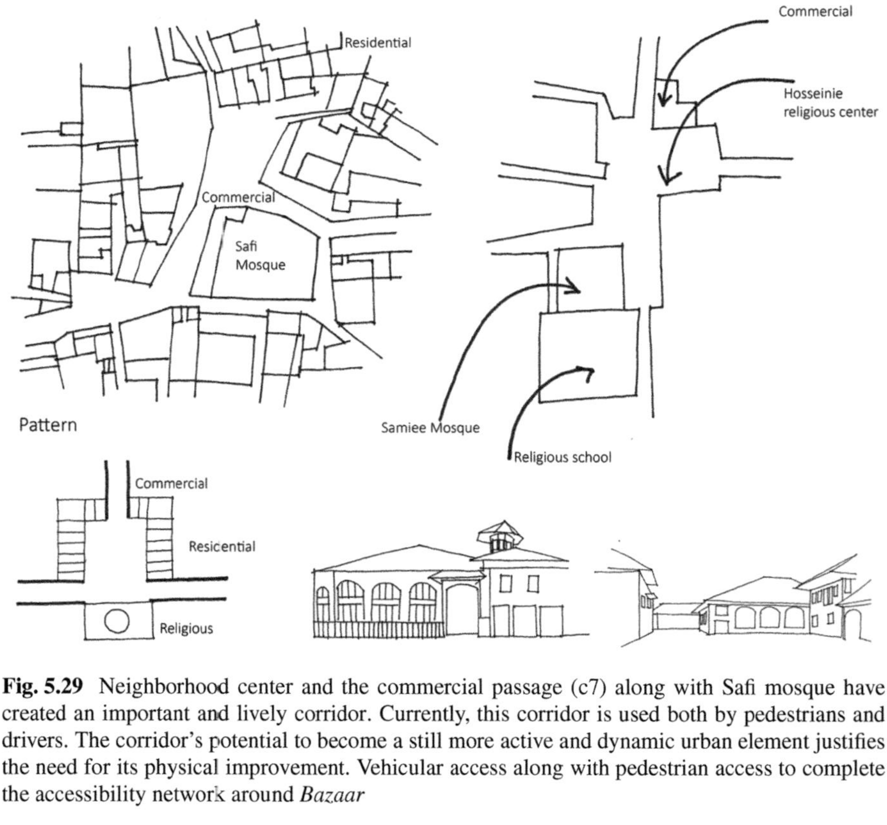

Fig. 5.29 Neighborhood center and the commercial passage (c7) along with Safi mosque have created an important and lively corridor. Currently, this corridor is used both by pedestrians and drivers. The corridor's potential to become a still more active and dynamic urban element justifies the need for its physical improvement. Vehicular access along with pedestrian access to complete the accessibility network around *Bazaar*

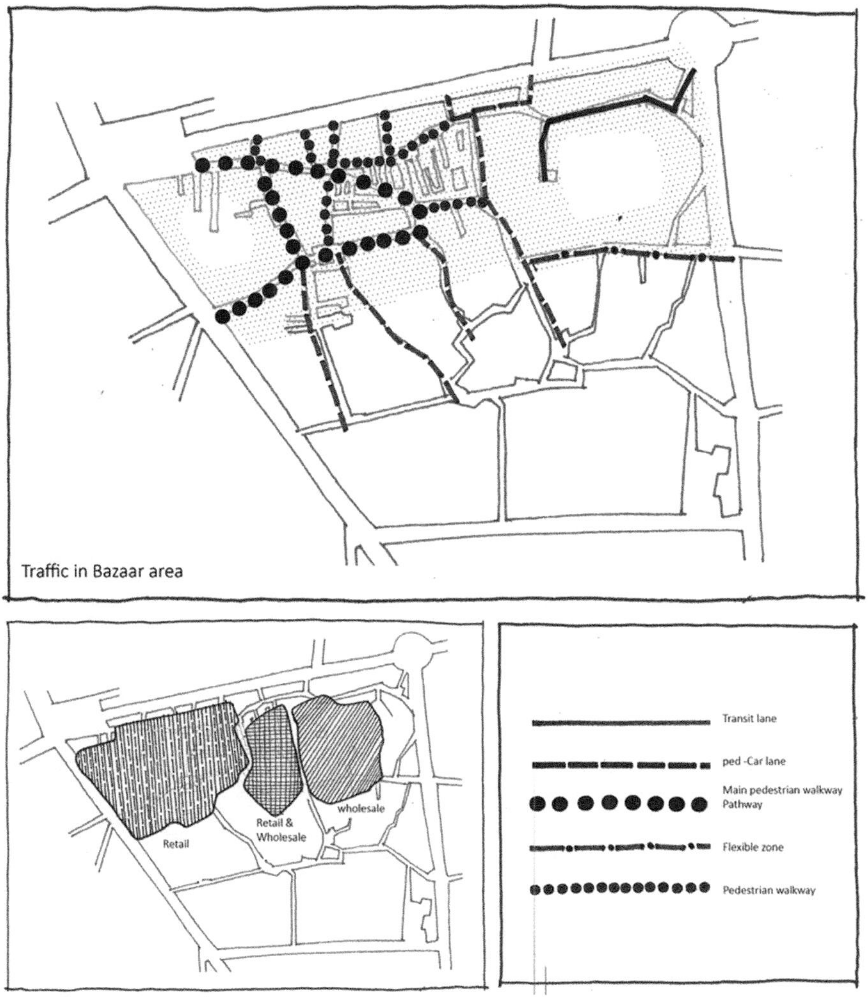

Fig. 5.30 Different traffic modes in the *Bazaar* area

neighborhood centers and nodes which include mosques and other religious centers and tombs, and also the neighborhood shops (Fig. 5.30).

The Bazaar complex, as the main and unique element of the city center and the destination of many people's daily trips for shopping, visit, and recreation, faces three basic requirements (Figs. 5.31 and 5.32).

Responding to those three requirements leads to a triangle shape, with three determinant points:

city square, Friday mosque, and the grand square.

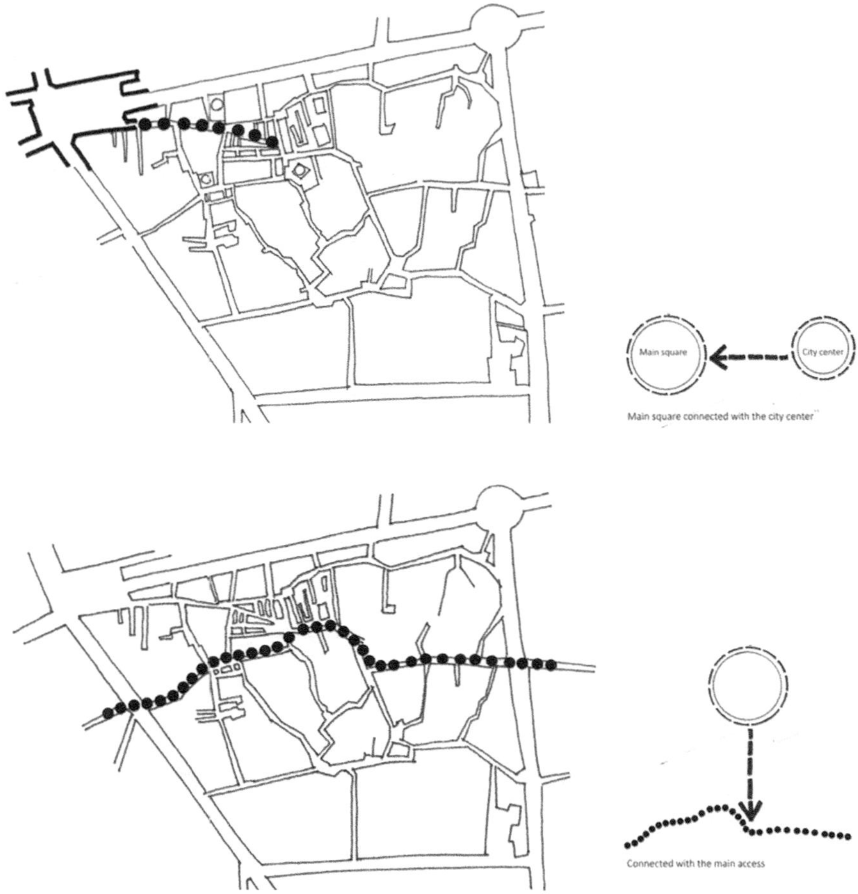

Fig. 5.31 Connection with the main city square and coordination with the city center (up) and connection with the city's main corridor (down)

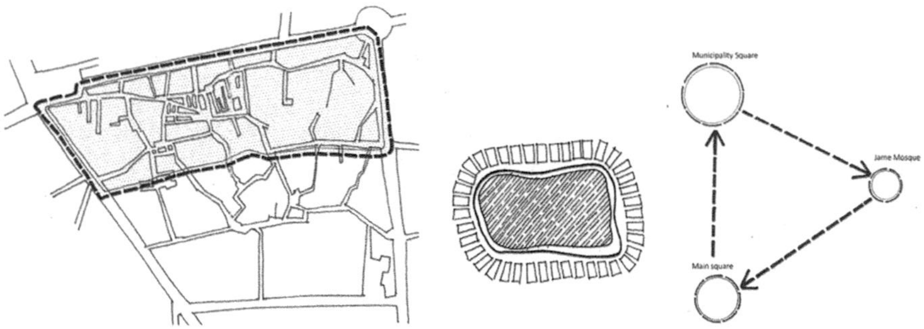

Fig. 5.32 The need for protecting the integrity of Bazaar as a unified complex

5.3 Vulnerability Analysis

The process of vulnerability analysis and design with the goal of vulnerability reduction due to seismic risk in the small case includes the following phases:

- Recognition and determination of damaging and damaged elements.
- Preparing earthquake scenario and determining the vulnerability zoning on its bases which design priorities could be known.
- Zoning.
- Specification—design implications.
- Design alternatives to mitigate vulnerability at the *Bazaar* scale.
- Reusing scenarios to evaluate design alternatives and choosing the optimum solution
- Complete the selected alternative.
- Loss and damage estimate based on isoseismic map of the city center: (1) very severe damage, (2) severe damage, (3) medium damage, (4) low damage.

5.3.1 Zoning Vulnerability of Buildings and Enclosed Spaces (Fig. 5.33)

- Zone 1, buildings with 75% damaging ratio, severe damaging physical type.
- Zone 2, buildings with ≥75% damaging ratio.
- Zone 3, buildings with ≤75% damaging ratio, and ≤50% severe damaging physical type.
- Zone 4, buildings with ≤75% and ≥50% damaging ratio.
- Zone 5, buildings with lower than 50% damaging ratio.
- Zone 6, the area with severe damaging due to mishaps in accessibility during disaster.
- Zone 7, the area with severe vulnerability due to fire risk.
- Zone 8, the area for basic needs storage for the disaster time (with high significance).

Fig. 5.33 Zoning vulnerability of buildings and closed spaces

5.3.2 *Damaging Factors and Vulnerable Elements (Fig. 5.34)*

Vulnerable elements during the disaster (the main activities are escape, refuge, and rescue). Supportive activities include first aid and emergency settlements.

- Vulnerability due to the lack of access to safe open spaces.
- Building vulnerability.
- Population vulnerability.
- Economic vulnerability.
- Vulnerability of critical goods and commodities.
- Vulnerability of goods due to fire.
- Vulnerability of buildings due to fire (Figs. 5.35 and 5.36).

Fig. 5.34 Earthquake damage in buildings, sidewalks, and passages

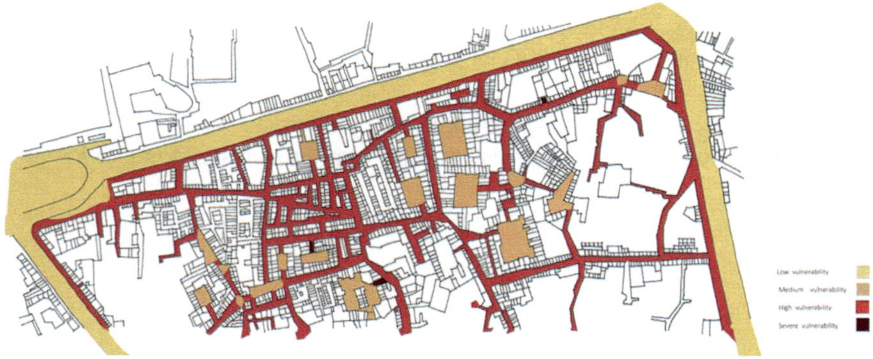

Fig. 5.35 Damage due to lack of access to open spaces

Fig. 5.36 Fire risk zoning

5.3.3 Building Vulnerability

Building vulnerability is determined on the basis of the following factors and using the specific guidelines: construction technique, roof structure, number of stories, construction quality, openings, overhangs, façade building materials; and also site characteristics such as soil, and land slope (Table 5.1).

5.3.3.1 Quantitative Analysis of Buildings Vulnerability

See Fig. 5.37.

Damage ratio

$$L_1 \times L_2 \times L_5 \times L_6 \times L_{11} \times [F_3 + F_4 + F_7 + F_9] = L_R$$

Table 5.1 Damage ratio in four different texture types

Lower damage, possibility of first aid and removing injured	*Caravanserai* type
Medium damage, average accessibility, possibility of moving less injured	Linear type
Higher damage, possibility of low penetration and accessibility, low settlement of injured	Grid type
Severe damage, very low accessibility, very low settlement of injured	Closed cluster type (passages)

On this basis, the damage ratio of buildings for 7, 8, and 9 Mercalli intensity will be in the following ranges: 0–25%, 25–50%, 50–75%, and 75% and higher

With the ratio lower than 25% minor, the building is usable, require repairment

With the ratio between 25 and 49% medium, the building is damage and possibility of repairment after evacuation

With the ratio between 50 and 75% severe damage, the building is compulsory evacuation and reconstruction required

With the ratio over 75% complete, the building is collapse and the possibility of death

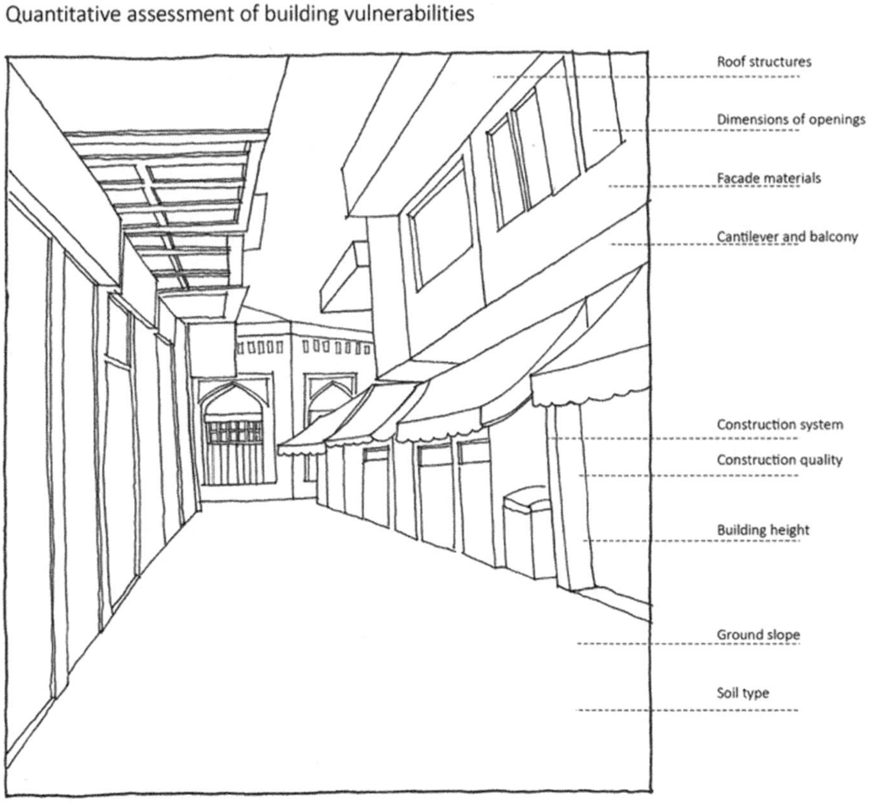

Fig. 5.37 Factors used in building vulnerability

5.3.3.2 Damaging elements During disaster (Main activities: escape, rescue, and refuge)

- Damaging buildings, damaged buildings will be also damaging.
- Damaging passages depend on the damaged or collapsed walls and buildings on the sides of the passage. The length of debris in the width of the passage due to the collapse of buildings along the passage equals half of the height of the buildings.

Fig. 5.38 Human vulnerability due to the lack of access to safe open spaces

5.3.4 Population Vulnerability

As a general rule, the higher the number of population in a surface, an increase in loss life possible.

Vulnerability due to the lack of access to safe open spaces. According to some criteria, "the distance run by a normal person would be 50 m in 30 s". Other factors such as damaged walls and population density would also affect vulnerability (Fig. 5.38).

5.3.5 Vulnerability of Buildings from Fire

Higher vertical density increases the chances of fire spreading, more open spaces will decrease spreading fire, more difference of building heights in adjacent buildings increases the possibility of spreading fire from lower buildings to the higher ones, and more openings will increase the possibility of fire spreading (Fig. 5.39).

Fig. 5.39 Building systems according to vulnerability against fire

5.3.6 Physical Types in the Bazaar Area and Their Damage Due to Earthquake

See Figs. 5.40, 5.41, 5.42, and 5.43.

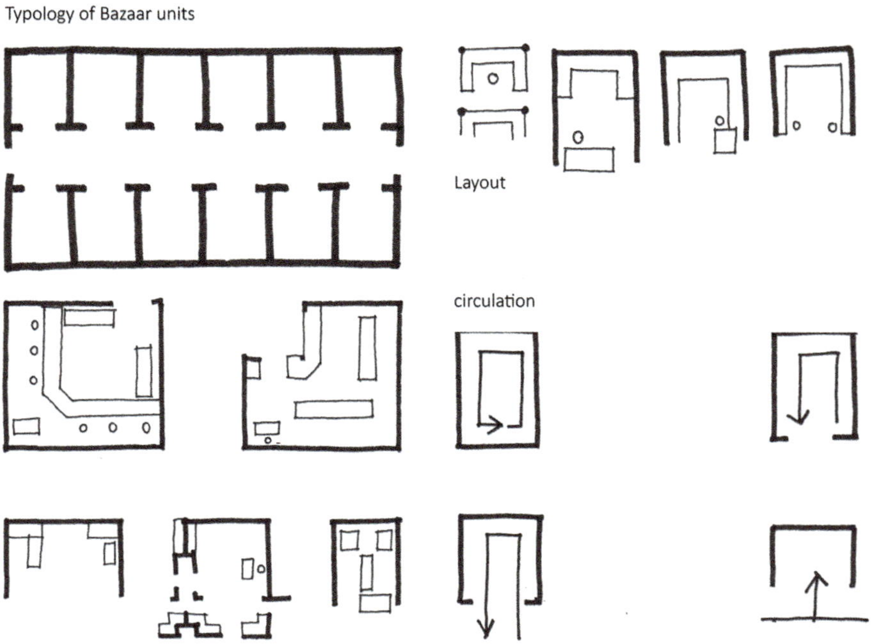

Fig. 5.40 Physical types in the Bazaar area and their damage due to earthquake

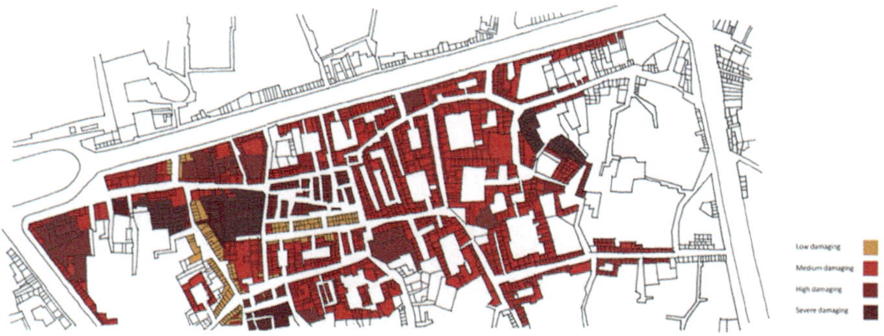

Fig. 5.41 Damages of the physical types (micro scale)

Fig. 5.42 Vulnerable construction systems due to earthquake

Fig. 5.43 Damage due to the collapse of passage walls

5.3.7 Damaging Non-structural Elements in the Passages and Open Spaces (Bahrainy et al. 1997a, b, c, d)

Non-structural elements include signs, powerlines, sloping roof covers, consules, façade materials, utility elements such as coolers, etc. (Figs. 5.44, 5.45, and 5.46; Table 5.2).

5.3.8 Inflicting Damages by Physical Types

The possibility of easy and fast connection between open and closed spaces leads to less damage. These factors are relevant: the depth of closed spaces, the kind of connection between closed and open spaces (with or without joint), the pattern of space use which indicates how the activities are taking place in the closed systems, long use along with lack of connection with open spaces, and non-structural elements in the closed space (Bahrainy et al. 1997a, b, c, d).

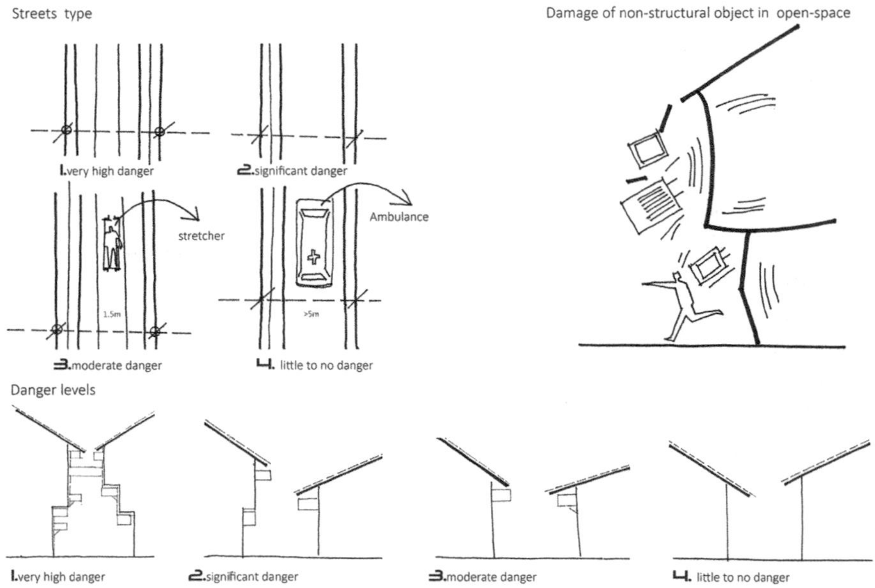

Fig. 5.44 Non-structural elements in four different types of roads

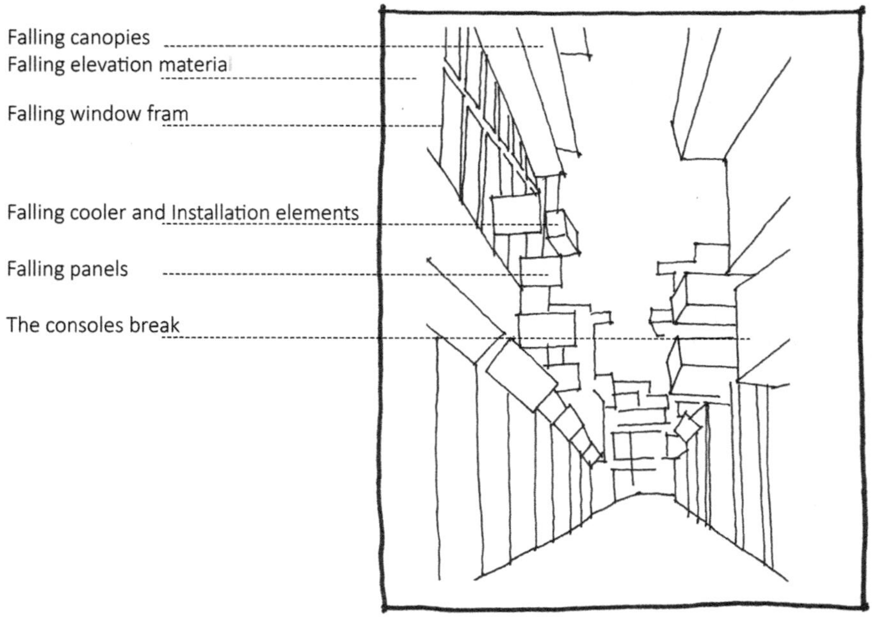

Fig. 5.45 Vulnerability of non-structural elements

Fig. 5.46 Damage of non-structural elements due to the passages walls

Table 5.2 Criteria for the degree of vulnerability of passages

Severe damaging	High-density signs, utility elements, dangerous covers, facades, and consules
High damaging	Medium density of non-structural elements and dangerous elements
Medium damaging	Medium density of non-structural elements, without dangerous consules,
Low damaging	Low density of non-structural elements, without dangerous consules

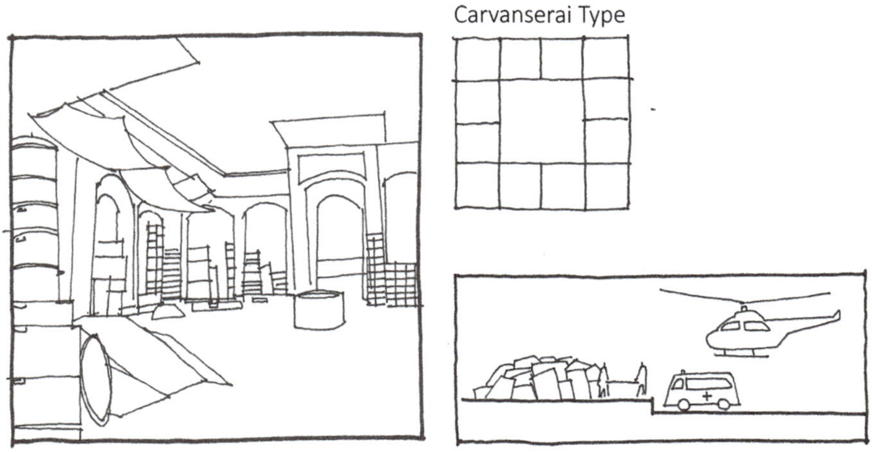

Fig. 5.47 The plan and a view to the caravanserai type

In the Bazaar area, four groups of elements cause damage, as follows (Figs. 5.47. 5.48, 5.49, and 5.50; Tables 5.3 and 5.4).

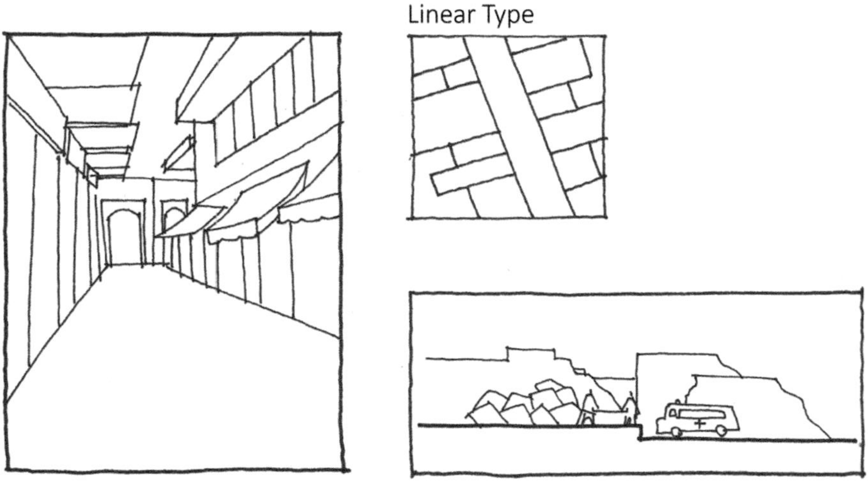

Fig. 5.48 The plan and a view of the linear type

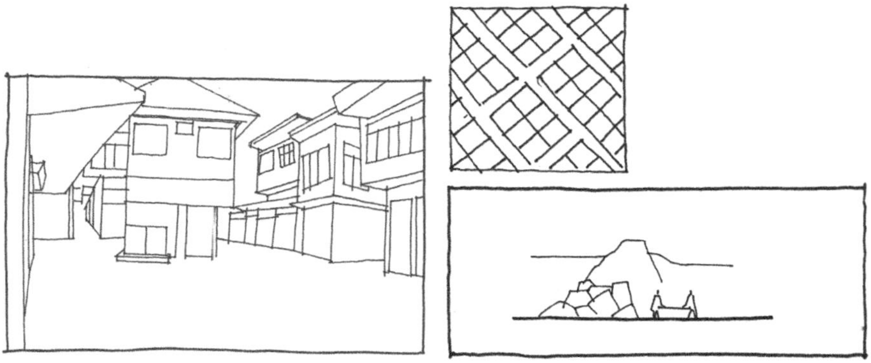

Fig. 5.49 The plan and a view of the grid type

5.3.9 *Vulnerability Zoning of Roads and Open Spaces (Bahrainy et al. 1997a, b, c, d)*

- Zone 1, roads type 1.
- Zone 2, roads type 2.
- Zone 3, roads type 3.
- Zone 4, the area with damaging potential due to accessibility mishaps during disaster.
- Zone 5, the area with severe potential of fire (Figs. 5.51 and 5.52).

Fig. 5.50 The plan and a view of the closed blocks type

Table 5.3 Severe, high, medium, and low damaging elements

Public baths, mosques, main offices, passages, and movie houses, which lack fast access of closed spaces to open spaces	Cause severe damage
Spaces which require complete presence in the space, but there is also the possibility of access to outside, such as banks	Cause High damage
The stores which their access is temporary	Cause Medium damage
Spaces which their access takes place in open or semi-open spaces, such as kiosks and fish stores	Low damaging

- Vulnerability of spaces that lack access to safe open space. The criteria in this regard: A normal person may go 50 m in 30 s. This has to be balanced with the vulnerability of road walls and population density.

  Damaging roads depend on the deteriorated walls of the road and the possibility of life loss and property damages. The possible length of falling debris of the destructed walls in the road equals half of the height of the walls (Fig. 5.53).

5.3.9.1 Vulnerable Factors During the Disaster 2: Main Activities (Emergency First Aid, Emergency Settlement)

Vulnerability of spaces which lack access to safe open spaces. These spaces are used for the transport and settlement of injured during the disaster (Figs. 5.54 and 5.55).

Table 5.4 Physical typology of the *Bazaar* area and their possible damage due to an earthquake (Bahrainy et al. 1997a, b, c, d)

Lower damage, possibility of first aid, and removing injured	*Caravanserai* type	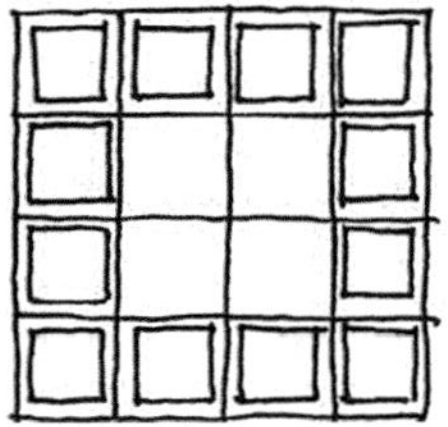
Medium damage, average accessibility, possibility of moving less injured	Linear type	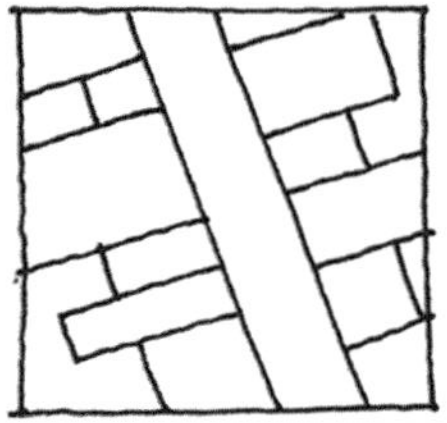
Higher damage, possibility of low penetration and accessibility, low settlement of injured	Grid type	

(continued)

Table 5.4 (continued)

Lower damage, possibility of first aid, and removing injured	*Caravanserai* type	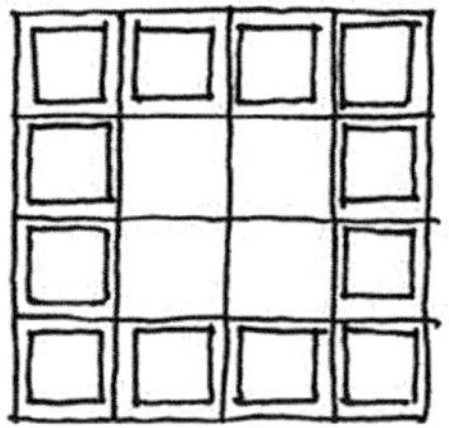
Severe damage, very low accessibility, very low settlement of injured	Closed cluster type (passages)	

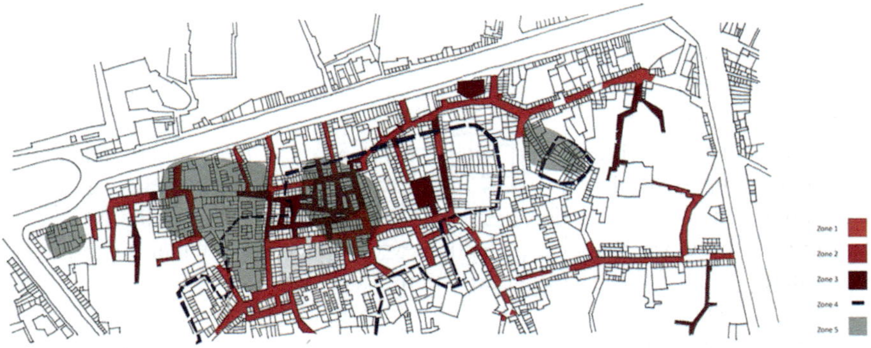

Fig. 5.51 Zoning vulnerability of roads and open spaces

Fig. 5.52 Zoning of buildings and closed spaces vulnerability

Fig. 5.53 Relationship between height of the wall and its width

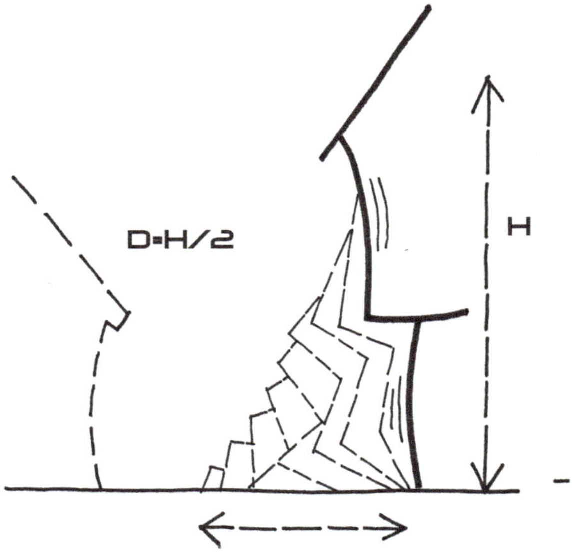

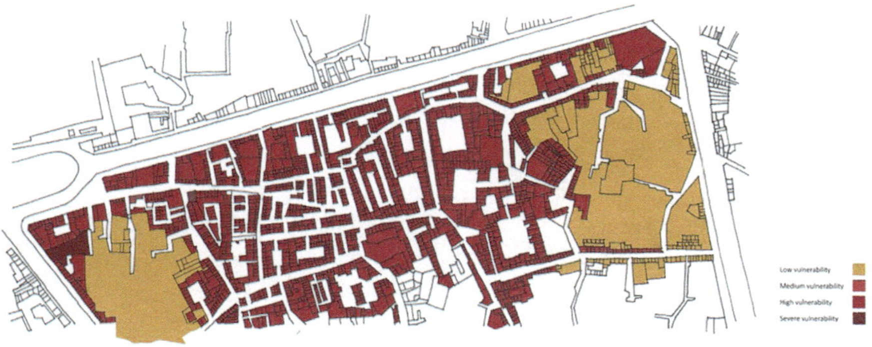

Fig. 5.54 Human vulnerability due to population density in the closed spaces

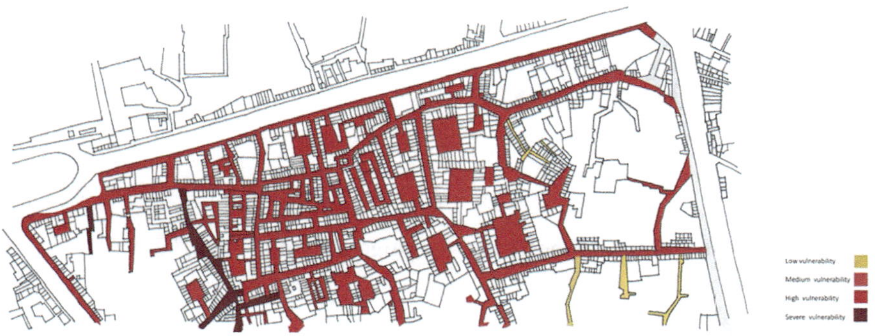

Fig. 5.55 Human vulnerability due to population density in the open spaces and passages

5.4 Design Ideas to Reduce Seismic Risk at Small Scale (Bahrainy et al. 1998)

5.4.1 City Structure

As stated before, the city of Rasht and particularly its center and the *Bazaar* areas, in spite of vulnerability against seismic risk, have significant cultural, economic, and social values which deserve to be taken into account in any planning and design activity. One of the primary goals in any intervention in the city structure is to provide physical cohesion and unity, not only in the Bazaar area, but the whole city (Figs. 5.56, 5.57, 5.58, and 5.59).

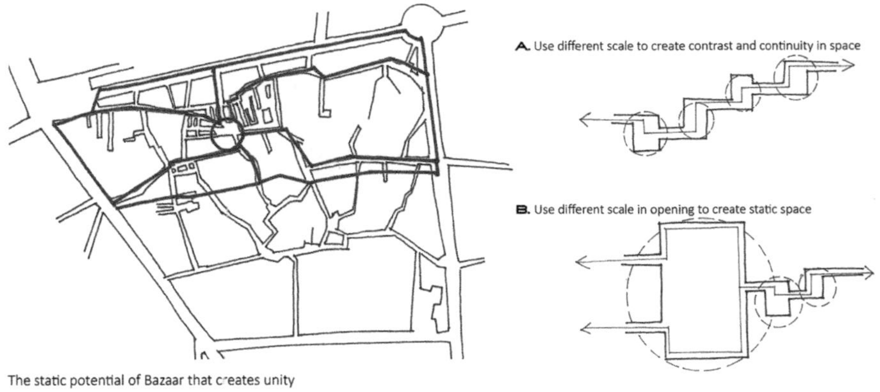

Fig. 5.56 The criteria and methods of physical cohesion and unity in the Rasht *Bazaar*

5.4.2 Operational Objectives for Roads and Open Spaces (Petrovski et al. 1998)

See Fig. 5.60.

Measures to be taken in zones 1 to 6 to mitigate seismic risk (Fig. 5.61).

- Zone 1, creating open space along with strengthening road walls
- Zone 2, provide access to open spaces as the first priority, and creating open space as the second priority
- Zone 3, provide access to existing open spaces as the first priority, and roads improvement, as the second priority
- Zone 4, strengthening roads, as the first priority, and road improvement, as the second
- Zone 5, removing the non-structural damaging elements, as the first priority
- Zone 6, provide accessibility for the time of disaster, as the first priority (Fig. 5.62).

5.4.3 Operational Objectives for the Closed Spaces

Following measures are recommended in different zones of the area (Fig. 5.63):

- Zone 1, urban renewal, as the first priority.
- Zone 2, urban renewal, as the second priority, partial repair, and improvement of physical type, as the first priority.
- Zone 3, repair and considering measures to facilitate access to existing and proposed open spaces outside the area and partial physical improvement.
- Zone 4, strengthening as the first priority.

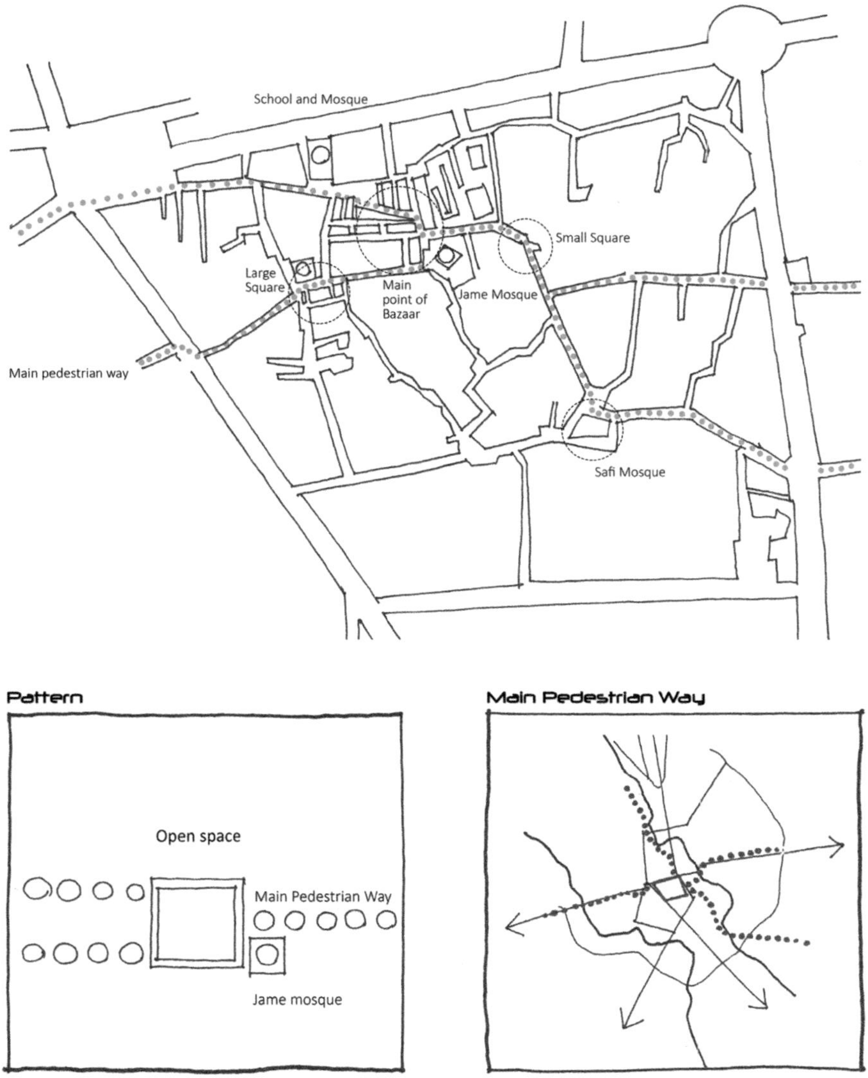

Fig. 5.57 Design concept based on physical factors

- Zone 5, strengthening as the first priority and urban renewal along with improvement of physical type, as the second priority.
- Zone 6, strengthening and partial improvement of physical type, as the third priority.
- Zone 7, the area has priority to provide measures for better access during disaster.
- Zone 8, the area is in need of appropriate measures to prevent breaking fire and spreading to other areas. Physical type improvement to prevent fire from spreading.

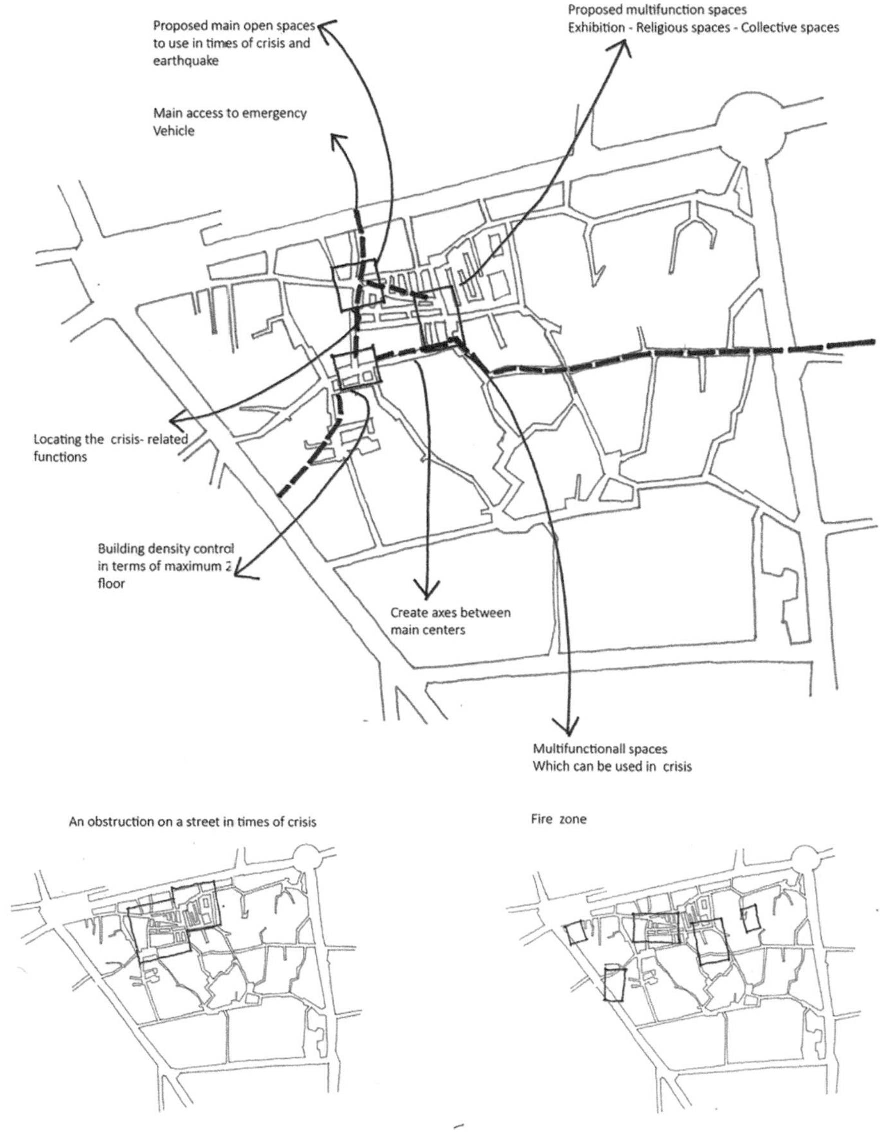

Fig. 5.58 General design concept to reduce seismic vulnerability

5.4.4 Measures to Be Taken in Different Physical Types to Reduce Seismic Risk

Some of the passages and Bazaar branches are of severe damaging type, with no appropriate access to open space. Due to the low depth of the enclosed spaces, any modification of this type of passages is not possible. A basic solution to reduce the

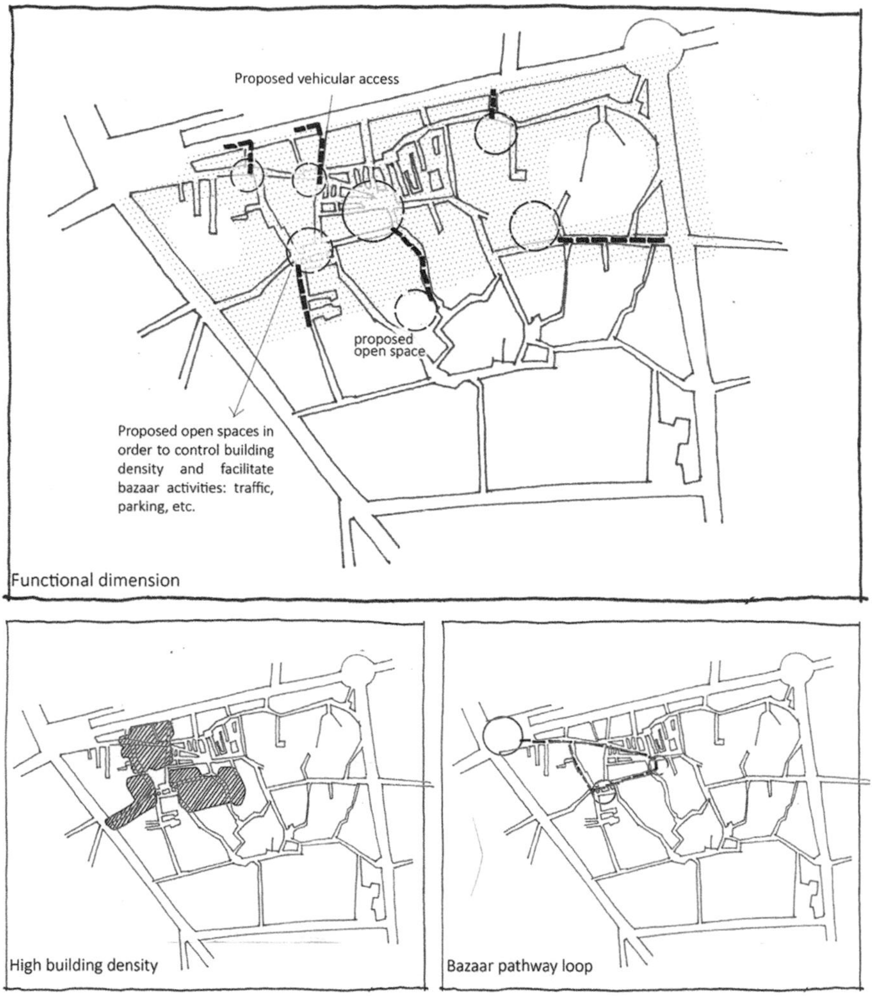

Fig. 5.59 Design concepts with regard to functional factors

vulnerability of these spaces is providing access from these spaces to the existing open spaces. In these cases, the existing open spaces should be recognized and the possibility of providing accessibility to those spaces and increasing the number of entrances and exits investigated.

Providing accessibility depends on two main factors:

- Securing space for movement and accessibility.
- Strengthening the movement space in order to prevent blockage in case of a disaster.

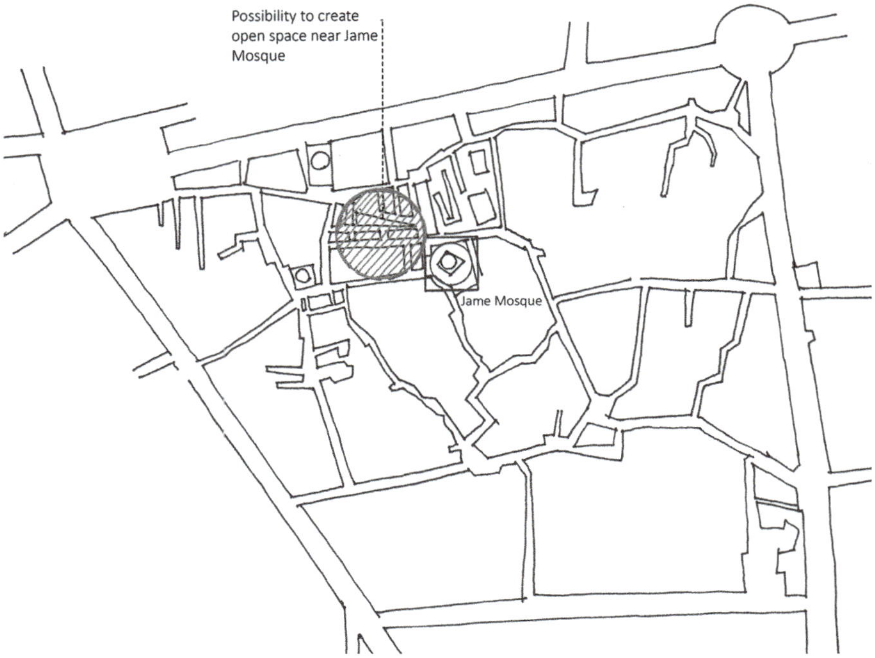

Fig. 5.60 Creating open space in the vicinity of the *Jame* mosque is critically important to serve the population using the mosque during and after a disaster

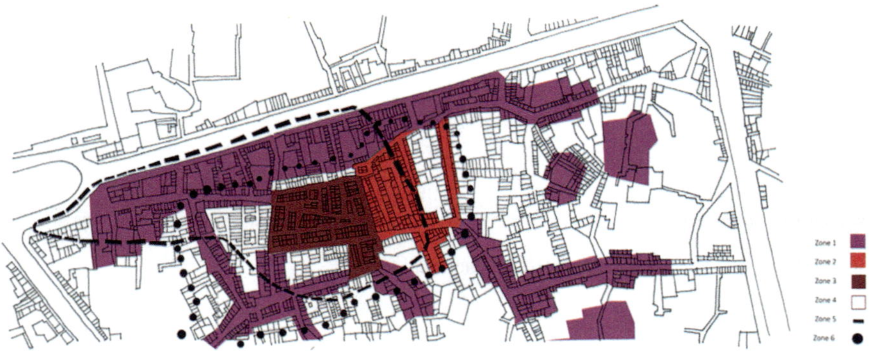

Fig. 5.61 Zoning specifications in passages and open spaces

Obviously providing space for the movement and accessibility requires the demolition of some parts of buildings which could take place on the basis of existing situation. Strengthening of accessibility networks and entrances are significant when they can play critical role in reducing vulnerability during a disaster.

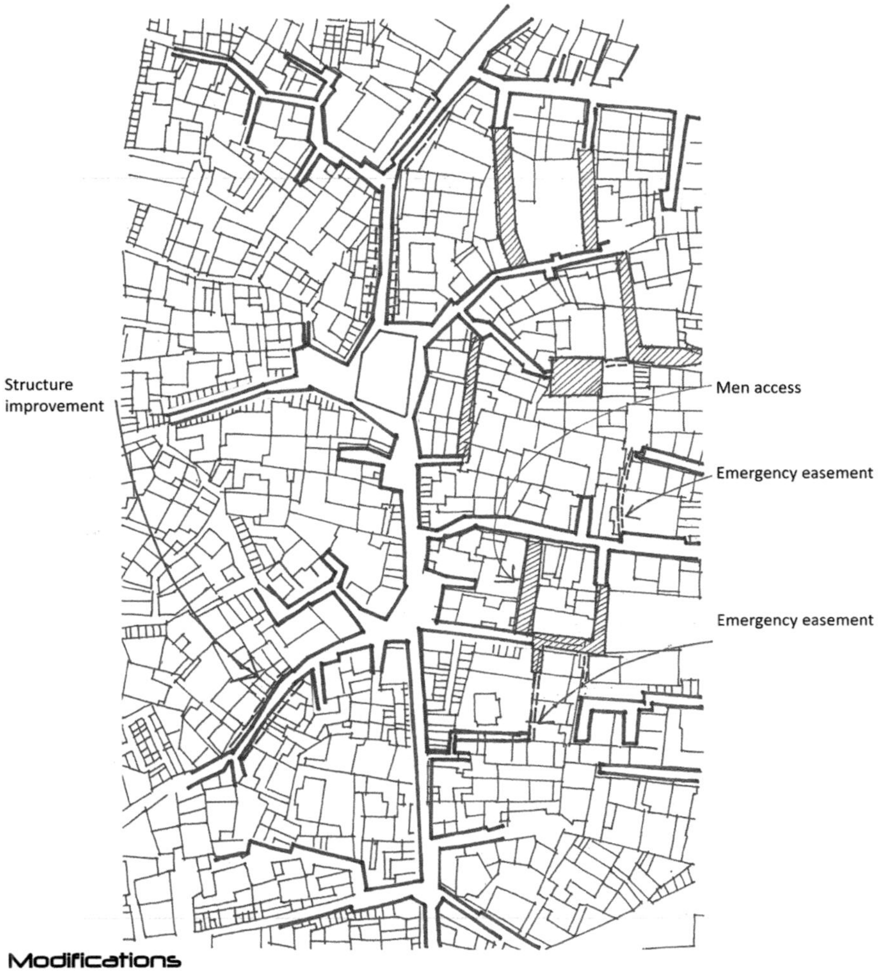

Fig. 5.62 Structural improvement, new access, emergency easement in the existing texture of the city of Rasht (after Banerjee 1993)

Some physical types may cause damages solely because the in-between spaces are covered. Removing the roofs and replacing it with very light material can turn these kinds of buildings into much less vulnerable ones (Figs. 5.64, 5.65, and 5.66).

Some physical types (see Fig. 5.67) have very deep open space and no proper access to open space, and these physical types seem dangerous. The proposed design includes a central open space which could be compensated for by increasing the number of story with a setback on each floor.

The complexes without access to open space, and with low depth of enclosed spaces (provide access to open space) (Fig. 5.68).

Fig. 5.63 Zoning of operational objectives for the "closed spaces"

Some physical types might face wall collapse, as a result which the passage will be completely blocked and any emergency access during the disaster will be stopped. Proposed design in these cases includes passage modification through vertical setback in the adjacent buildings to reduce risk and its closure in case of building collapse (see Fig. 5.69).

5.4.5 Analysis and Design of a Central Open Space (Square) as an Example, on the Basis of Seismic Risk Reduction

See Figs. 5.70, 5.71, and 5.72.

Compatibility with climate and ecology are taken into account by providing west–east orientation and open space between buildings for ventilation (Figs. 5.73, 5.74, and 5.75).

5.4.6 Improvements in a Dense Building Block to Reduce Seismic Risk

Developing dense building blocks with vertical setbacks and spatial access through several entrances and exits to the adjacent external open spaces.

Existing condition: Building complexes with dangerous and low depth passages and without access to safe open space.

The proposed improvement includes:

- Improving the existing open spaces to be used during disaster.
- Entrances need to be strong enough to be appropriate for use during a disaster.
- Make the use of existing open spaces possible by providing proper entrance and exit spaces.

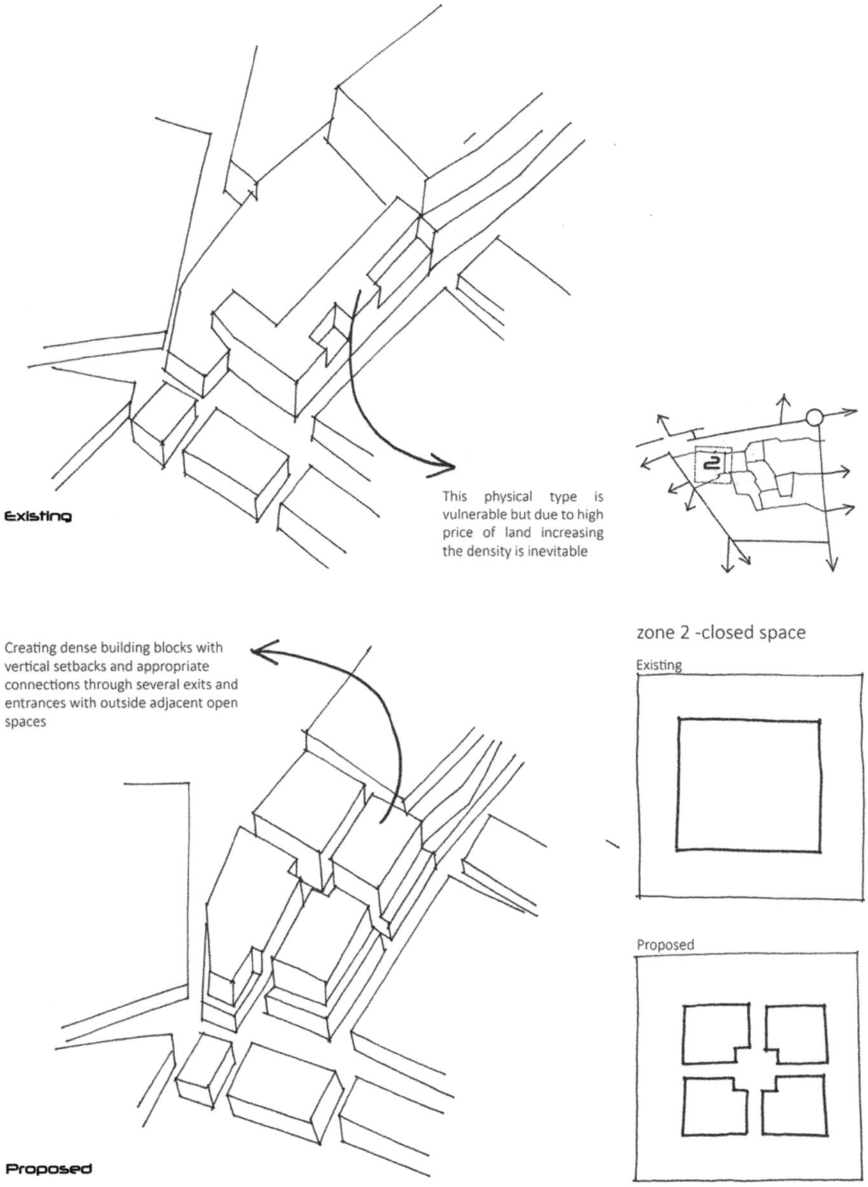

Fig. 5.64 Modification of physical type, medium scale

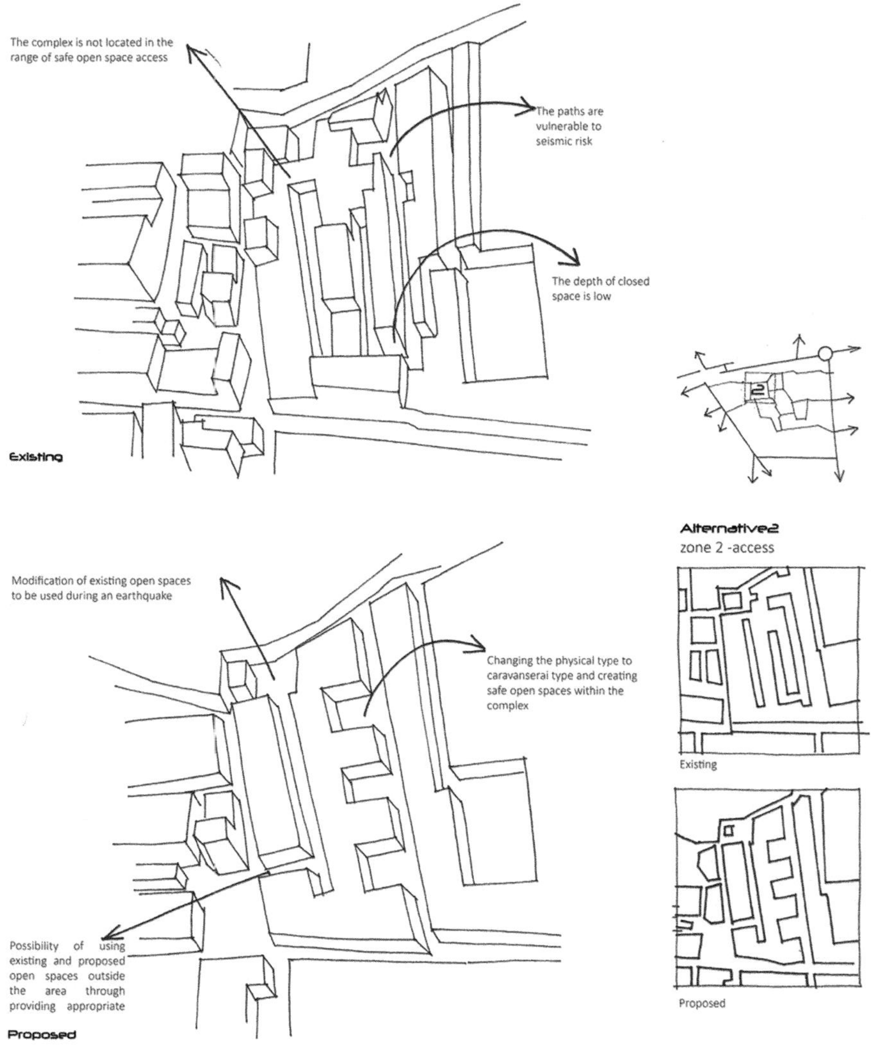

Fig. 5.65 Modification of the physical type at the medium scale in the complexes with dangerous passages and with low depth and without access to safe open space. 2nd alternative

- Strengthening buildings.
- In the case of buildings with damage ratio of less than 50%, or of low risk physical type, with damage ratio of above 50% and lower than 75% strengthening and repairment measures may be applied. These measures vary based on the kind of construction system which has been used (Figs. 5.76 and 5.77).

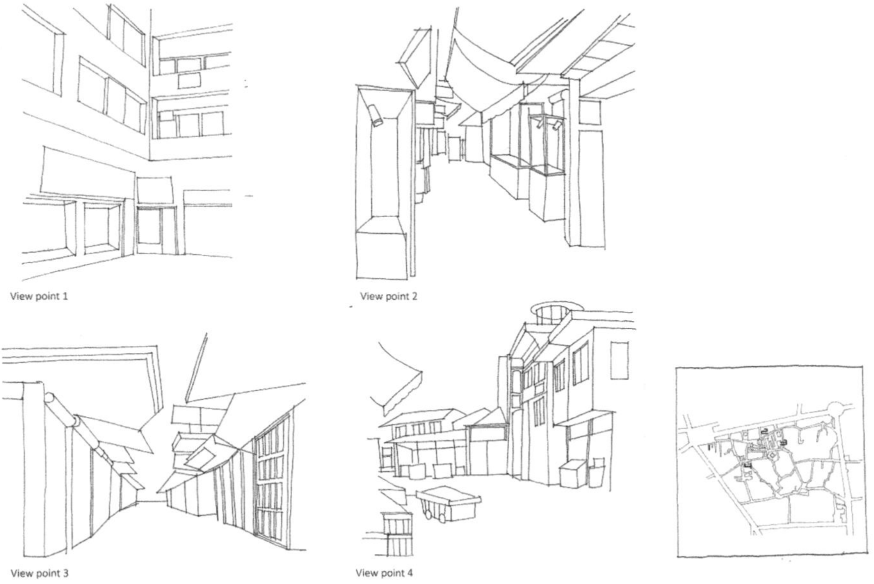

Fig. 5.66 Neighborhood center and the commercial passage (c7) along with Safi mosque have created an important and lively corridor. Currently, this corridor is used both by pedestrians and drivers. The corridor's potential to become a still more active and dynamic urban element justifies the need for its physical improvement. Vehicular access along with pedestrian access to complete the accessibility network around *Bazaar*

5.5 Conclusion

It is believed that the most effective and commonly used scale in urban design is small scale. Urban design through its methods and techniques try to manipulate physical environment in order to achieve certain goals and particularly reducing seismic risks in urban areas. Manipulating the physical environment encompasses a broad set of activities in a wide range of physical settings. "Manipulation" may include direct physical design and construction, regulatory measures to guide physical changes over time, economic and community development efforts, regional growth strategies, and institutional measures such as funding programs that ultimately result in physical actions.

An analysis of the existing conditions, including: location of the area in the city, typology of the city structure (large scale), typology of the *Bazaar* structure, (macro, medium, and micro scales), area characteristics, and general patterns of the physical elements is required. The process of vulnerability analysis and design with the goal of vulnerability reduction due to seismic risk in the small case includes seven phases which begins with recognition and determination of damaging and damaged elements, and ends with loss and damage estimate based on isoseismic map of the

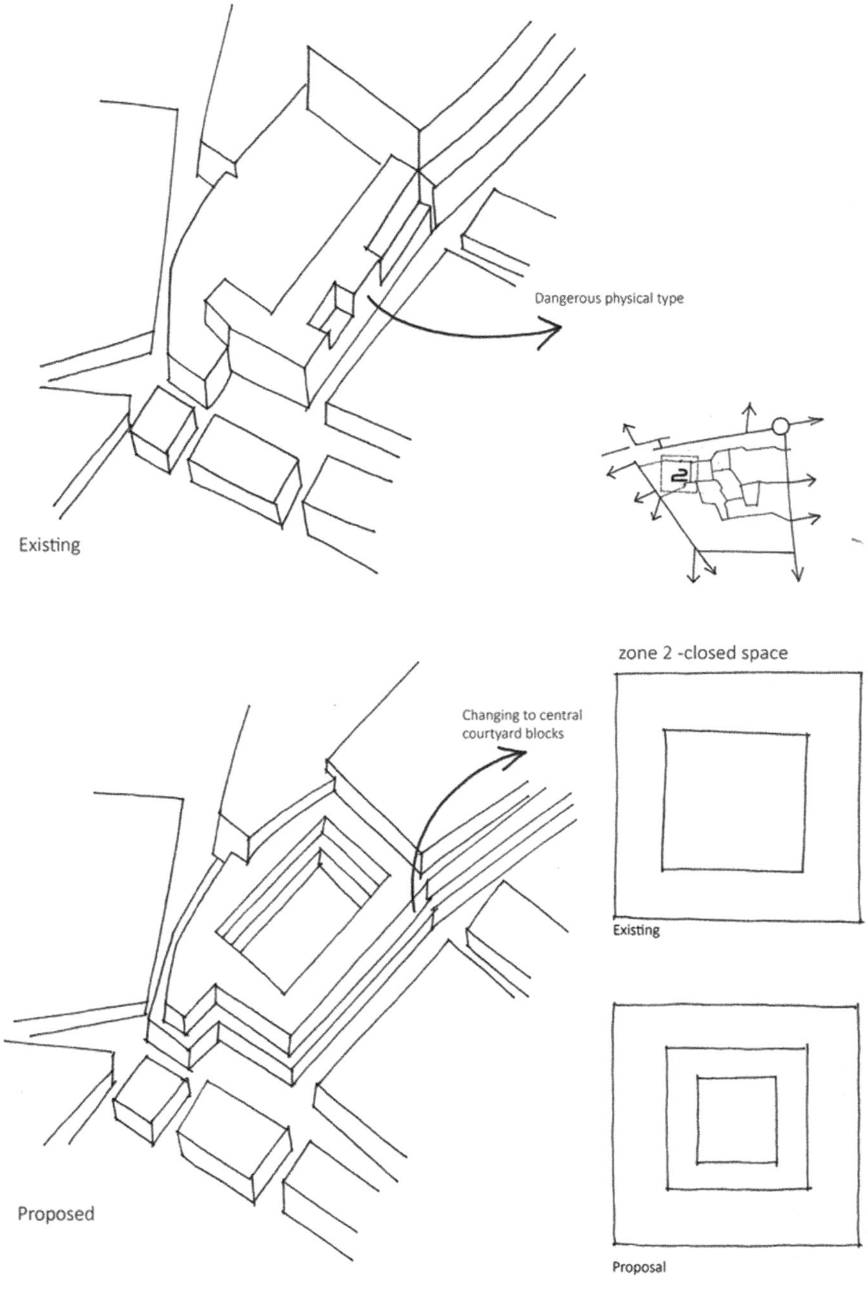

Fig. 5.67 Existing and proposed physical type, closed space

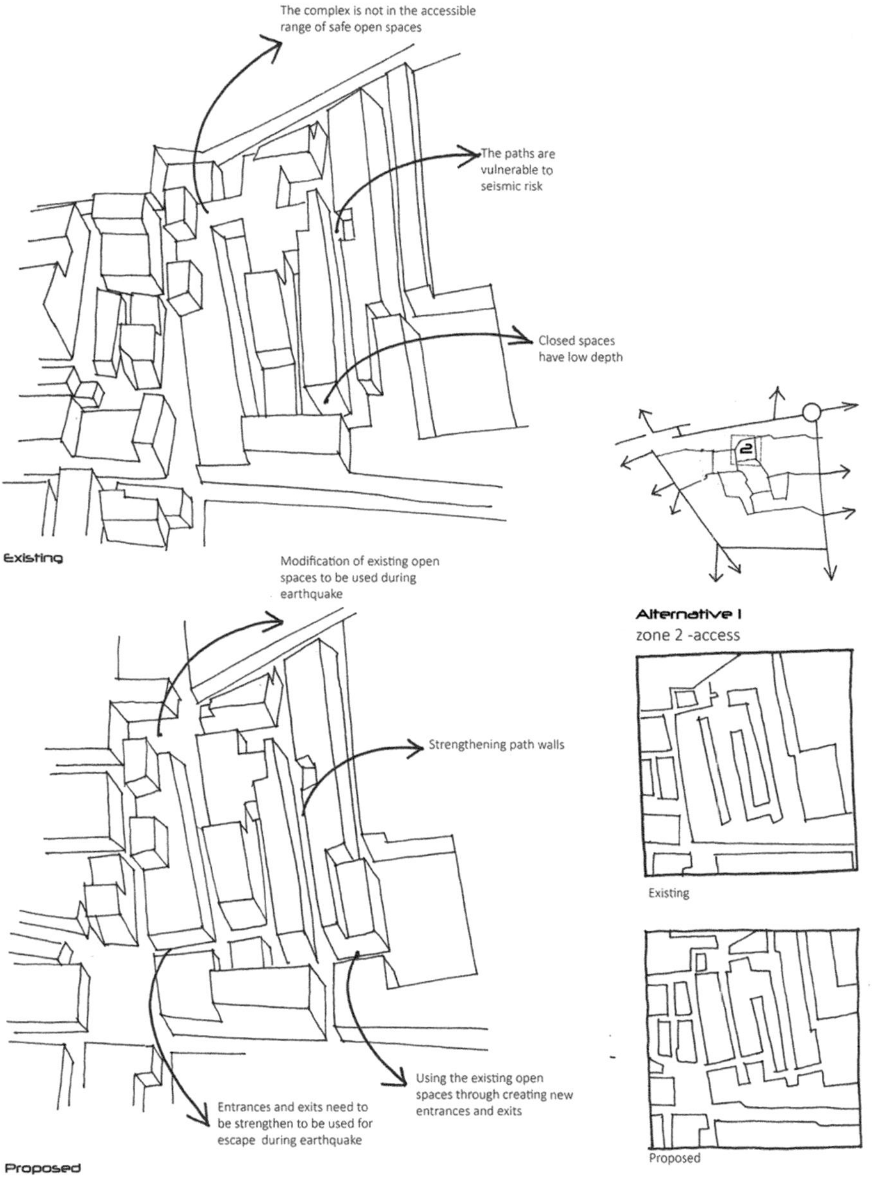

Fig. 5.68 Existing and proposed modifications in complexes with low depth dangerous passages and without safe access to open space, 1st alternative

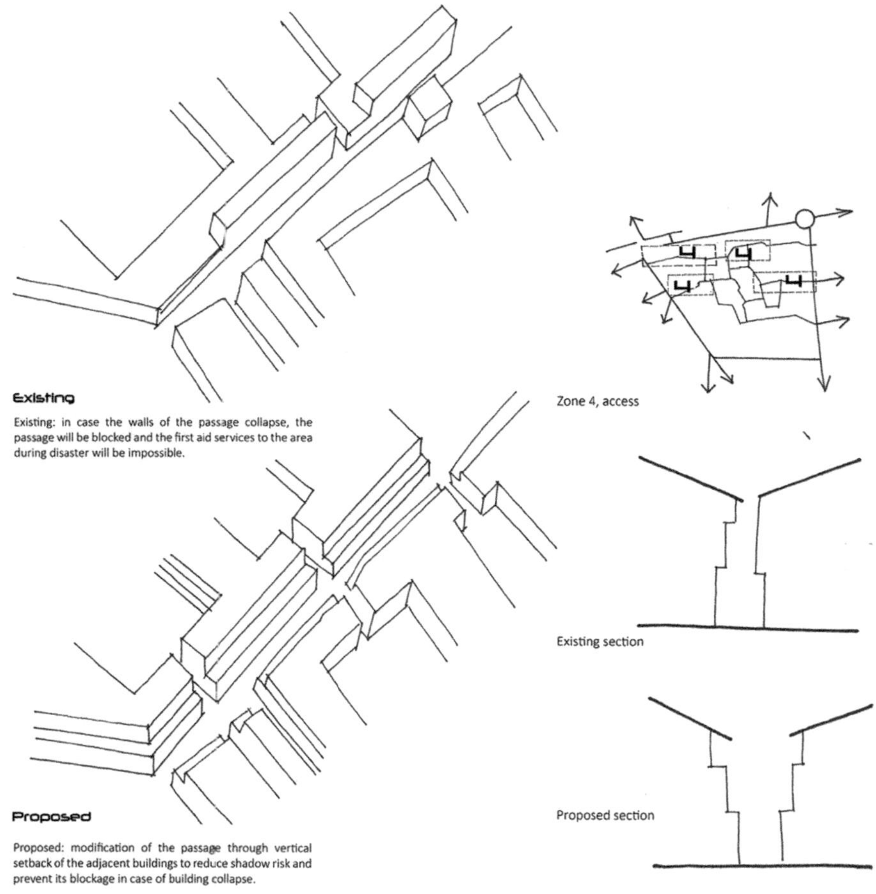

Fig. 5.69 Improvement of the main passages: entrance and exit, and modification of the passage through vertical setback of the adjacent buildings to reduce shadow risk and prevent its blockage in case of building collapse

city center: (1) very severe damage, (2) severe damage, (3) medium damage, and (4) low damage.

Bazaar activities are reflected in diverse physical types. These types may be differentiated through diversity of aggregation patterns and juxtaposition of essential components and smaller scale of the texture and the type of commercial activities. The dominant types are as follows: central type (*caravanserai*), grid type, linear type, and dense and integrated blocks. Each type has its strengths, as well as weaknesses, in general and particularly with regard to seismic risk.

Building vulnerability is determined on the basis of the following factors and using the specific guidelines: construction technique, roof structure, number of stories,

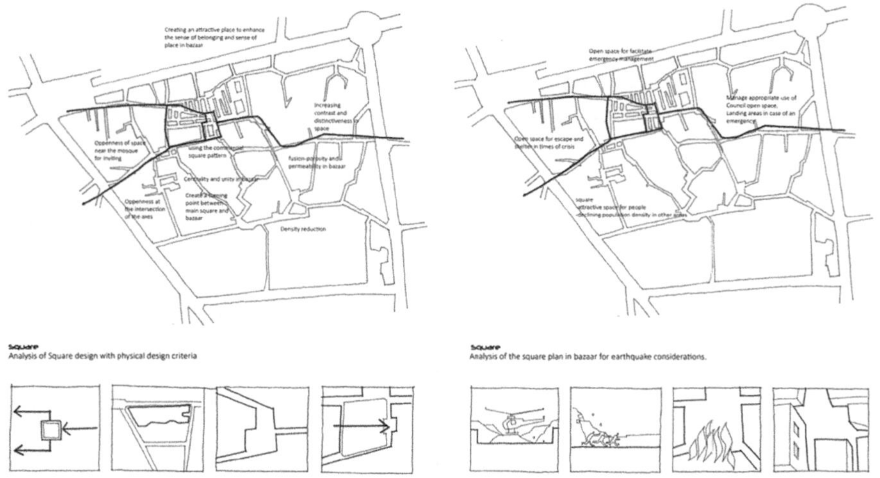

Fig. 5.70 Analysis of square design on the basis of seismic risk reduction

construction quality, openings, overhangs, façade building materials; and also site characteristics such as soil, and land slope.

A quantitative analysis of buildings vulnerability, population vulnerability, vulnerability of buildings from fire, vulnerability of non-structural elements in the passages and open spaces, and vulnerability of roads and open spaces are carried out, on their basis which, design ideas to reduce seismic risk at small scale are presented. Operational objectives for roads and open spaces and also for the closed spaces are formulated.

And finally, as an example, analysis and design of a central open space (square) are presented on the basis of seismic risk reduction.

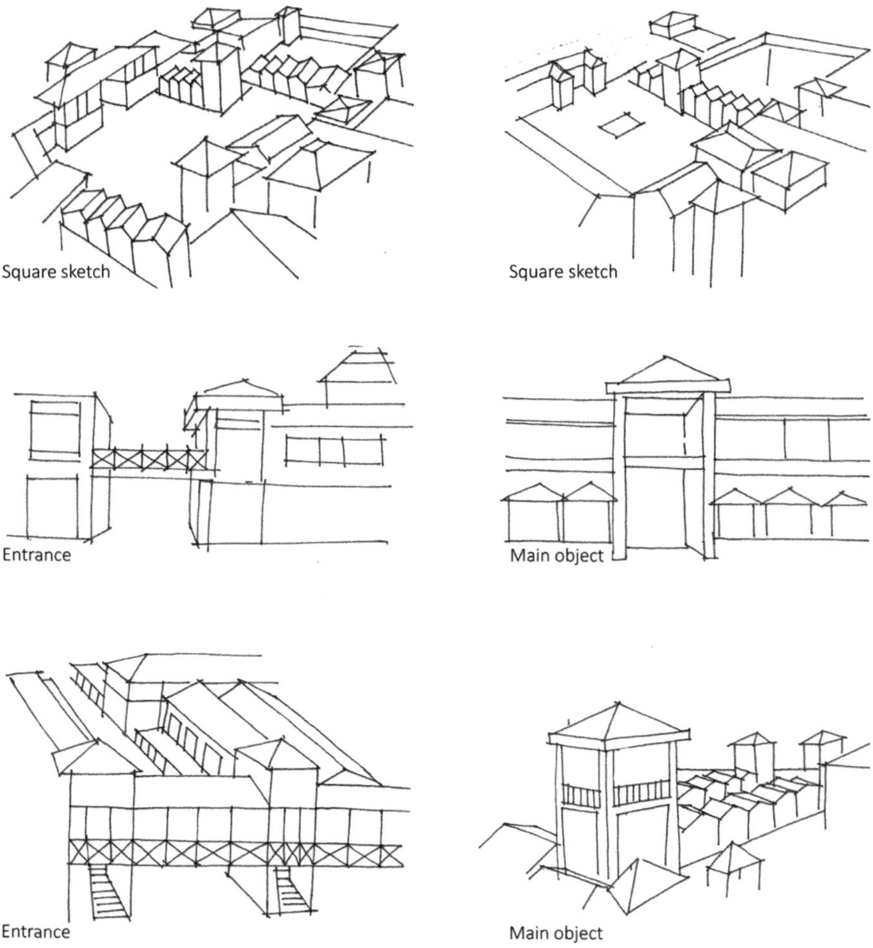

Fig. 5.71 Design ideas for the main square, based on seismic risk reduction

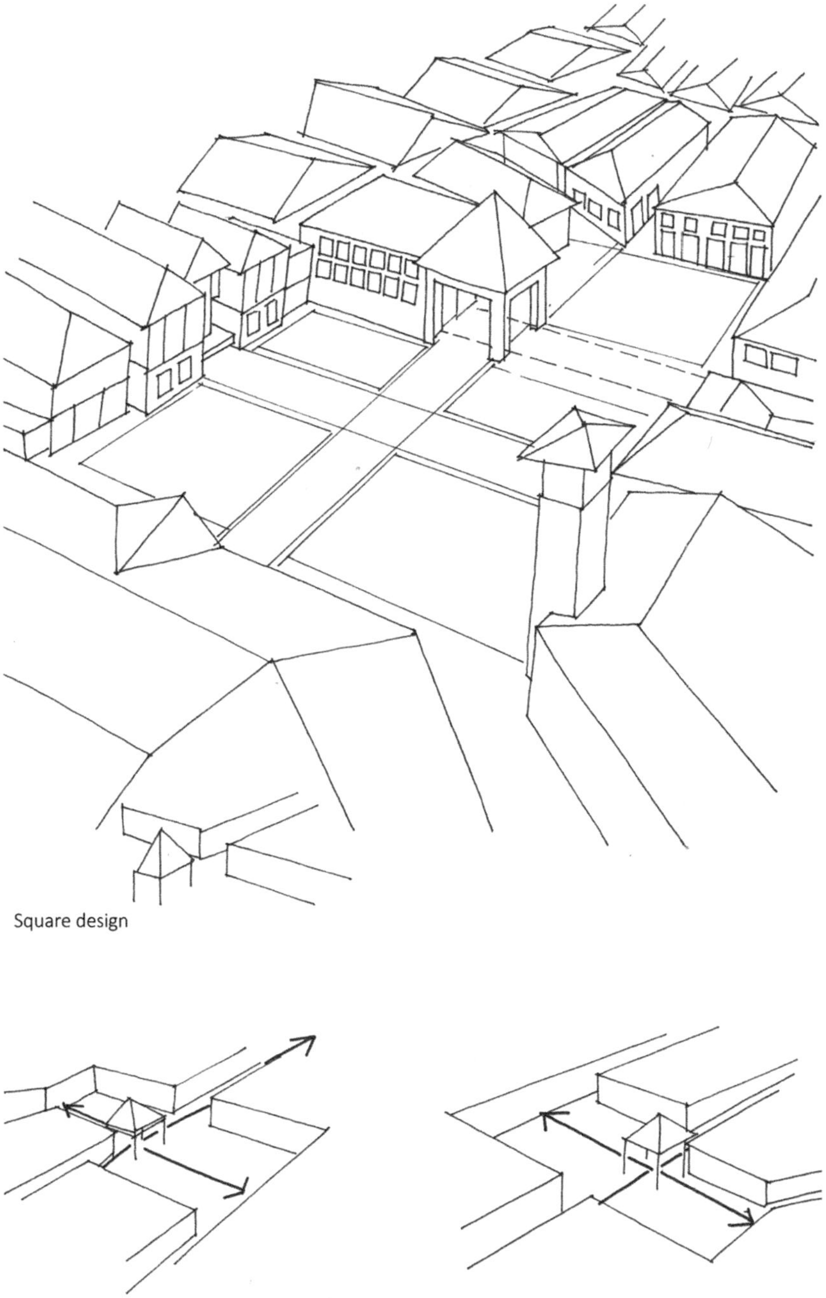

Fig. 5.72 The preliminary design for the square: final alternative. Entrance to a point which is the center of the complex and defines axes which provides order and meaning to the area

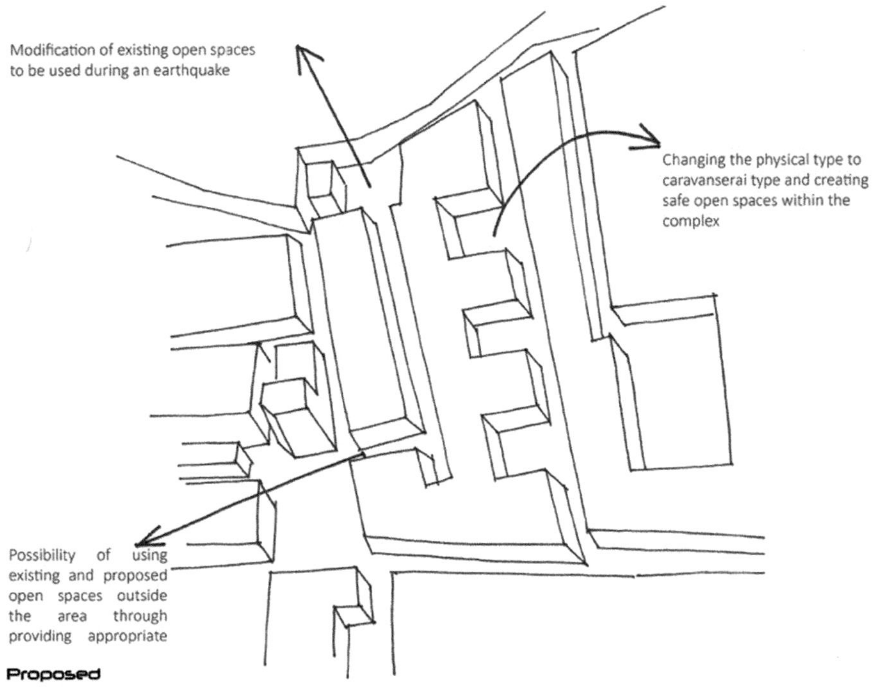

Fig. 5.73 Modification of existing open spaces and changing the physical type

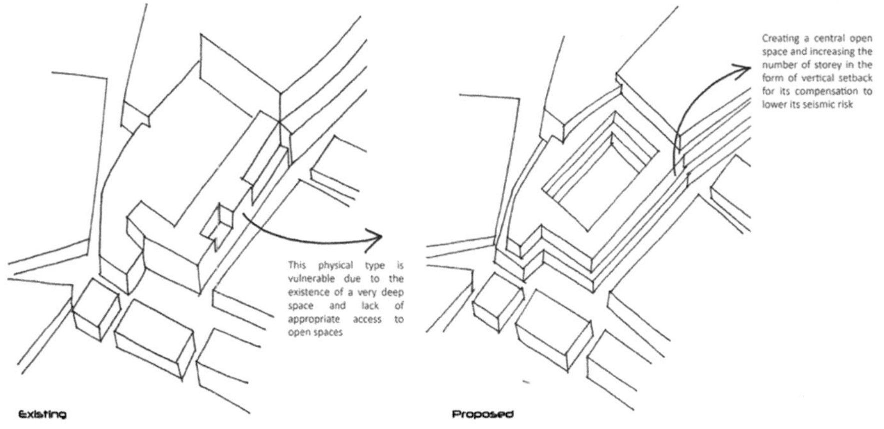

Fig. 5.74 Creating a central open space and increasing the number of story in the form of vertical setback for its compensation to lower its seismic risk

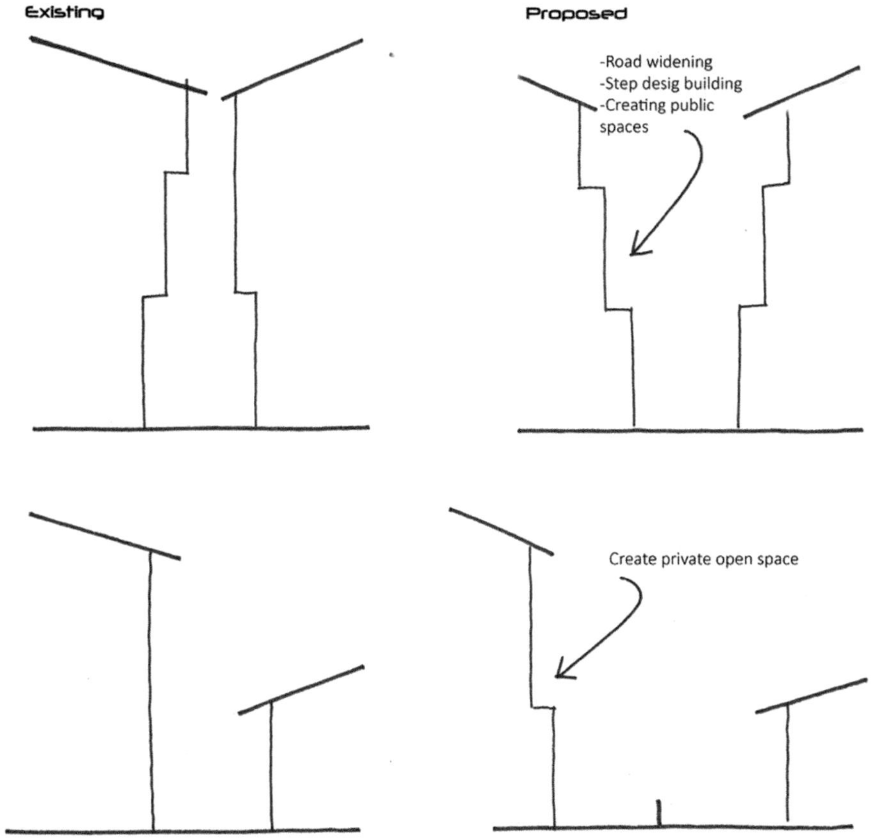

Fig. 5.75 Access modification through widening of passage and vertical setback

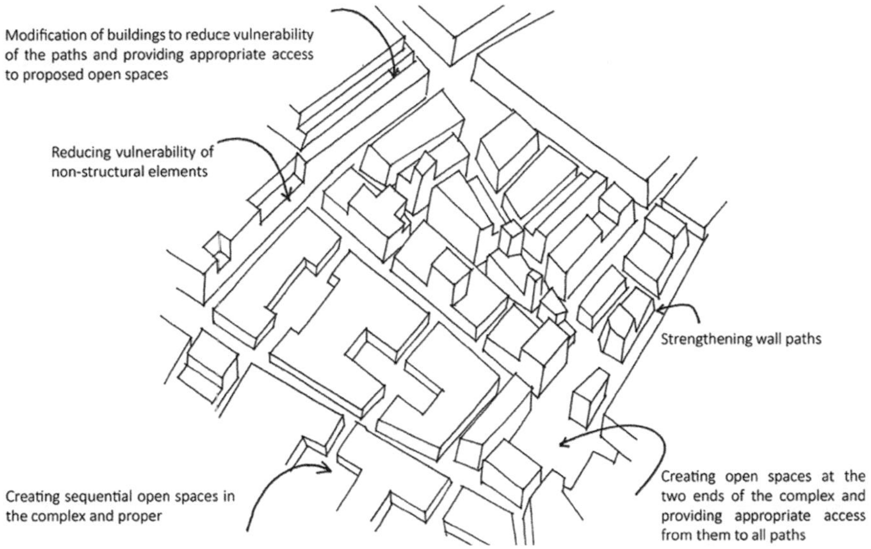

Fig. 5.76 Due to the existence of a deep space without appropriate access to open space, this physical type is dangerous. By creating central open space and compensating it with increasing the number of story in the form of vertical setback, this physical type will turn into less vulnerable

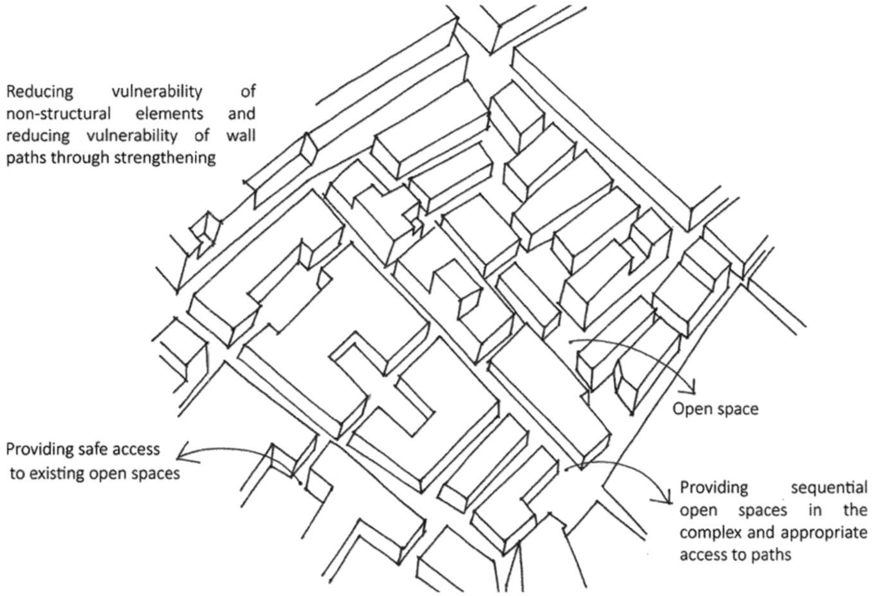

Fig. 5.77 Provide open space, safe access and strengthen wall paths in the dense block

References

Bahrainy and Bakhtiar (2016) Toward integrative theory of urban design. Springer

Bahrainy H et al (1997a) Housing typology in the city of Rasht for reducing of vulnerability. National Report, vol 33, UNDP-UNCHS (Habitat) Project IRA1901004, Housing Foundation of I.R., NDPC (Translated in English), Tehran, August 1997a

Bahrainy H et al (1997b) Open spaces in the city of Rasht. National Report, vol 39, UNDP-UNCHS (Habitat) Project IRA1901004, Housing Foundation of I.R., NDPC (Translated in English), Tehran, June 1997b

Bahrainy H et al (1997c) Planning and design of specific land uses in the city of Rasht for reducing of earthquake vulnerability. National Report, vol 52, UNDP-UNCHS (Habitat) Project IRA1901004, Housing Foundation of I.R., NDPC (Translated in English), Tehran, November 1997c

Bahrainy H et al (1997d) Planning and design of transportation network in the city of Rasht. National Report, UNDP - UNCHS (Habitat) Project IRA1901004, Housing Foundation of I.R., NDPC (Translated in English), Tehran, August 1997d

Bahrainy H et al (1998) Synthesis on redesign of the city of Rasht. National Report, UNDP-UNCHS (Habitat) Project IRA1901004, Housing Foundation of I.R., NDPC (Translated in English), Tehran, December 1998

Banerjee T (1993) Urban form and seismic vulnerability. Earthquake protection through urban design and planning. International Consultant Report, UNDP-UNCHS (Habitat) Project IRA1901004, Housing Foundation of I.R., Tehran, October

Kazemi M et al (1997) Socio-economic structure of the city of Rasht. National Report, Volume 34, UNDP-UNCHS (Habitat) Project IRA1901004, Housing Foundation of I.R., NDPC (Translated in English), Tehran, January 1997

Majidi H et al (1997) Central part of the city of Rasht. National Report, Volume 47, UNDP-UNCHS (Habitat) Project IRA1901004, Housing Foundation of I.R., NDPC (Translated in English), Tehran, May 1997

Music V (1993) Earthquakes and management of urban environment. Urban planning and design as a part of preventive strategies", International Consultant Report, UNDP-UNCHS (Habitat) Project IRA1901004, Housing Foundation of I.R., Tehran, October 1993

Music V (1998) Alternative development scenarios of the city of Rasht and its functional urban region. International Consultant Report, UNDP-UNCHS (Habitat) Project IRA1901004, Housing Foundation of I.R., Tehran, December

National Consultant Report (1993) UNDP-UNCHS (Habitat) Project IRA1901004. Housing Foundation of I.R, Tehran

Owen J (2019) URBAN DESIGN: a definition, approach and conceptual framework makers architecture & urban design. Makersarch.com, September

Petrovski JT, Bahrainy H, Music V (eds) (1998) Guidelines for reduction of seismic risk in regional and urban planning, Volume VII of the Guidelines for Earthquake Disaster Management, UNDP-UNCHS (Habitat) and Housing Foundation of I. R., Tehran-Skopje

Chapter 6
Creating Open Spaces, Modification, and Improvement of Access Network: The Case of Central Area of the City of *Rasht*

Abstract Open spaces and access neworks, as the most significant elements of urban space, play a critical role in every phase of earthquake diasters, from escape, refuge, relief, rescue, first aid, emegency, and temporary settlements. Urban space is defined here as the container of the average daily circuit which is composed of urban activity systems. The aggregate of the daily circuit of urban activity systems is the average of a person's daily, weekly, or yearly routine of activities in the urban environment. In normal situations, these activities include such things as recreational activities, shopping, driving to work, going to church, walking to school, and visiting. Repetitive patterns of such activities are called urban activity patterns, examples of which are shopping patterns, home to work commuting patterns, recreational patterns, and so on. When a disaster strikes, these activities will be interrupted and urban spaces play quite different functions. Application of seismic risk mitigation strategies in dealing with urban spaces in the event of a disaster can have more tangible results in smaller scales. Areas of the cities with high risk or vulnerability such as city cores, historical sections, mountain slopes, or areas with land instability potentials are examples for sector scale application of seismic risk mitigation. One should, however, bear in mind that changing the existing fabric of a city, particularly in the dense and traditional core of the cities, to secure the safety and well-being of residents is a complicated and difficult task. Attempts should be made to make immediate and short-term modifications in such dense, built-up areas realistically and limited to the most crucial elements, i.e., those with significant positive consequences. This will ensure smooth implementation of recommendations at this level. From an urban design point of view, project scale is the lowest possible level one can deal with, with short-term and specific objectives and easy-to-implement proposals. It could be applied to the old section of the Central Business District of traditional cities where the interwoven fabric, deteriorated structures, overpopulated passages, high density of buildings, and substandard safety against earthquake and fire make them the most vulnerable sections of the cities.

Keywords Open space · Modification · Improvement · Access network · *Rasht* · Central area · Design ideas · Specifications · Passages · Non-structural · Fire

H. Bahrainy and A. Bakhtiar, *Urban Design in Seismic-Prone Regions*,
The Urban Book Series, https://doi.org/10.1007/978-3-031-08321-1_6

6.1 Introduction

> The key to the successful integration of disaster risk planning and urban design lies in a shift of thinking that sees a city's open spaces as a "second city": a network of open space designed not only to contribute significantly to the quality everyday urban life, but with the latent capacity to act as essential life support and an agent of recovery in the event of an earthquake (Allan and Bryant 2014).

Open spaces and access neworks, as the most significant elements of urban space, play a critical role in every phase of earthquake diasters, from escape, refuge, relief, rescue, first aid, emegency, and temporary settlements.

Urban space is defined here as the container of the average daily circuit which is composed of urban activity systems. The aggregate of the daily circuit of urban activity systems is the average of a person's daily, weekly, or yearly routine of activities in the urban environment. In normal situations, these activities include such things as recreational activities, shopping, driving to work, going to church, walking to school, and visiting. Repetitive patterns of such activities are called urban activity patterns, examples of which are shopping patterns, home to work commuting patterns, recreational patterns, and so on (Bahrainy and Bakhtiar 2016). When a disaster strikes, these activities will be interrupted and urban spaces play quite different functions.

Application of seismic risk mitigation strategies in dealing with urban spaces in the event of a disaster can have more tangible results in smaller scales. Areas of the cities with high risk or vulnerability such as city cores, historical sections, mountain slopes, or areas with land instability potentials are examples for sector scale application of seismic risk mitigation. One should, however, bear in mind that changing the existing fabric of a city, particularly in the dense and traditional core of the cities, to secure the safety and well-being of residents is a complicated and difficult task. Attempts should be made to make immediate and short-term modifications in such dense, built-up areas realistically and limited to the most crucial elements, i.e., those with significant positive consequences. This will ensure smooth implementation of recommendations at this level.

From an urban design point of view, project scale is the lowest possible level one can deal with, with short-term and specific objectives and easy-to-implement proposals. It could be applied to the old section of the Central Business District of traditional cities where the interwoven fabric, deteriorated structures, overpopulated passages, high density of buildings, and substandard safety against earthquake and fire make them the most vulnerable sections of the cities (Bahrainy 1998).

Central area of the city of *Rasht* has a complex grid texture with the advantage of possible accessibility, but at the same time, it includes dense paths without access to open space which result in its high vulnerability. Low depth of stores and relatively high population density of open and closed spaces will increase the presence of population in the paths and open spaces. The high density of non-structural elements in the paths will practically increase the risk even more.

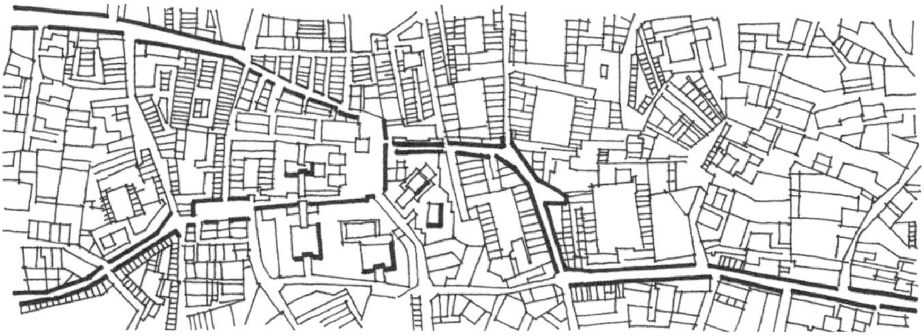

Fig. 6.1 The proposed design for passages, open spaces, and the central square in the Bazaar area of city of *Rash*

The goals of the area design to reduce vulnerability against seismic risk are as follows (see Fig. 6.1):

- The need for creating the main open space in the central core, next to the Jame Mosque, to hold the commercial activities and, at the same time, to reduce population density.
- The need for the modification and improvement of the main passages to provide accessibility during disaster.
- Creation or improvement of emergency access network and facilities in the vicinity to be used in case of disaster.
- Demolition and reconstruction of highly vulnerable buildings due to seismic risk.
- Creating open and semi-open spaces in order to connect closed spaces to make speedy evacuation during an earthquake possible.
- Change and modification of dangerous buildings, such as mosques and schools, with high density of population.
- Providing spaces with commercial activities to be flexible so that their uses could be changed in case of emergency (Fig. 6.2) (Petrovsky and Bahrainy 1992; Bahrainy 1998).

6.1.1 Green and Open Space

In normal conditions, these spaces are used for recreation, sport activities, and air purification, but during a disaster, they will be used for escape, rescue, and relief and, immediately after the earthquake, will be used for emergency and temporary settlement. This becomes more important in the old and decayed textures of the cities which are usually very compact but lack such spaces. Easy, fast, and safe access to these spaces during and after earthquake is urgently needed (Kazemi et al. 1997).

Fig. 6.2 Several views of the existing retail stores in the Bazaar area which need modification to reduce their vulnerability against seismic risk

The shortage of public services and their inappropriate distribution based on population distribution has resulted in a number of problems. The concentration of the city's major land uses at the dense center of the city where the streets are incapable of responding to the passage of people coming to the area from the other parts, while creating day-to-day urban mobility problems, causes many deaths and impedes relief operations in a crisis situation owing to the excessive concentration of population at one point.

Presence of areas prone to liquefaction in the urban area is a great hazard regardless of which construction works have been pursued in the past, especially at the city's center where major land uses are settled and the texture is old, the thoroughfares narrow, the street sides unsafe, and the density high (Bahrainy 1992).

6.2 Design Ideas at the Block Scale to Reduce Seismic Risk in the Bazaar Area

6.2.1 Operational Objectives in Roads and Open Spaces (Bahrainy et al. 1998; Kazemi et al. 1997)

In this regard, following changes are proposed in different zones (see Chap. 5):

- Zone 1, creating open space along with strengthening road walls.
- Zone 2, provide access to open spaces as the first priority and creating open space as the second priority.
- Zone 3, provide access to existing open spaces as the first priority and roads improvement as the second priority.
- Zone 4, strengthening roads, as the first priority, and road improvement, as the second.
- Zone 5, removing the non-structural damaging elements, as the first priority.
- Zone 6, provide accessibility for the time of disaster, as the first priority.

6.2.2 The Central Area with Grid Texture, Zone 1 Paths, Alternative 1 (Majidi 1997)

See Fig. 6.3.

6.2.3 Design Ideas at the Block Scale to Reduce Seismic Risk in the Bazaar Area, Alternative 2

See Fig. 6.4.

6.2.4 Design Ideas at the Block Scale to Reduce Seismic Risk in the Bazaar Area, Alternative 3

See Fig. 6.5.

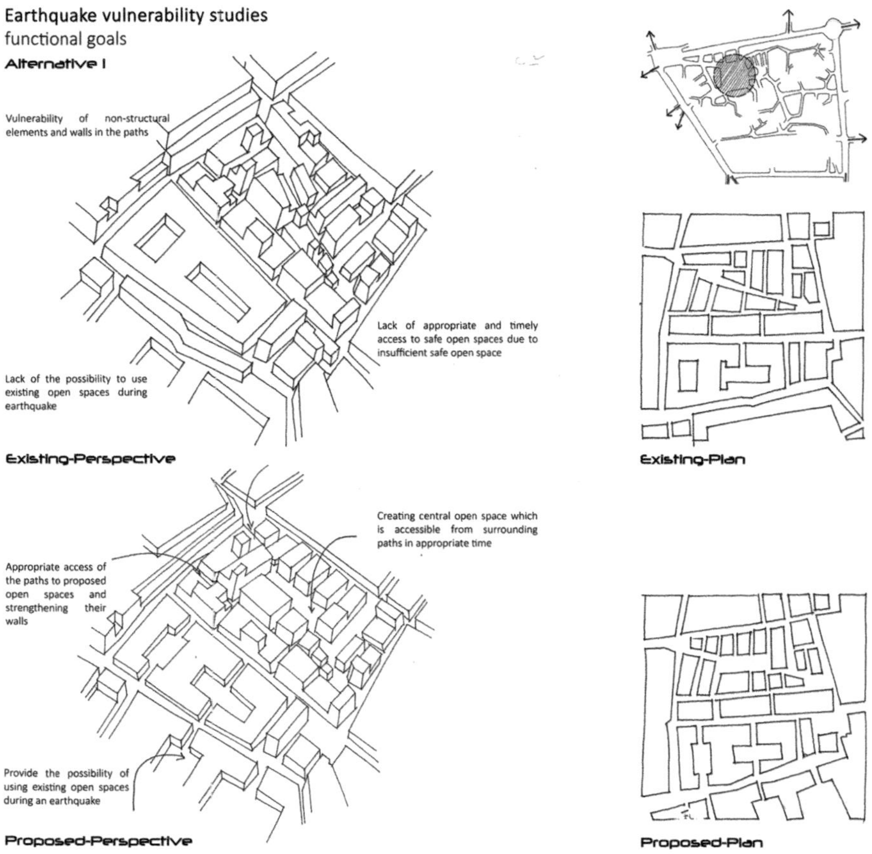

Fig. 6.3 Existing and proposed perspectives and plans for a block in the Bazaar area, alternative 1

6.3 Design Solution

Solution to reduce seismic risk and vulnerability of the elements at risk in the studied block will include the following characteristics: lower density, open spaces, wide access networks, light roofs, etc. (Bahrainy et al. 1996; Music 1993; Majidi 1997) (Figs. 6.6 and 6.7).

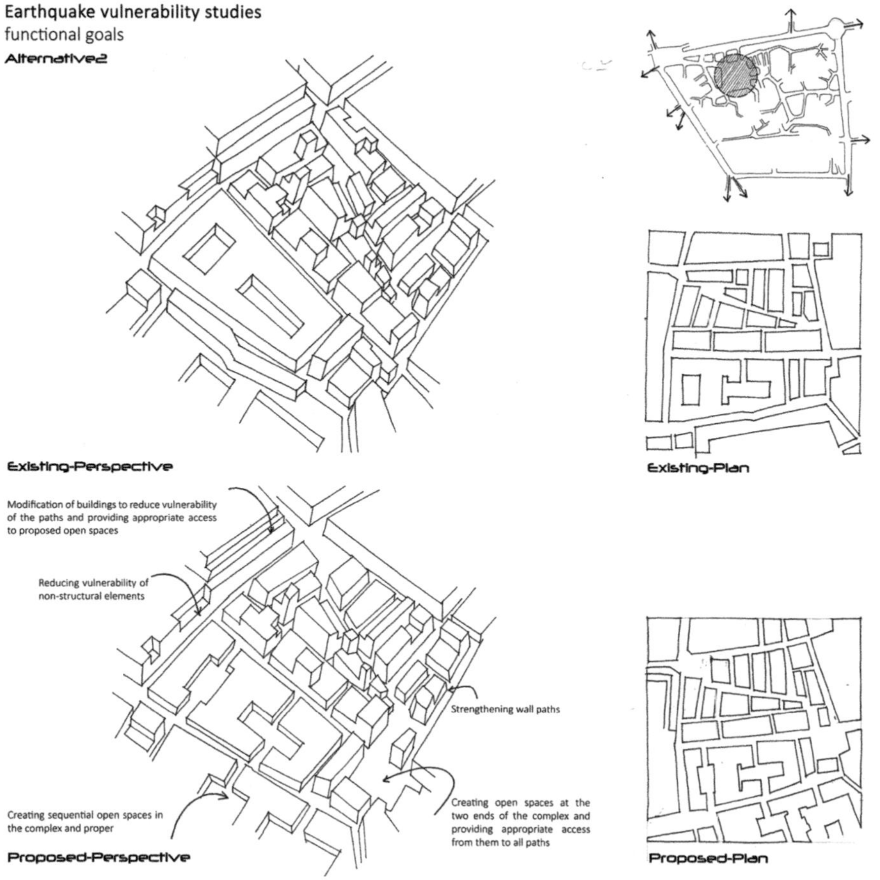

Fig. 6.4 Existing and proposed perspective and plan for a block in the Bazaar area, alternative 2

6.4 Seismic Vulnerability—Specifications

6.4.1 Modification of the Main Roads: Exits and Entrances, Zone 4

See Fig. 6.8.

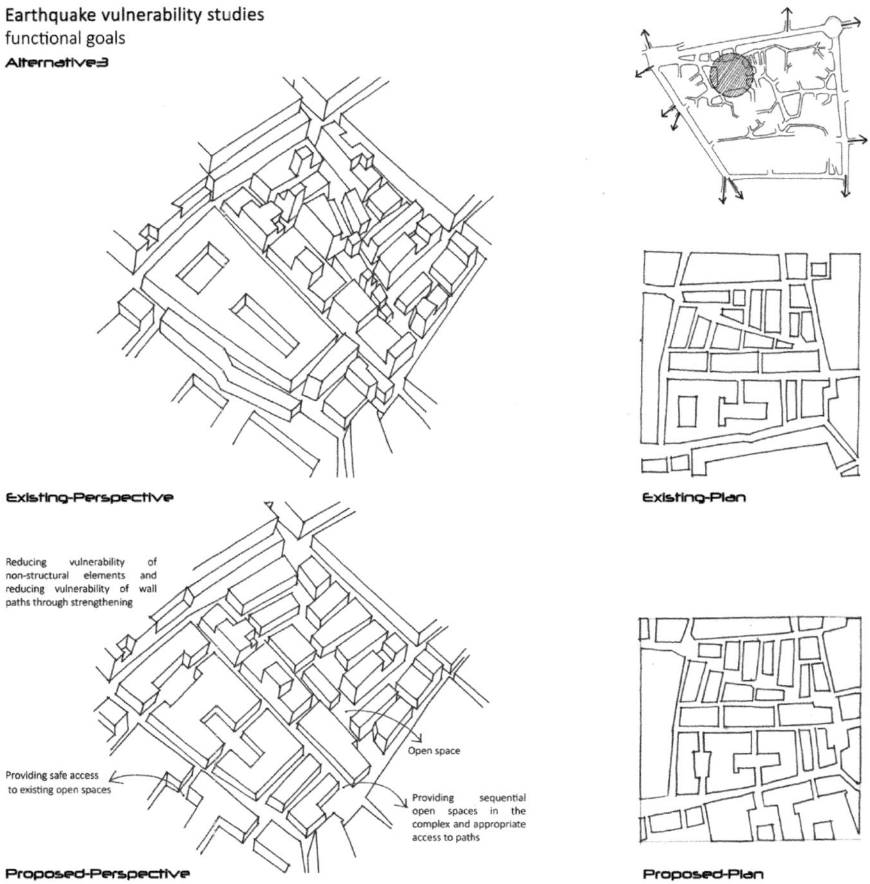

Fig. 6.5 Existing and proposed perspective and plan for a block in the Bazaar area, alternative 3

6.4.2 Areas at Risk with Low Depth and Lack of Access to Safe Open space, Zone 2, Access Network Alternative 1

See Fig. 6.9.

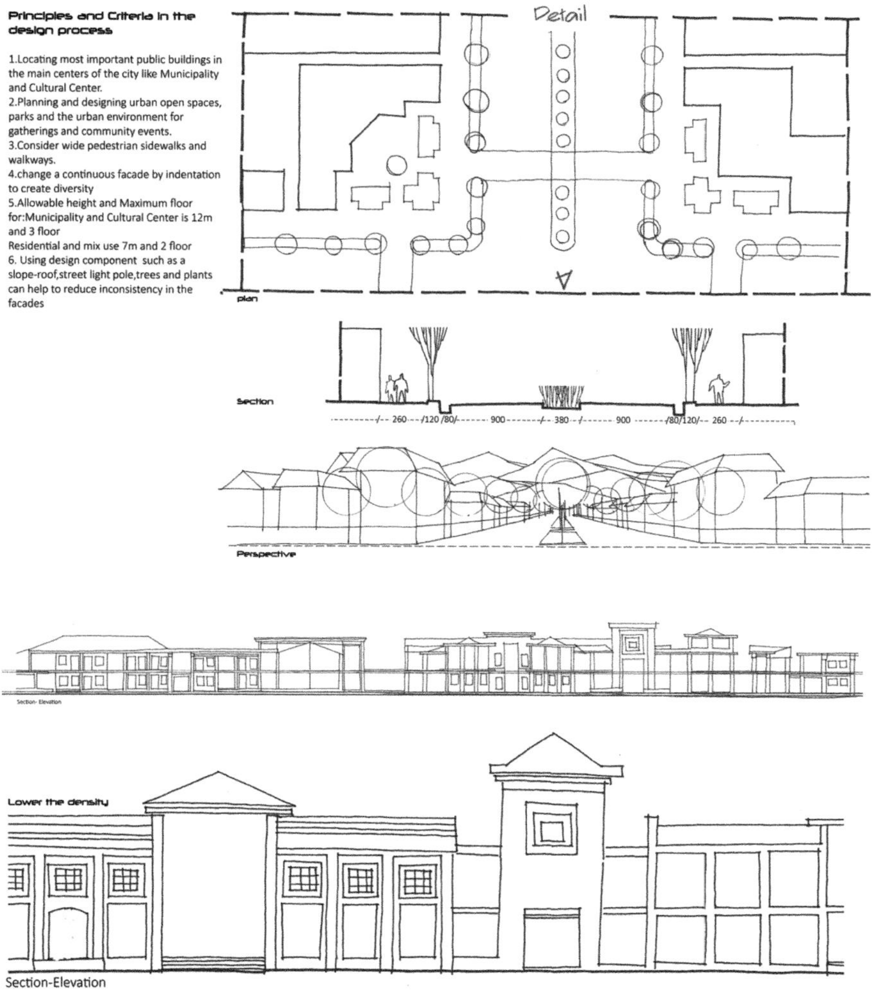

Fig. 6.6 Proposed plan, perspective and façade of the studied block in the Bazaar area to reduce seismic risk

6.4.3 Modification of Physical Types at Medium Scale in the Complexes with Dangerous and Low Depth Passages and Lack of Access to Safe Open Space, Alternative 2, Zone 2

See Fig. 6.10.

Fig. 6.7 Birds eye view of the proposed design for the studied block in the Bazaar area

6.4.4 Modification of Physical Type, Medium Scale, Zone 2, Enclosed Spaces

See Fig. 6.11.

6.4.5 Modifications of Physical Types, Medium Scale, Zone 2, Enclosed Spaces

See Fig. 6.12.

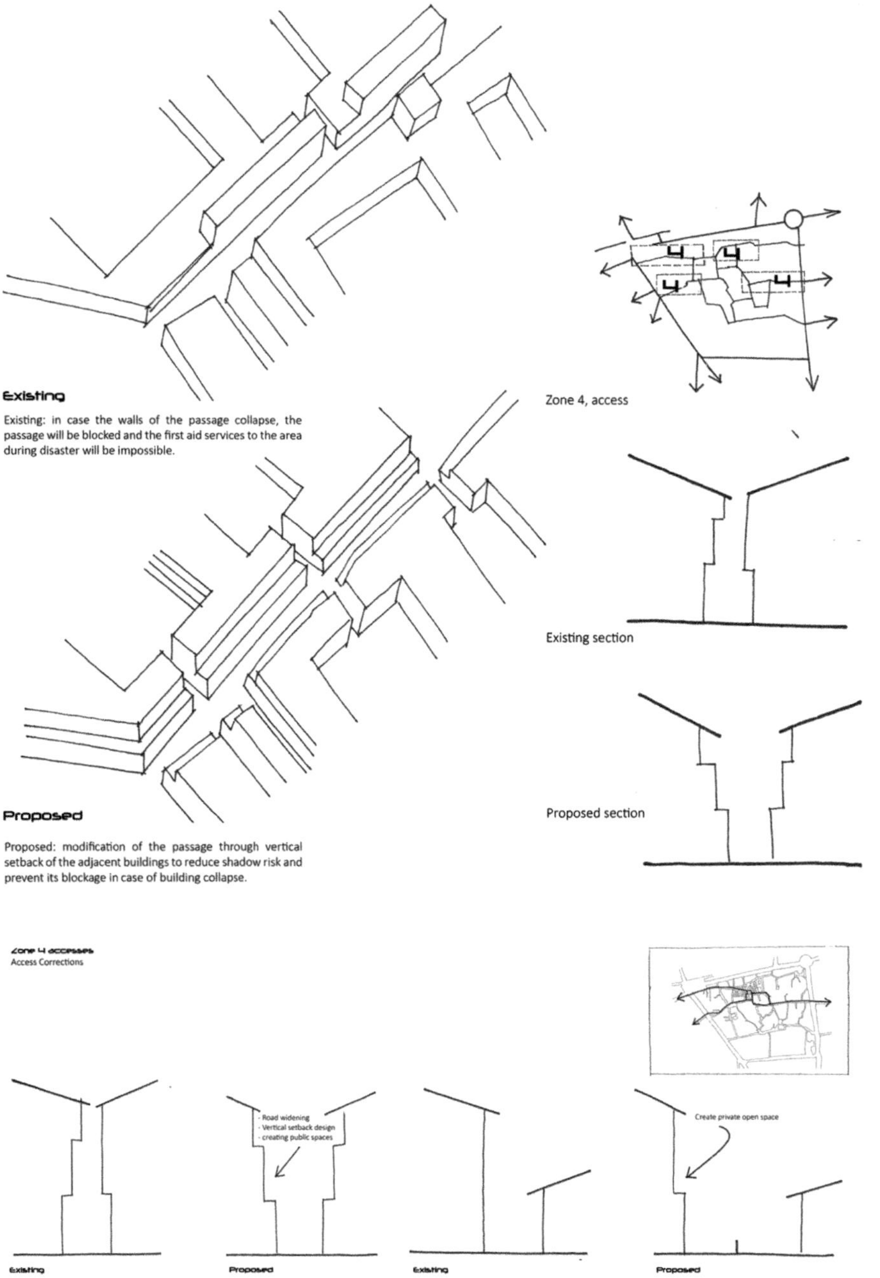

Fig. 6.8 Existing and proposed roads. In case the walls collapse, the roads will be completely blocked which prevents first aids to reach to the area. Modification of the road by vertical setback of the adjacent buildings to reduce shadow risk and prevent road blockage in case of buildings collapse

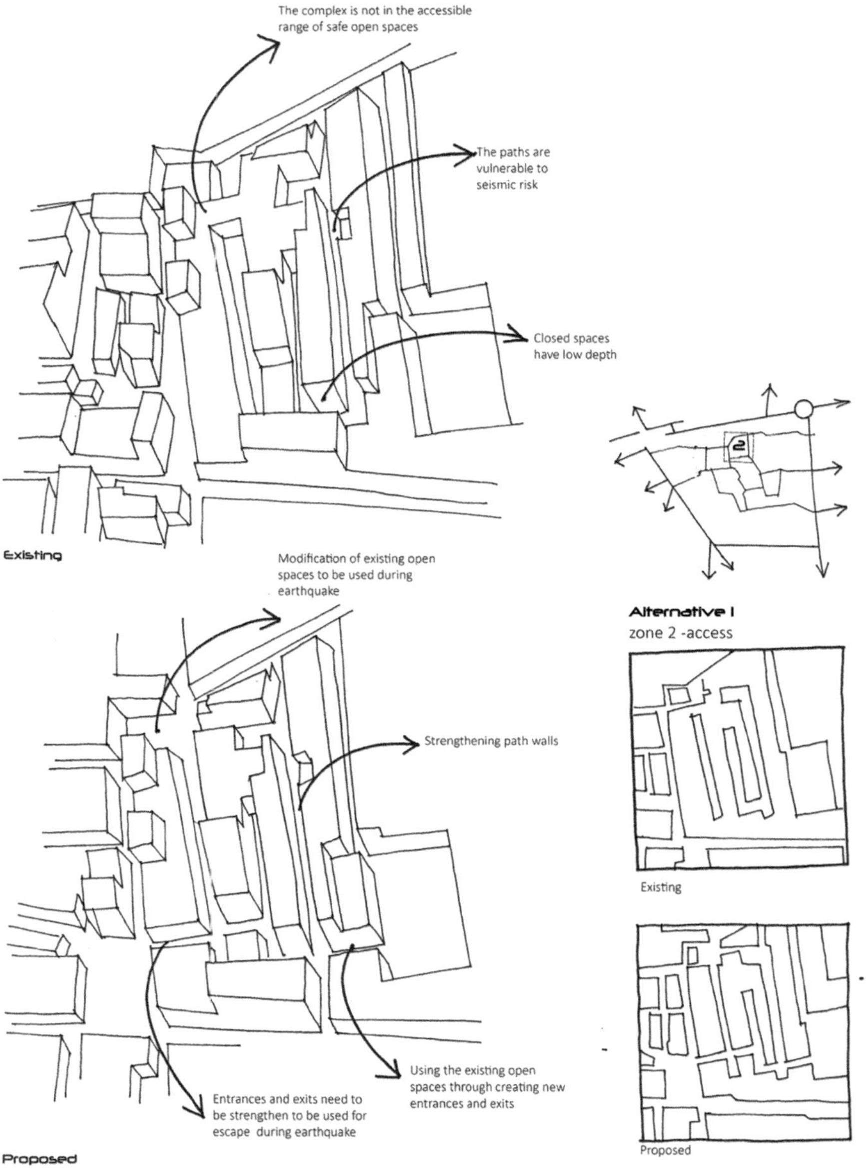

Fig. 6.9 Existing and proposed access network. Passages are at risk, the depth of enclosed spaces is low, and the complex is not in the range of safe open spaces. Strengthening passage walls, modification of existing open spaces for the use during disaster, and entrances should be strong enough to be safely used for escape during disaster. Also, existing open spaces should become usable through providing exit and entrance spaces. The proposed design includes a central open space which could be compensated for by increasing the number of story with a setback on each floor

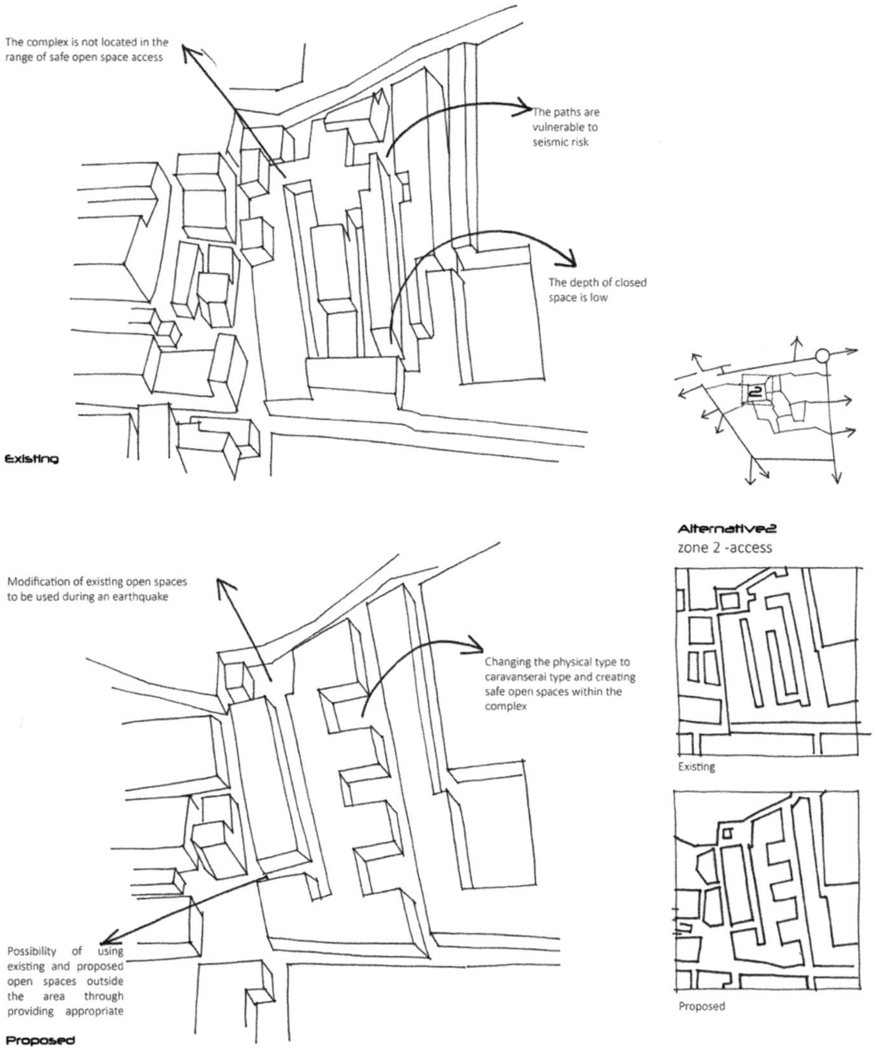

Fig. 6.10 Existing and proposed access network. Passages are at risk, the complex is not in the range of access to safe open spaces, and the depth of closed spaces is low. Changing the physical type to caravanserai and creating safe open spaces within the complex, improving the existing open spaces to be easily usable during disaster, and the possibility of using the existing and proposed open spaces beyond the complex through providing appropriate access spaces

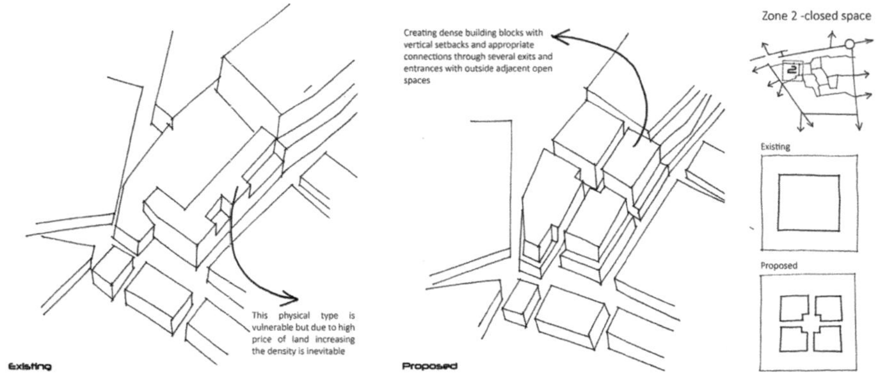

Fig. 6.11 Existing and proposed access network. The physical type is at risk, and due to the high price of land, increase in density is inevitable. Creating dense building blocks with vertical setback and appropriate connection with adjacent open spaces through several exits and entrances

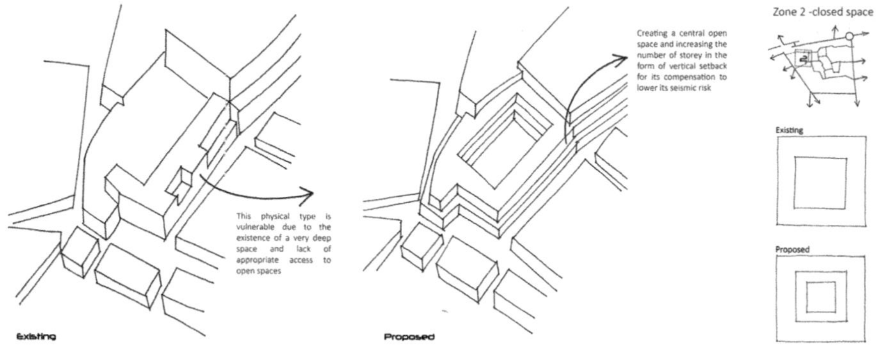

Fig. 6.12 Existing and proposed access network. The type is at risk due to the high depth and lack of appropriate access to open spaces. It can be changed to a lower risk through central open space and compensate it by increasing the number of story with a vertical setback

6.4.6 Modifications of Physical Types, Microscale, Open Space Design for Rescue, Relief, and Emergency Settlement

General principles were suggested (Bahrainy 1993).

- Using mosques for the preparation and provision center during earthquake.
- Considering the inclined and slope roofs in this climate and also the use of various unstable materials, such as tile, which could cause damage, and even when the building remains safe against earthquake, these unstable elements may fall and cause human injuries and even death. This is especially critical in case of roads and

passages, where people would escape to take refuge there. So it is recommended to fix these elements and prevent them from falling.

- The schools with large court are appropriate for emergency settlement.
- Deteriorated buildings with high vulnerability should be demolished and turned into open space.
- As the buildings with high vulnerability will collapse in less than 10 s, the ones with average vulnerability in less than 20 s, and the ones with low vulnerability in less than 30 s, residents will have this 30 s to flee and find refuge.
- Put the open space for emergency settlement at the beginning of the cul-du-sacs and alleys to have a better control of the people entering and leaving the area and increase the safety level of each space.

The spaces with large open spaces (private, semi-private, and governmental) should be left to disaster management authorities during disaster. This would provide security for their properties, but will be also used for certain needs, such as helicopter landing, collecting supporting materials, and rescue.

The general design concept in relation to reducing seismic risk are as follows:

- Providing multi-function spaces, such as exhibitions, religious spaces, and small squares.
- Proposing main open spaces to be used during the disaster.
- The main corridors to be used by the vehicular traffic during disaster.
- Limiting the density to two stories along the passages with high vulnerability.
- Assigning crises-related land uses around the proposed open spaces, such as police and relief stations, mosques, pharmacies, and multi-functional spaces.
- Connection possibility between three centers.
- Providing multi-functional spaces around open spaces makes it possible to change their activities to disaster-related activities, such as rescue and relief and essential needs, during disaster.
- Vulnerable zones for fire.
- The area facing serious difficulties for accessibility during disaster because of blocked passages.
- Improvement of the passages and main pedestrian paths in the *Bazaar* area to promote spatial cohesion.
- Improvement and modification of passages to provide accessibility to the area during a disaster.
- Creating and improving of fringe roads to provide vehicular access to the warehouses and parking lots and secure the safety of local pedestrian passages.
- Per capita space needed for refuge is considered 1_{sm} and for emergency settlement 5_{sm}.
- Crowded spaces filled with strangers make them unsafe and therefore prevent concerned residents to settle in these spaces. That is why privacy of the residents should be considered in planning for emergency settlements. Smaller spaces close to people's residencies are not only easy and fast to access, but it is also safe because people know each other.
- Practical guideline for reconstruction is as follows:

H_1 and H_2= Height of the building on both sides
D = passage width (recommended)

$$D - \frac{h1 + h2}{2} > 3$$

Three meters is the minimum width for walking and driving rescue people.

6.5 Modification of Passages, Selected Cases

Some of the passages and Bazaar branches are of severe damaging type, with no appropriate access to open space. Due to the low depth of the enclosed spaces, any modification of this type of passages is not possible. A basic solution to reduce the vulnerability of these spaces is to provide access from these spaces to the existing open spaces. In these cases, the existing open spaces should be recognized and the possibility of providing accessibility to those spaces and increasing the number of entrances and exits investigated (Kazemi et al. 1997).

Providing accessibility depends on two main factors:

- Securing space for movement and accessibility.
- Strengthening the movement space in order to prevent blockage in case of a disaster.

Obviously, providing space for the movement and accessibility requires the demolition of some parts of buildings which could take place on the basis of existing situation. Strengthening of accessibility networks and entrances are significant when they can play critical role in reducing vulnerability during a disaster, such as:

- The buildings which have the possibility of access to open space and are expected to be used for rescue and relief during a disaster and therefore have entrance and exit to open spaces.
- The passages which have access to open spaces and are intended to serve as rescue and relief during disaster.
- Physical improvement: Some physical types may cause damages solely because the in-between spaces are covered. Removing the roofs and replacing it with very light material can turn these kinds of buildings into much less vulnerable ones (Bahrainy et al. 1997b) (Figs. 6.13, 6.14, 6.15, 6.16, 6.17, 6.18, 6.19, 6.20 and 6.21).

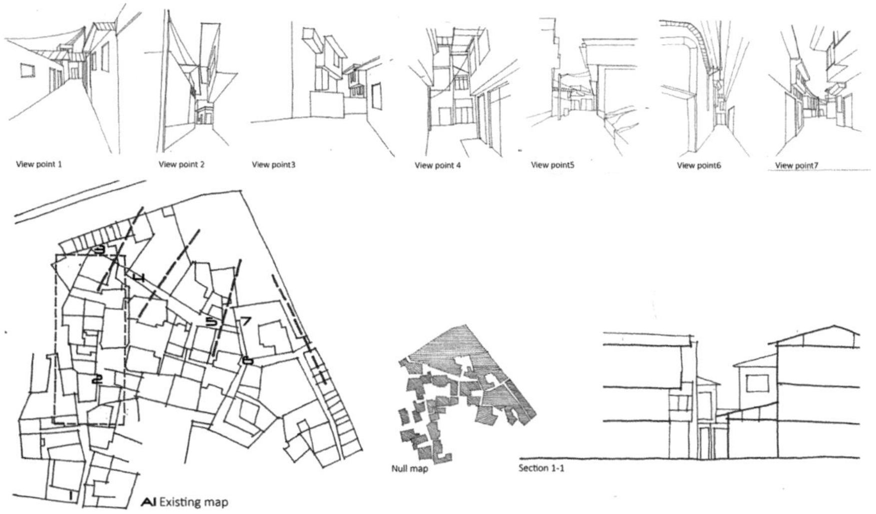

Fig. 6.13 A1 is a narrow pedestrian passage next to the municipality square which connects two streets of alamalhoda and Imam Khomeini for pedestrians. The passage was blocked during the *Manjil* earthquake of 1990 and has not been cleared ever since. Part of its walls have been extremely deteriorated which could cause serious risk in case of another earthquake

The proposed alternative includes the following:

- Improving the existing open spaces to be used during disaster.
- Entrances need to be strong enough to be appropriate for use during a disaster.
- Make use of existing open spaces possible by providing proper entrance and exit spaces.
- Strengthening buildings.
- In the case of buildings with damage ratio of less than 50%, or of low risk physical type, with damage ratio of above 50% and lower than 75%, strengthening and repairment measures may be applied. These measures vary based on the kind of construction system that has been used (Fig. 6.22).

6.5.1 Operational Objectives for the Closed Spaces in Zones 1 to 8 (for Zones, See Chap. 5)

- Zone 1: Priority is with renewal.
- Zone 2: Renewal has the second priority, and restoration and local modification have the first priority.
- Zone 3: Restoration and ways to facilitate access to existing and proposed open spaces beyond the complex.
- Zone 4: Strengthening has the first priority.

- Zone 5: Strengthening has the first priority, and renewal along with local modification has the second priority.
- Zone 6: Strengthening and local modification have the third priority.
- Zone 7: The area has the priority for ways better access during disaster.
- Zone 8: The area needs to find ways to prevent breaking and spreading fire. Also building modification to prevent fire from spreading.

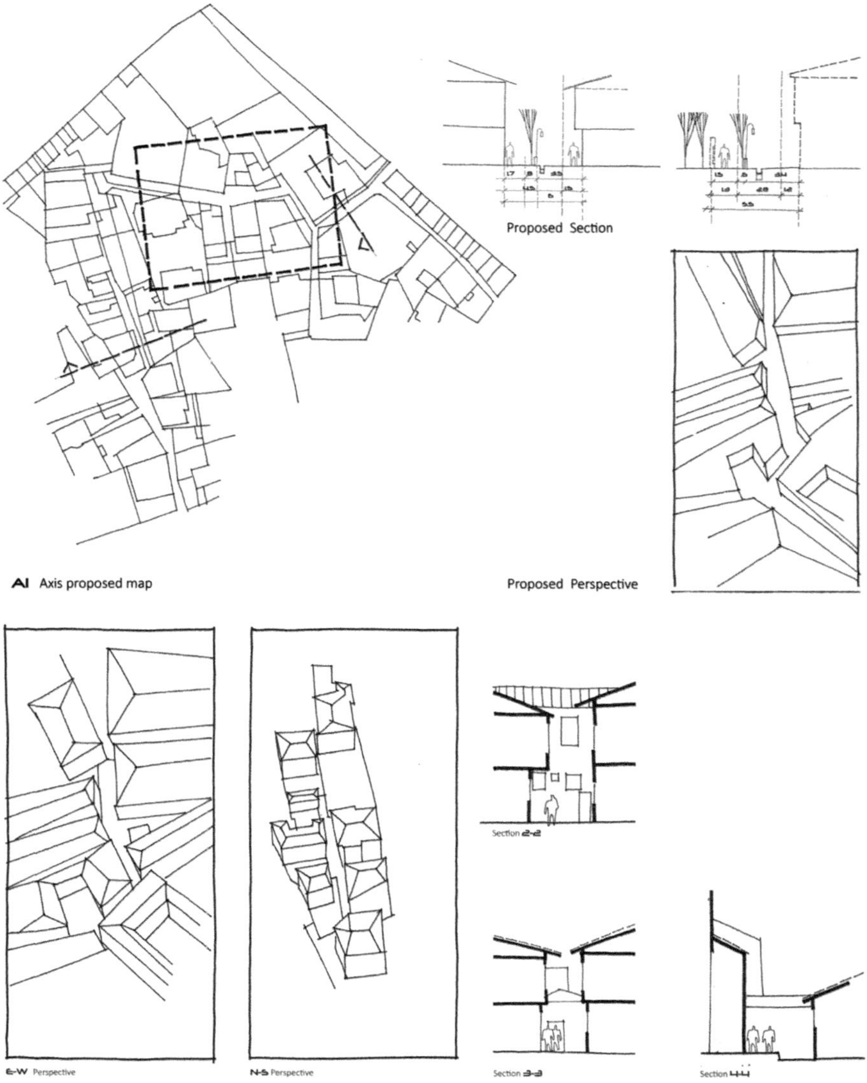

Fig. 6.14 The need to open up the passage in the proposed design for the area was clearly felt. Although this improvement was initially intended to provide pedestrian access, motor vehicle access was also possible in case of emergency. The deteriorated and vacant buildings adjacent to the passage provide an appropriate possibility for widening of the passage

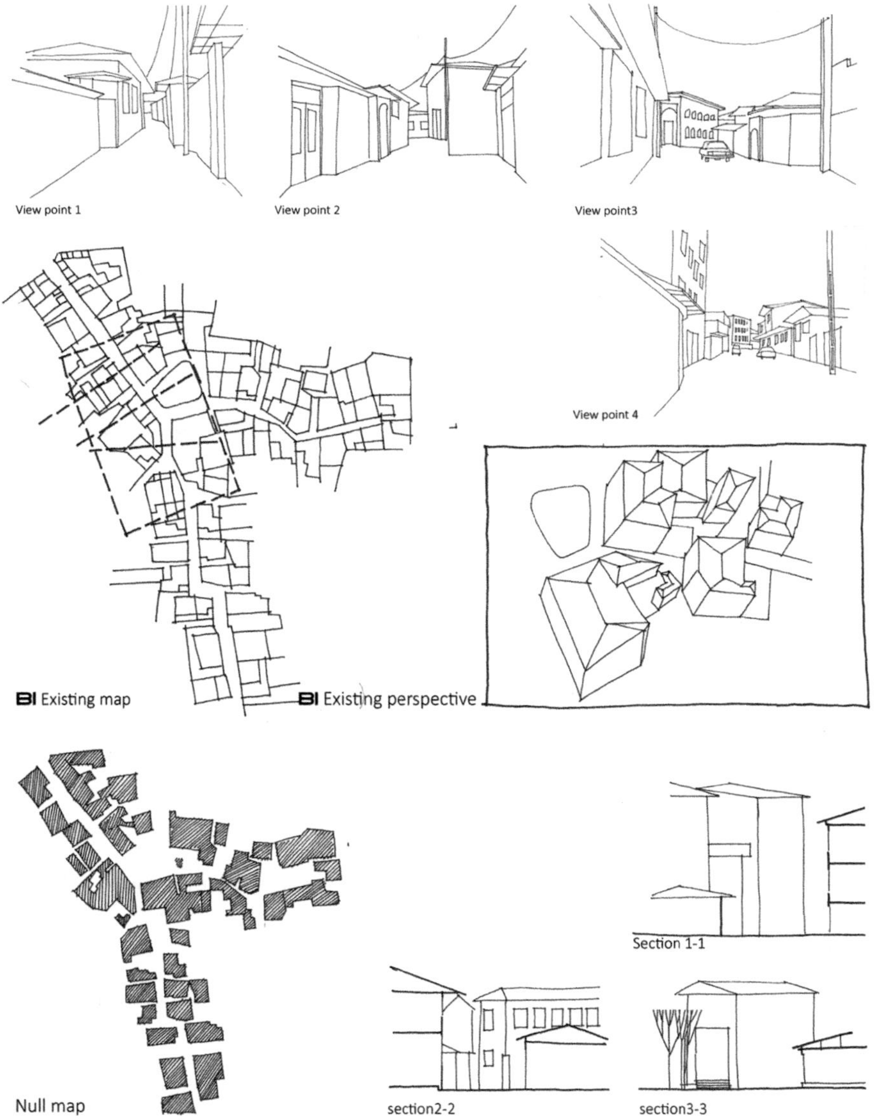

Fig. 6.15 The passage B1 is one of the important passages in the central area. The center of this passage is in southern section, and in spite of modern urban changes, particularly heavy traffic, life still continues

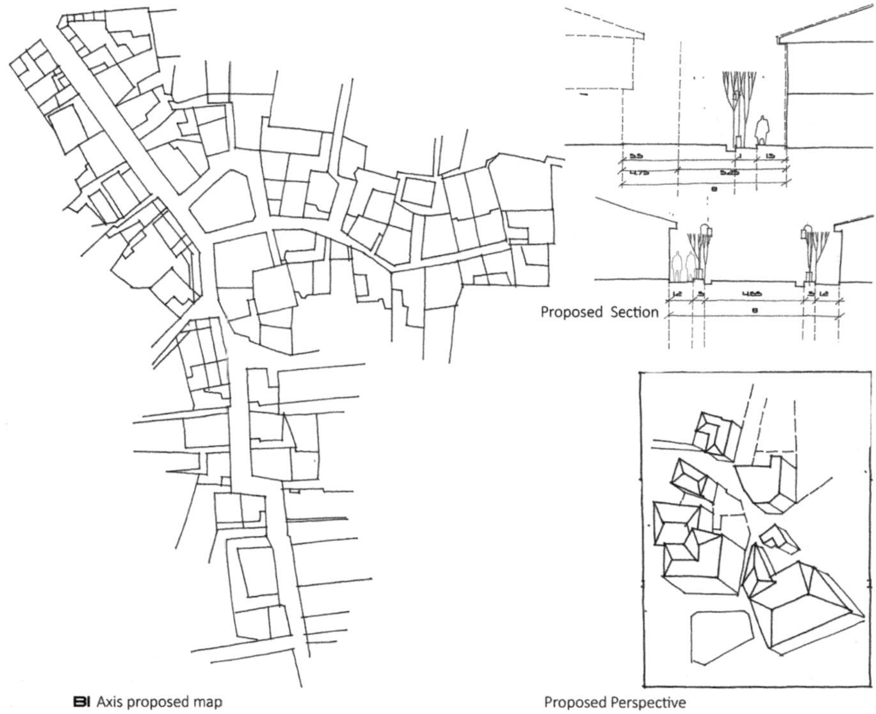

Fig. 6.15 (continued)

6.6 Selected Site to Analyze and Propose Open Spaces (Bahrainy 1993; Bahrainy et al. 1997b)

The city has many different types of open spaces, one of which has been selected, as example, for detail analysis and recommended creation and/or improvements to mitigate seismic risk in this public space. The selected case includes the following:

- The location of open space in the city. This space is located in the northern part of the central section of the city with three separate spaces which are considered as open space in this dense area of the city. They are in fact part of the passages in the organic texture which have more width in these spaces. These spaces are in the old commercial zone of the city where the main *Bazaar*, as the most important element for the city, is located.
- The area lacks adequate open space.
- Area texture. Irregular texture and organic form, along with fine grain and high density, with very few open spaces, and one- and two-story buildings.

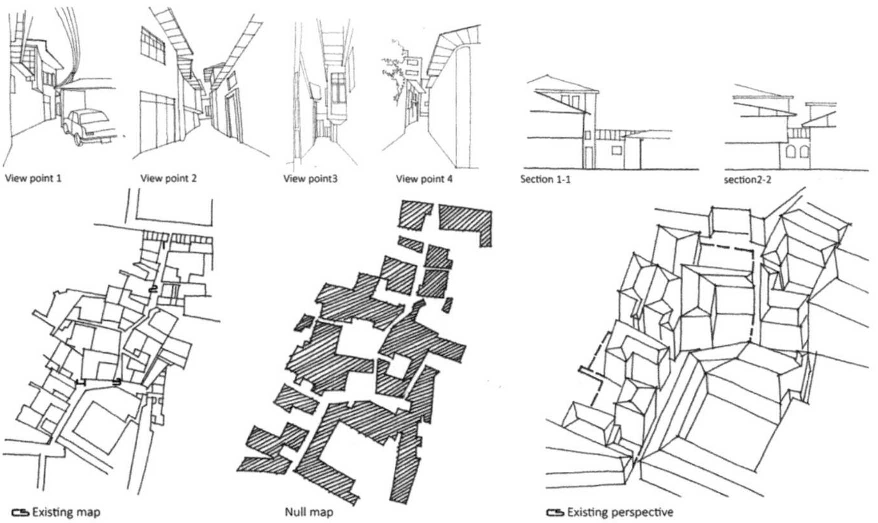

Fig. 6.16 C5 is a relatively secondary passage next to the neighborhood. Most of the current land uses along the passage are residential with considerable deterioration. There are a few commercial uses at both ends which create heavy pedestrian traffic in the passage. A religious school in the passage had added to its recognition and provided a sense of identity

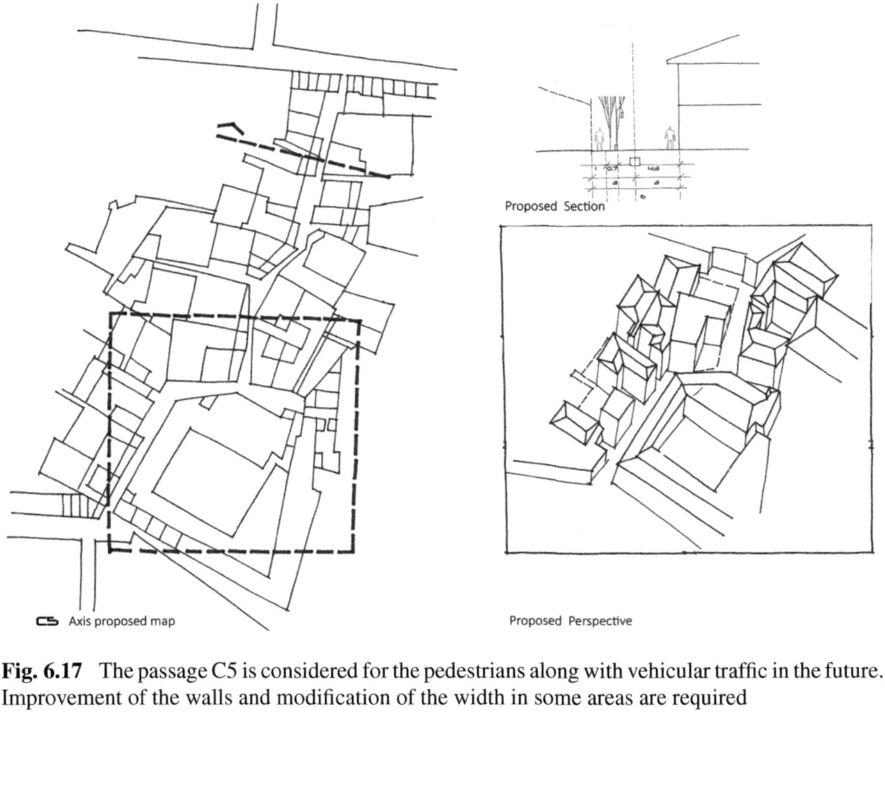

Fig. 6.17 The passage C5 is considered for the pedestrians along with vehicular traffic in the future. Improvement of the walls and modification of the width in some areas are required

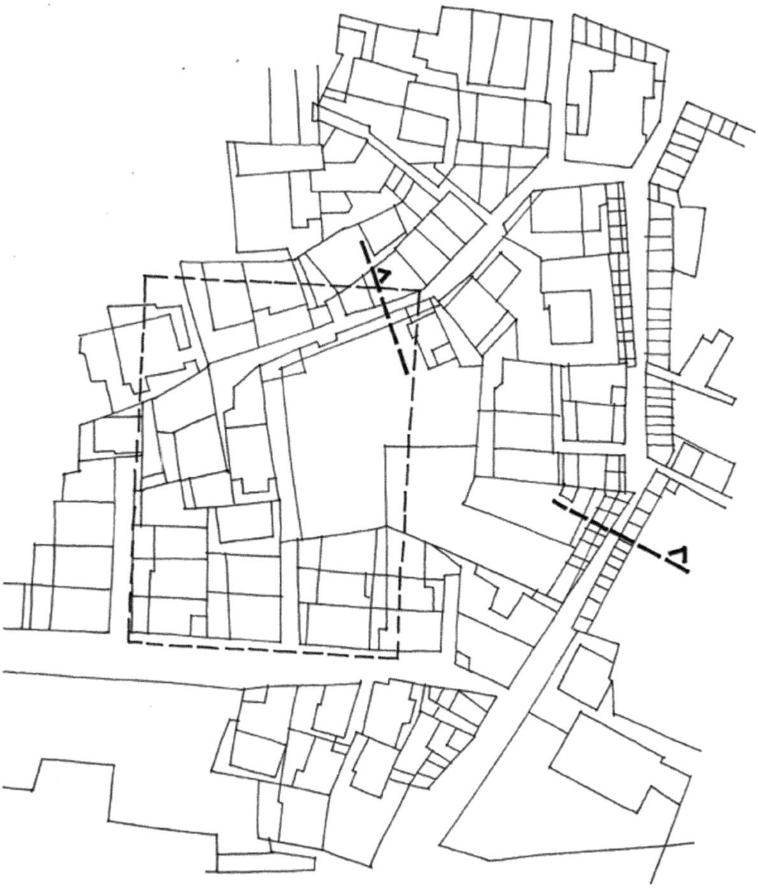

C7 Axis proposed map

Fig. 6.18 Widening the passage C7 to connect it to the adjacent passages during normal time and particularly during disaster, considering the effective factors causing the vulnerability of adjacent buildings along the passages. Demolition of part of the building due to its age, height, the quality of land use. and other vulnerability factors

- The status of existing buildings, due to the old age, buildings have a low quality and weak strength. Old and abundant storehouses make the area quite vulnerable during earthquake and also fire.
- Land uses, the dominant use of the area is commercial, and residential uses are mixed with commercial ones in some parts.

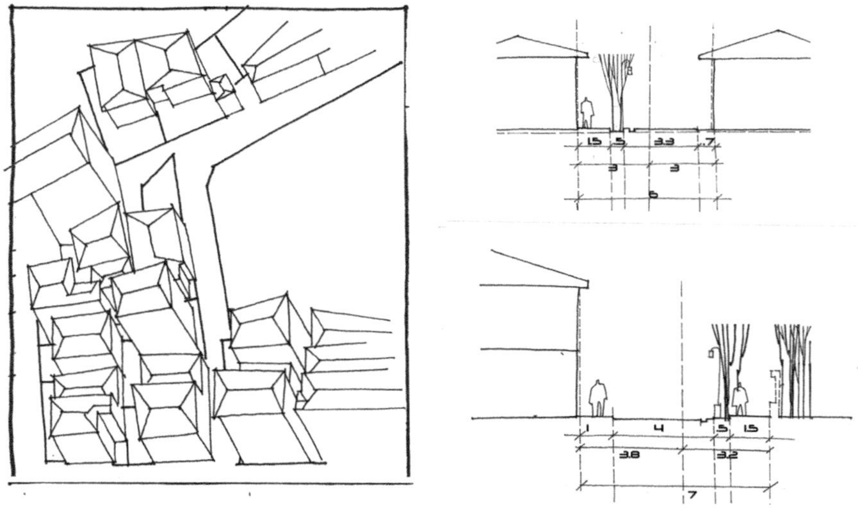

C7 Proposed Perspective

Proposed Section

Fig. 6.19 Part of C7 passage. Physical improvement and spatial arrangement need to be carried out through proposed plans. These measures intend to create a desirable urban image and also provide the possibility of appropriate pedestrian and vehicular connection especially during a disaster. Another problem is controlling the density of the passage

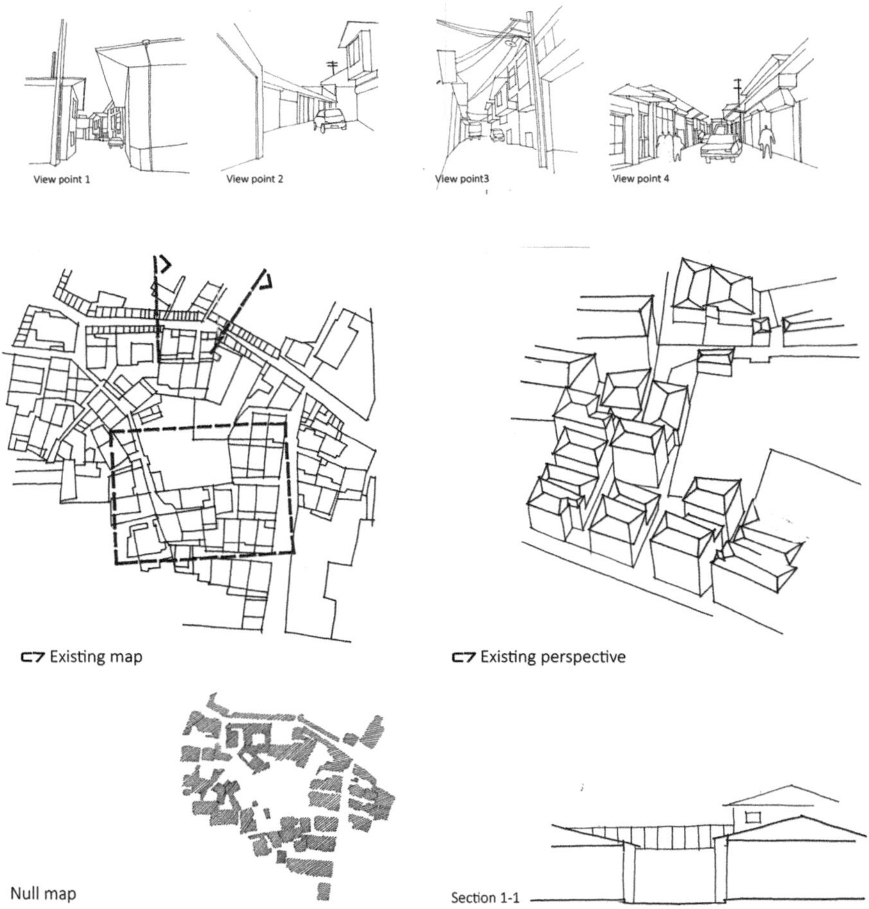

Fig. 6.20 Neighborhood center and the commercial passage (c7) along with Safi mosque have created an important and lively corridor. Currently, this corridor is used both by pedestrians and drivers. The corridor's potential to become a still more active and dynamic urban element justifies the need for its physical improvement. Vehicular access along with pedestrian access to complete the accessibility network around Bazaar

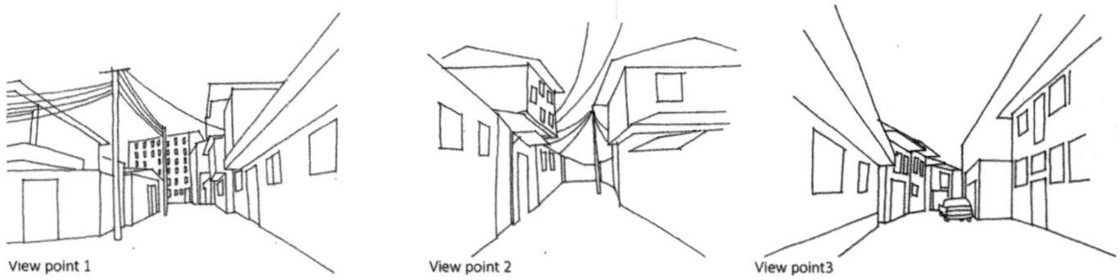

Fig. 6.21 D1 is one of the main passages of Hajiabad neighborhood. The dominant land uses of this passage are residential and have a relatively better quality. However, some walls and the width of the secondary branches face difficulties

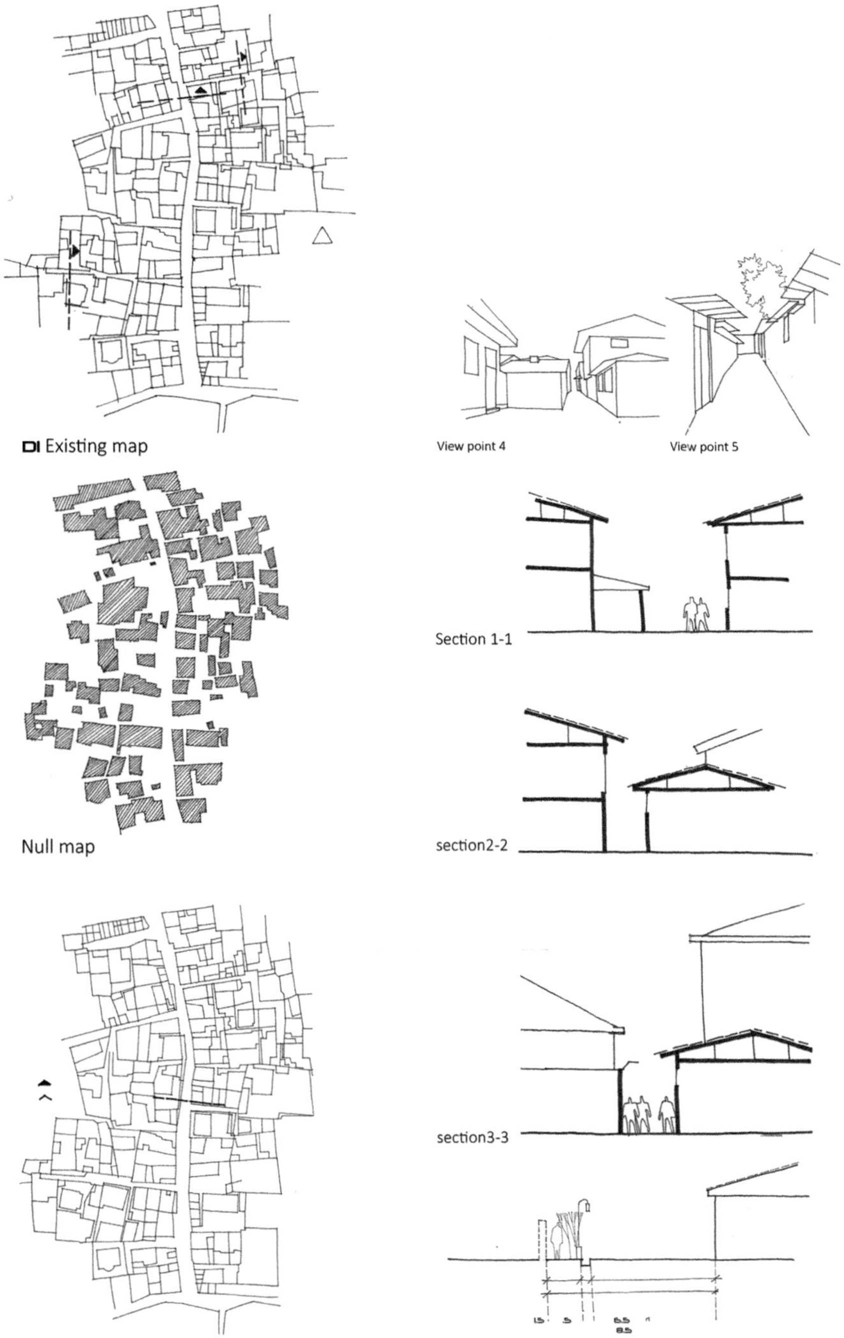

Fig. 6.21 (continued)

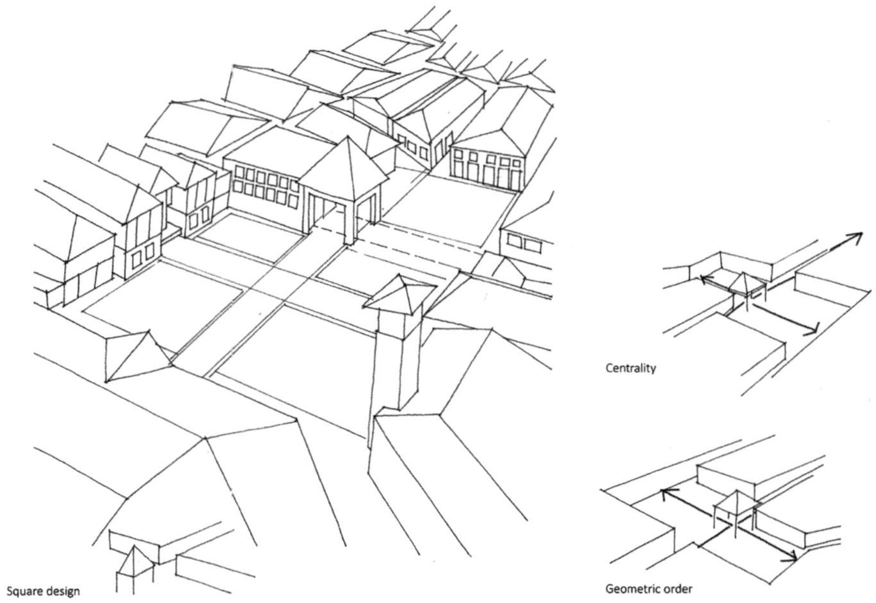

Fig. 6.22 The preliminary design for the square: final alternative. Entrance to a point which is the center of the complex and defines axes which provides order and meaning to the area

1)Building height
2)Risk affected areas (of Access network) due to Building destruction
3)Risk affected areas (of Access network) due to wall destruction
1. Height stories
3m
4m
5m
6m
7m
8m
9m
2. Risk affected areas (of Access network) due to Building destruction
3. Risk affected areas (of Access network) due to wall destruction
Vulnerability probability of low
Vulnerability probability of average
Vulnerability probability of high
4) Vulnerability degree
5) Existing lifelines
6) Land uses
4. Vulnerability degree of buildings surrounding the open space
Vulnerability probability of zero
Vulnerability probability of very low
Vulnerability probability of low
Vulnerability probability of average
Vulnerability probability of high
Vulnerability probability of very high
5. Existing lifeline
Ground slope
Hydrant
Sewage canal
6. Land-use
Residential
Commercial
Residential and commercial building
Office
Unused land
Religious, Education, Culture, Sport, Health, Public service, Military,Industry Industrial,Infrastructural, Green space
7)Wall material
8) quality of wall construction
9) wall type
7. Wall material
Brick
Hollow cement blocks façade
Stone
Concert
Clay
Wooden
Brick facade
Ceramic or stone facade
Cement façade
Steel frame
8. Quality of wall construction
Old
Normal
Newly built
9. Wall type
Yard wall
Fence
Building
10) Structural system
11)wall height
12) roof covering
10. Structure system
Load-bearing wall
Steel frame
Concert
11. Wall height
2m
3m
4m
5m
8m
9m
9. Roof covering
Galvanized Iron
Flat roof, Straw-covered sheets, Asbestos corrugated sheet,Asbestos roof shingle, Clay tiles, Galvanized plate

- Transportation network, the passages and alleys are irregular, narrow, and winding, so that only pedestrians can use them. The inadequate width of alleys along with old and deteriorated walls leads to the blockage during the earthquake.
- Urban utilities and facilities, the area has a sewage system, but lacks natural gas pipeline.
- Assessing the role of the space during disaster, since most of the activities take place during the day, so these spaces can be used for rescue and relief. For this purpose, visibility plays a critical role, because the users are passerbys and not familiar with the area. Considering open space along the movement paths is appropriate.

For this analysis, several factors in the site have been taken into considerations: building height, risk affected areas (of access network) due to building destruction, risk affected areas (of access network) due to wall destruction, structural system, wall height, roof covering, wall material, quality of wall construction, and wall type (See Figs. 1–12) (Bahrainy 1997a, c; Kazemi 1997).

The results of the site analysis is shown in Fig. 6.23.

The dense and crowded section of the old cities with narrow alleys and deteriorated buildings (Fig. 6.24) is prone to seismic activities, which not only cause life loss and injuries during earthquake, but make rescue and relief operations impossible because of passage closure. These passageways, therefore, are of great need of abundant small

Fig. 6.23 Existing and proposed open space in the selected site, based on the analysis

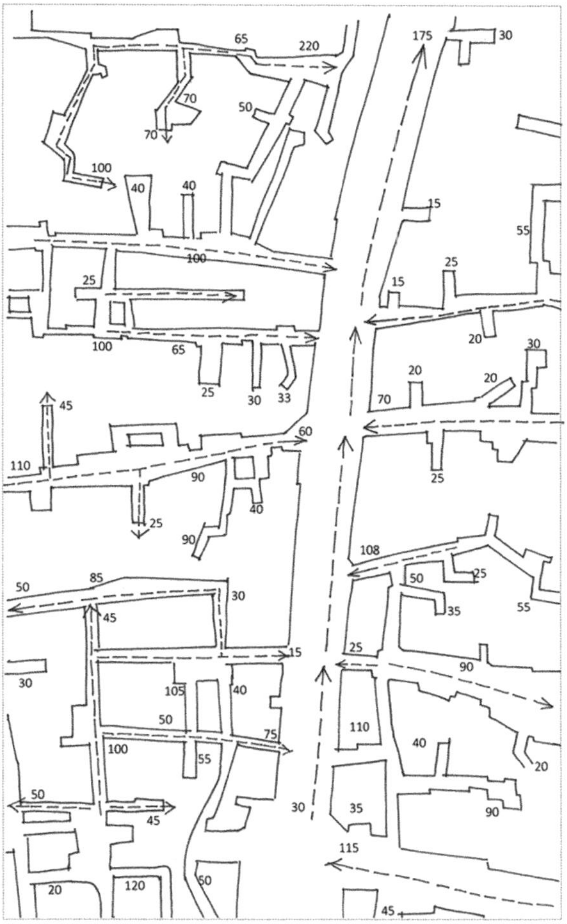

Fig. 6.24 The existing dense texture

open spaces close to where population live and work to act (Fig. 6.25) (Bahrainy et al. 1997c).

6.7 Vulnerability of Non-structural Elements in the Passages and Open Spaces

Non-structural elements include signs, powerlines, sloping roof covers, consules, façade materials, and utility elements such as coolers. The damage these elements cause in case of an earthquake ranges from severe to low (see Table 6.1 and Figs. 6.26 and 6.27).

Damaging passages depend on the deteriorated walls of the road and the possibility of life and property damages. The possible length of falling debris of the destructed walls in the road equals half of the height of the walls (Fig. 6.28).

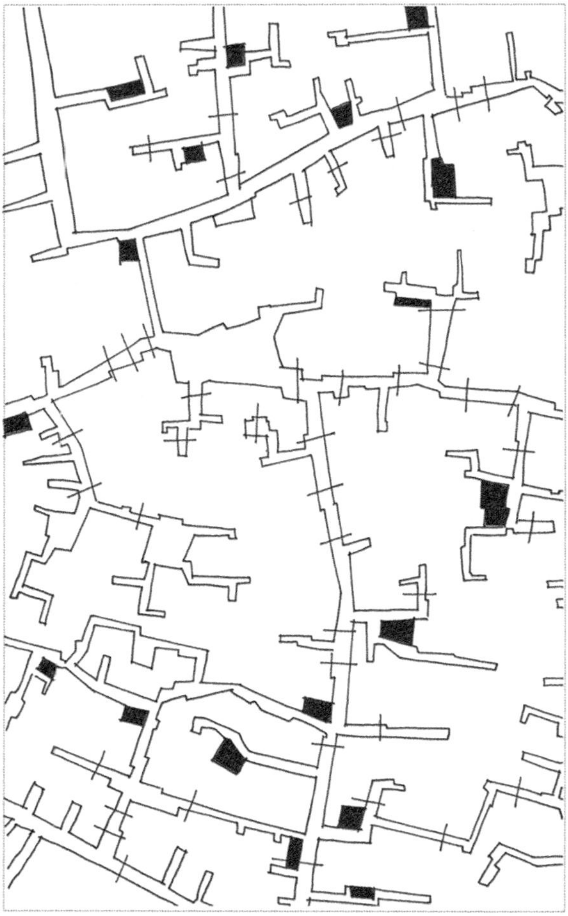

Fig. 6.25 Providing open space next in the old section of traditional to the residential units (right) city (left)

Table 6.1 Elements causing damage from severe to low

Severe damaging	High density signs, utility elements, dangerous covers, facades, and consules
High damaging	Medium density of non-structural elements and dangerous elements
Medium damaging	Medium density of non-structural elements, without dangerous consules
Low damaging	low density of non-structural elements, without dangerous consules

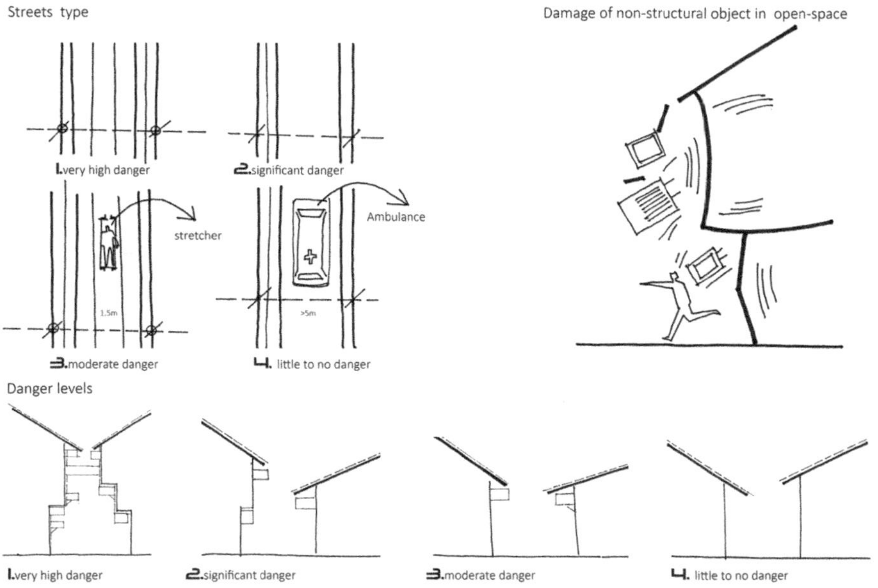

Fig. 6.26 Damages caused by non-structural elements

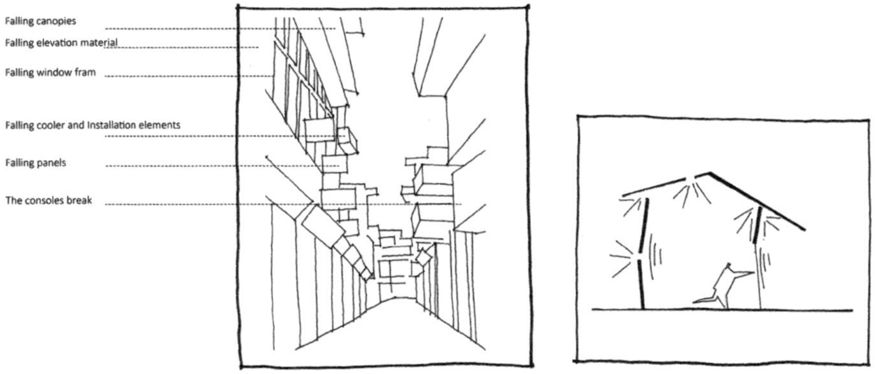

Fig. 6.27 Damages caused by physical type (microscale)

6.8 Vulnerability Against Fire

- Higher vertical density increases the chances of fire spreading, more open spaces will decrease spreading fire, more difference of building heights in adjacent buildings increases the possibility of spreading fire from lower buildings to the higher ones, and more openings will increase the possibility of fire spreads (Banerjee 1993; Bahrainy et al. 1997b) (Figs. 6.29 and 6.30).

Fig. 6.28 Earthquake damage in buildings and in streets

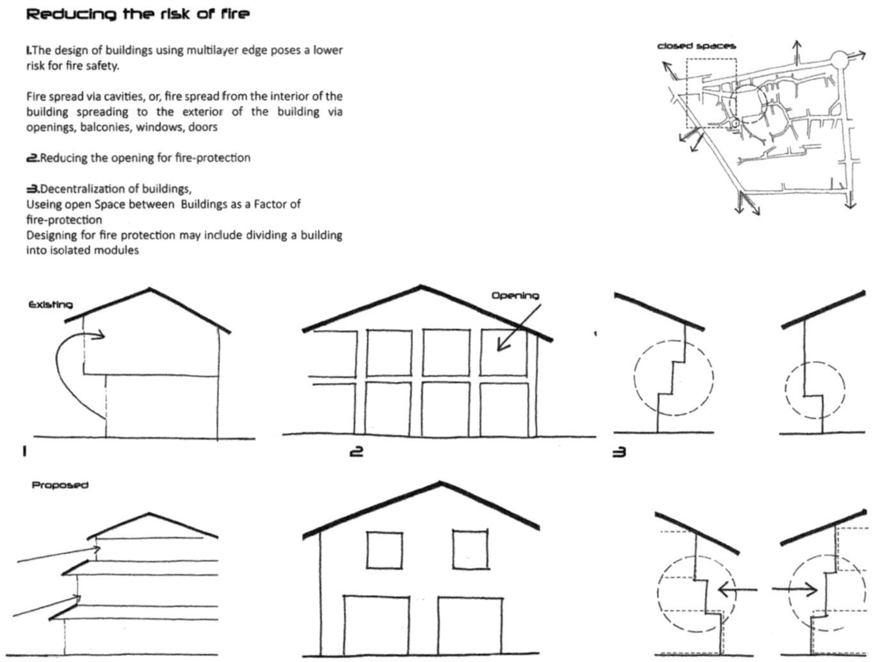

Fig. 6.29 Building form to reduce fire risk

6.9 Earthquake Vulnerability, Specifications, the Central Area with Grid Layout

Various reinforcing methods of passages walls, zone 1 (see Chap. 5, p.) (Figs. 6.31, 6.32, 6.33 and 6.34).

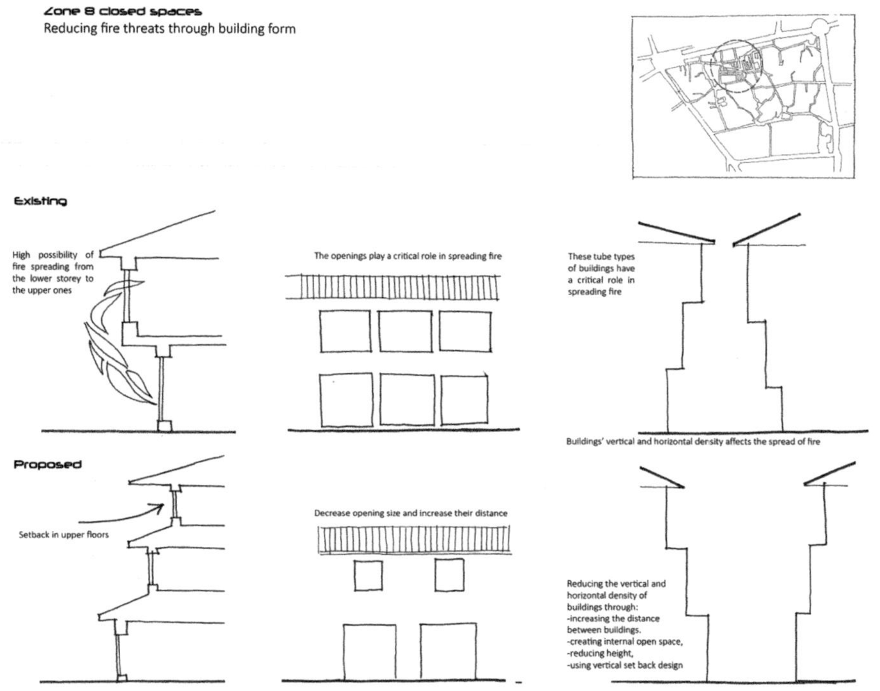

Fig. 6.30 Building form to reduce fire risk, cross section and facade

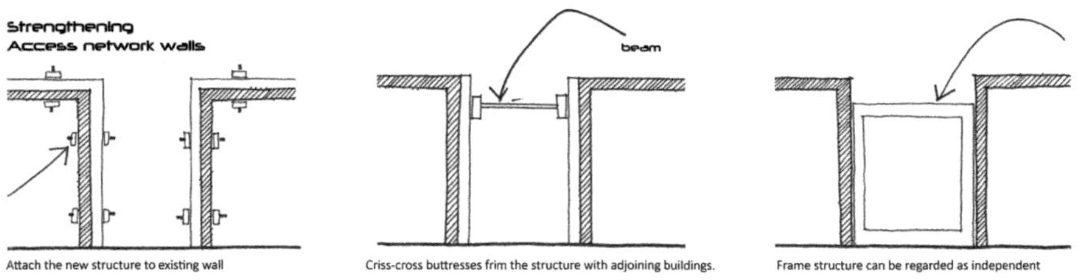

Fig. 6.31 Strengthening access network walls

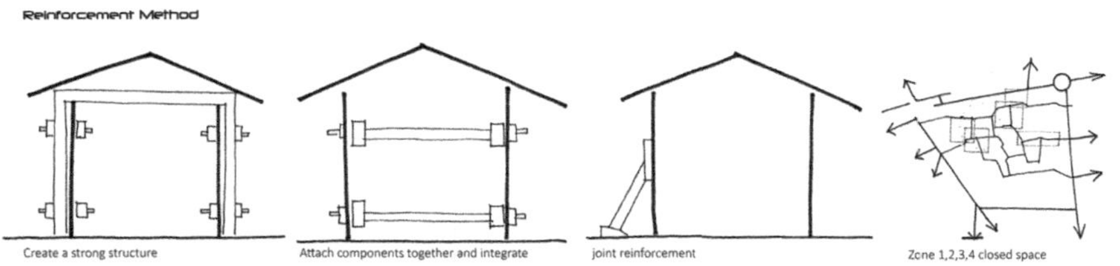

Fig. 6.32 Reinforcement methods in the buildings with the steel-frame structure zones 1, 2, 3, and 4 (closed spaces) (see Chap. 5)

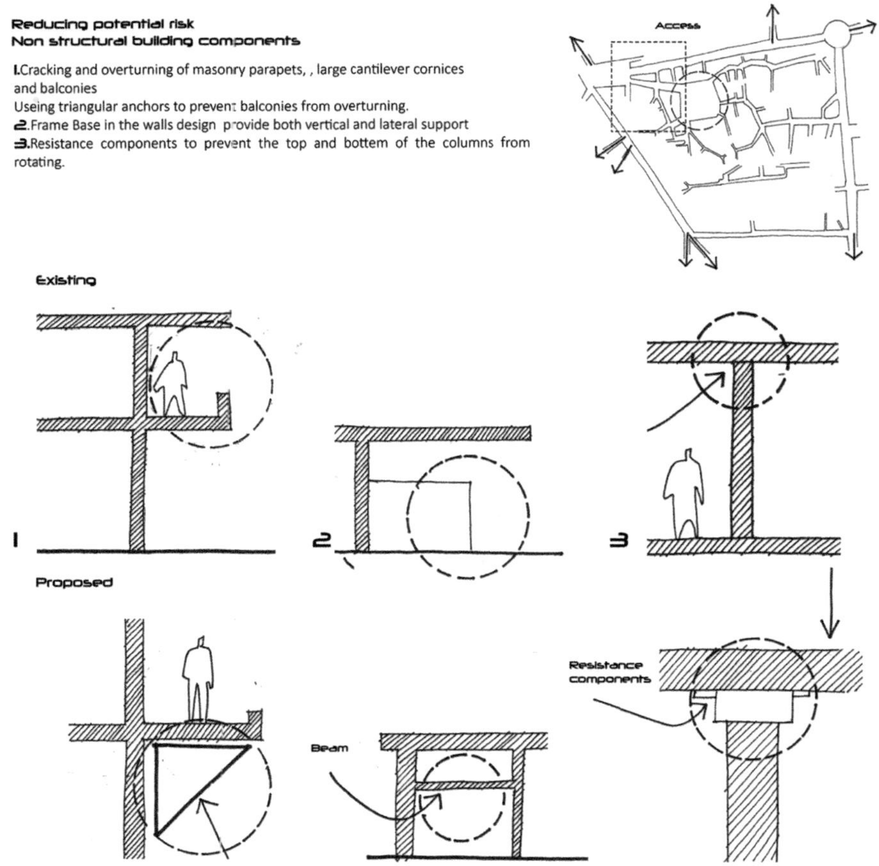

Fig. 6.33 Reducing potential risk through modification of non-structural components

6.9.1 Strengthening Methods for Buildings with Load-Bearing Walls, Zones 1, 2, 3, 4, 5, and 6. Enclosed Spaces

See Fig. 6.35.

Zones 1, 2, 3, 4, 5, and 6 (closed spaces) (see Chap. 5).

6.9.2 Strengthening Methods in Steel-Frame Buildings, Zones 2, 3, 4, 5, and 6 (See Chap. 5)

See Fig. 6.36.

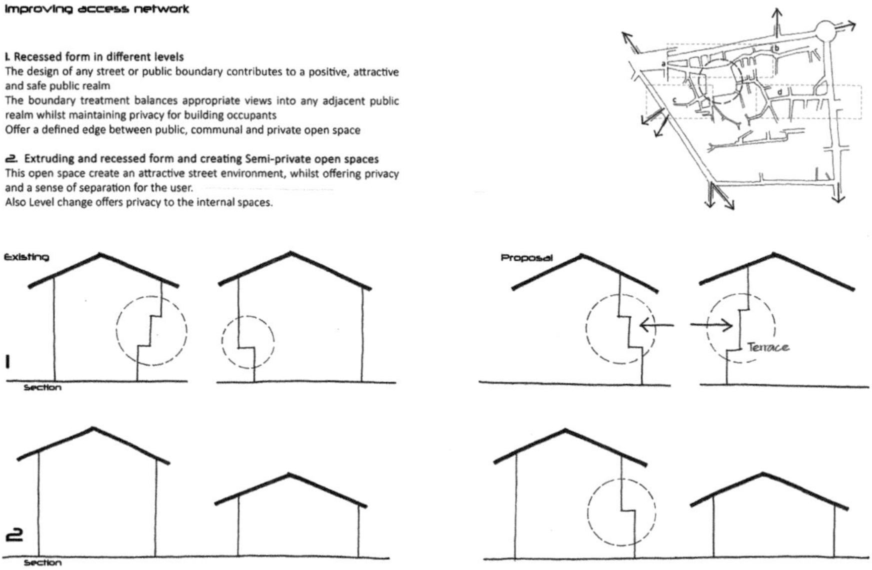

Fig. 6.34 Building form to improve access network

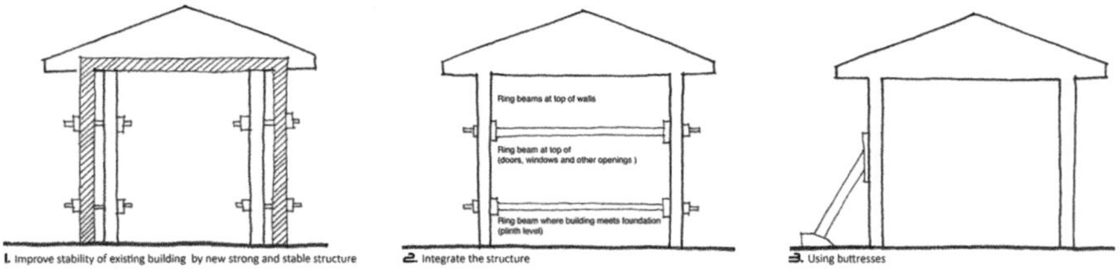

Fig. 6.35 Reinforcement methods in the buildings with the steel-frame structure

Open spaces and access networks, as the most significant elements of urban space, play a critical role in every phase of earthquake disasters, from escape, refuge, relief, rescue, first aid, emergency, and temporary settlements. Application of seismic risk mitigation strategies in dealing with urban spaces in the event of a disaster can have more tangible results in smaller scales. Areas of the cities with high risk or vulnerability such as city cores, historical sections, mountain slopes, or areas with land instability potentials are examples for sector scale application of seismic risk mitigation.

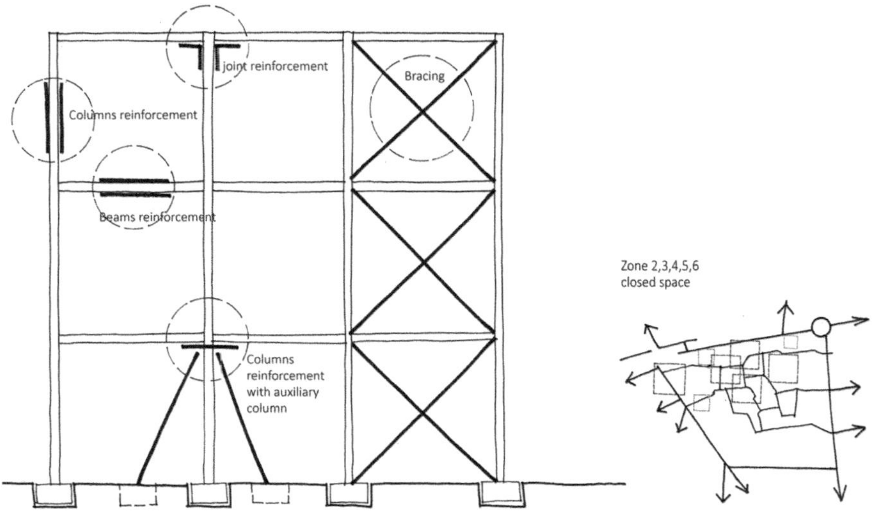

Fig. 6.36 Reinforcement methods in steel structure buildings of closed spaces, zones 2, 3, 4, 5, and 6

From an urban design point of view, project scale is the lowest possible level one can deal with, with short-term and specific objectives and easy-to-implement proposals. It could be applied to the old section of the Central Business District of traditional cities where the interwoven fabric, deteriorated structures, overpopulated passages, high density of buildings, and substandard safety against earthquake and fire make them the most vulnerable sections of the cities.

The goals of the area design to reduce vulnerability against seismic risk are as follows:

- The need for creating the main open space in the central core.
- The need for the modification and improvement of the main passages.
- Creation or improvement of emergency access network and facilities.
- Demolition and reconstruction of highly vulnerable buildings due to seismic risk.
- Creating open and semi-open spaces.
- Change and modification of dangerous buildings.
- Providing open spaces next to commercial activities to be flexible.

In normal conditions, green and open spaces, are used for recreation, sport activities, and air purification, but during a disaster, they will be used for escape, rescue, and relief and, immediately after the earthquake, will be used for emergency and temporary settlement.

Design ideas at the block scale to reduce seismic risk in the Bazaar area and the operational objectives in roads and open spaces include the following:

- Creating open space, along with strengthening road walls.
- Provide access to open spaces as the first priority and creating open space as the second.
- Provide access to existing open spaces as the first priority and roads improvement as the second.
- Strengthening roads, as the first priority, and road improvement, as the second.
- Removing the non-structural damaging elements, as the first priority.
- Provide accessibility for the time of disaster, as the first priority.

Design alternatives are presented for the paths in the central area with grid texture, also design solutions are suggested to reduce seismic risk, and vulnerability of the elements at risk in the studied block will include the following characteristics: lower density, open spaces, wide access networks, light roofs, etc.

Specifications are needed to practically reduce seismic vulnerability in the roads and open spaces; examples are as follows:

Modification of the main roads: exits and entrances, and areas at risk with low depth and lack of access to safe open space, modification of physical types at medium scale in the complexes with dangerous and low depth passages and lack of access to safe open space, modification of physical type, medium scale, modifications of physical types, microscale and open space design for rescue, relief, and emergency settlement.

Modification of passages are suggested for selected cases which are of severe risk type, and with no appropriate access to open space. Also, a few sites are selelcted for analysis and proposing open spaces. And finally, vulnerability of non-structural elements in the passages and open spaces, as well as vulnerability against fire, is analyzed and appropriate design solutions are recommended.

References

Allan, Bryany PM (2014) The attributes of resilience. Int J Disaster Resilience Built Environ 5(2):109–129

Bahrainy H (national director) (1992) Natural disaster management. UNDP-UNCH (Habitat) Project IRA 90//004, Assistance in the implementation of Post earthquake Reconstruction Program, 8 vols

Bahrainy H (1993) Cities and earthquake. Paper presented to the International symposium: experiences, prevention, protection, Iraklion, Greece, 6–8 May

Bahrainy H (1998) Urban planning and design in seismic-prone region (the case of Rasht in Northern Iran). American Society of Civil Engineering, Urban Planning and Development Division. 124(4): December

Bahrainy H, Bakhtiar A (2016) Toward Integrative Theory of Urban Design, Springer

Bahrainy H et al (1996) Urban design in the seismic-prone regions. In: Urban design of the city of Rostam Abaad. Research Report. Housing Foundation of Islamic Republic of Iran

Bahrainy H et al (1997a) Open spaces in the city of Rasht. National Report, vol 39, UNDP—UNCHS (Habitat) Project IRAl90l004, Housing Foundation of IR, NDPC (Translated in English), Tehran, June

Bahrainy H et al (1997b) Planning and design of specific land uses in the city of Rasht for reducing of earthquake vulnerability. National Report, vol 52, UNDP—UNCHS (Habitat) Project IRA1901004, Housing Foundation of IR, NDPC (Translated in English), Tehran, November

Bahrainy H et al (1997c) Planning and design of transportation network in the city of Rasht. National Report, UNDP—UNCHS (Habitat) Project IRA1901004, Housing Foundation of IR, NDPC (Translated in English), Tehran, August

Bahrainy H et al (1998) Synthesis on redesign of the city of Rasht. National Report, UNDP—UNCHS (Habitat) Project IRA1901004, Housing Foundation of IR, NDPC (Translated in English), Tehran, December

Banerjee T (1993) Urban form and seismic vulnerability. Earthquake protection through urban design and planning. International Consultant Report, UNDP—UNCHS (Habitat) Project IRA1901004, Housing Foundation of IR, Tehran, October

Kazemi M et al (1997). Socio-economic structure of the city of Rasht. National Report, vol 34, UNDP—UNCHS (Habitat) Project IRA1901004, Housing Foundation of IR, NDPC (Translated in English), Tehran, January

Majidi H et al (1997) Central part of the city of Rasht. National Report, vol 47, UNDP—UNCHS (Habitat) Project IRA1901004, Housing Foundation of IR, NDPC (Translated in English), Tehran, May

Music V (1993) Earthquakes and management of urban environment. Urban planning and design as a part of preventive strategies. International Consultant Report, UNDP—UNCHS (Habitat) Project IRA1901004, Housing Foundation of IR, Tehran, October

Music V (1998) Alternative development scenarios of the city of Rasht and its functional urban region. International Consultant Report, UNDP—UNCHS (Habitat) Project IRA1901004, Housing Foundation of IR, Tehran, December

National Consultant Report (1993) UNDP—UNCHS (Habitat) Project IRA1901004, Housing Foundation of IR, Tehran, October.

Petrovsky J, H Bahrainy (1992) Analysis and development of empirical vulnerability functions. Res Rep UNDP-UNCHS Project IRA/90/004

Petrovski JT, Bahrainy H, Music V (eds) (1998) Guidelines for reduction of seismic risk in regional and urban planning, vol. VII of the Guidelines for Earthquake Disaster Management, UNDP-UNCHS (Habitat) and Housing Foundation of IR, Tehran—Skopje

Chapter 7
General Guidelines for Urban Design in Seismic-Prone Regions

Abstract What can urban design do to practically reduce the vulnerability of the built environment? Here, in this chapter some guidelines are visually illustrated to help users easily understand and apply the ideas in practice. These include the following: Earthquake Scenario, evacuation times from multi-story buildings for different age groups and handicaps (since time is a determinant factor in earthquake management), and also some structural recommendations. Secondary damages, such as fire, explosions, flooding, subsidence, landslide, psychological trauma of shake, disorientation uncertainty, and grief, are all concerned issues that are presented graphically to show the risks involved when disasters are not dealt with in the right way. General principles include, for example: system redundancy, restorability, serviceability, and evacuation potential, etc. Specific principles include, for example: shorter distances between different land uses, network hierarchy, shorter distances between critical land uses during disaster, fast access to important land uses, more possible control, more possibility of escape and evacuation through the network, adequate speed, shorter length and more safety of the network, higher ratio of open space over built-up areas, and more independent network for each district, etc.

Keywords Guidelines · Urban design · Seismic-prone regions · Information needed · General principles · Specific principles · Regulatory measures · Location decisions · Optimum distribution · Evacuation

What can urban design do to practically reduce the vulnerability of the built environment? Here, in this chapter some guidelines are visually illustrated to help users easily understand and apply the ideas in practice. These include the following: Earthquake Scenario, evacuation times from multi-story buildings for different age groups and handicaps (since time is a determinant factor in earthquake management), and also some structural recommendations. Secondary damages, such as fire, explosions, flooding, subsidence, landslide, psychological trauma of shake, disorientation uncertainty, and grief, are all concerned issues that are presented graphically to show the risks involved when disasters are not dealt with in the right way.

More specifically, as for issues dealing with general location or site selection for different land uses, following factors should be taken into consideration:

H. Bahrainy and A. Bakhtiar, *Urban Design in Seismic-Prone Regions*,
The Urban Book Series, https://doi.org/10.1007/978-3-031-08321-1_7

7.1 Information Needed for the City Scale Design in the Seismic-Prone Regions

7.1.1 Basic Information

- Seismicity map.
- Hazard Assessment, Risk Assessment, Vulnerability Assessment (Fig. 7.1).

- Seismic zonation/microzonation
- Geological condition/soil stability/soil dynamics
  - Landslide.
  - Rockfall.
  - Liquefaction.
  - Land slope/topography

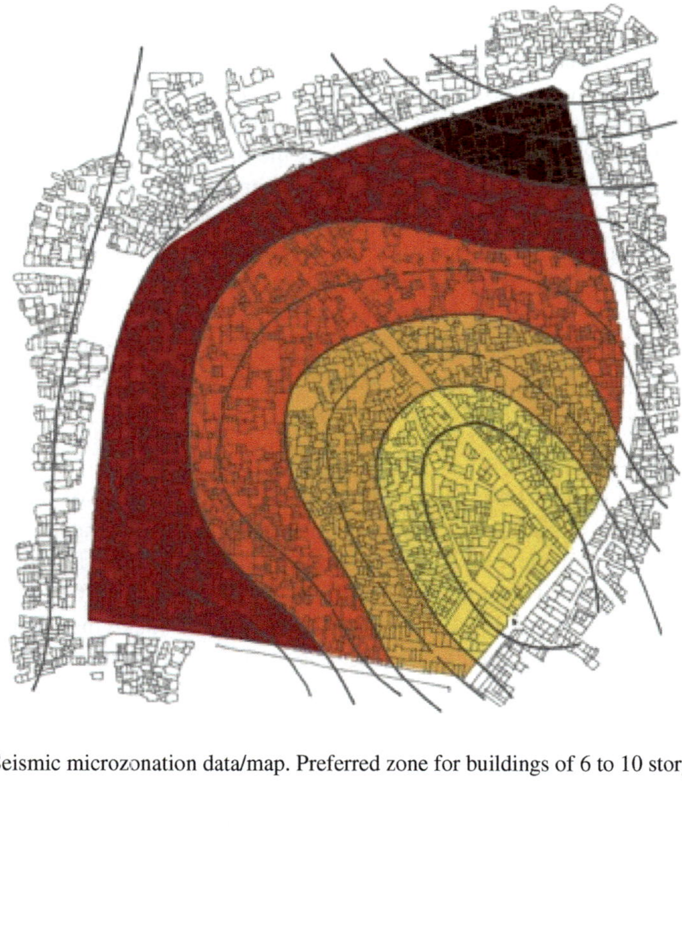

Fig. 7.1 Seismic microzonation data/map. Preferred zone for buildings of 6 to 10 story

7.1.2 *Information Needed for Planning and Design Purposes*

- Land uses (sensitive, vulnerable, and damaging).
- Construction quality.
- Buildings' age and depreciation.
- Construction techniques.
- Building materials.
- Building form.
- Texture (grain, open, and green space, etc.).
- Density (Height) of buildings.
- Accessibility network (hierarchy, bridges, parking, traffic volume, alternatives, etc.).
- Urban utilities (water, sewage, electricity, communication, etc.).
- Urban facilities (health centers, fire extinguisher, security facilities, etc.).

7.2 Urban Design Principles in the Seismic-Prone Region

In any urban planning and design activity in the seismic-prone regions, the following principles need to be taken into account.

7.2.1 *General Principles (Bahrainy 1998)*

- System Redundancy.
- Restorability.
- Serviceability.
- Evacuation Potential.
- Hazard Potential.
- Trauma Potential.
- Occupancy.
- Coping ability.
- Critical residents.

7.2.2 *Specific Principles*

- Shorter distances between different land uses.
- Network hierarchy.
- Shorter distances between critical land uses during disaster.
- Fast access to important land uses.
- More possible control.

- More possibility of escape and evacuation through the network.
- Adequate speed, shorter length, and more safety of the network.
- Less traffic in the network.
- Lower traffic (users) intensity.
- Higher ratio of open space over built-up areas.
- More direct connection between important uses.
- More connection between different parts of the network.
- Less the number of cul-de-sacs in the network.
- More vehicular access to the network.
- Less the number of intersections.
- More independent network for each district.
- The lower the ratio between the length and width of passages.
- The more the ratio between the width to half of the walls' height of the passages.
- The more seismic codes are applied in the construction of utilities and facilities.

7.3 Urban Design as Regularity Measures to Reduce Seismic Risk

Science, technology, and regulatory measures can direct their activities toward reducing seismic risk in human settlements and reduce life loss and property damage in case a disaster strikes. Most of urban design activities focus on regularity measures which include mitigation, preparedness, emergency response, and recover measures.

7.4 Time of Earthquake

Although property damage might not change, life loss and injuries depend to a large extent on the time earthquake occurs (Fig. 7.2).

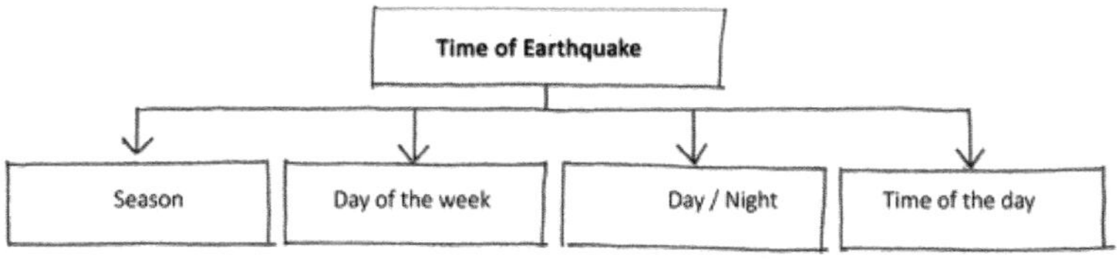

Fig. 7.2 The time that earthquake strikes has a determinant factor in the number of losses and injuries. Worst season would be winter time; worst day of the week is any week day and nighttime (Bahrainy 2005)

7.5 Location Decisions

Deciding the right location for any specific urban development project could prevent a series of disasters in the future. Seismic zonation and especially microzonation is a required tool for this purpose (see Chap. 2) (Figs. 7.3 and 7.4).

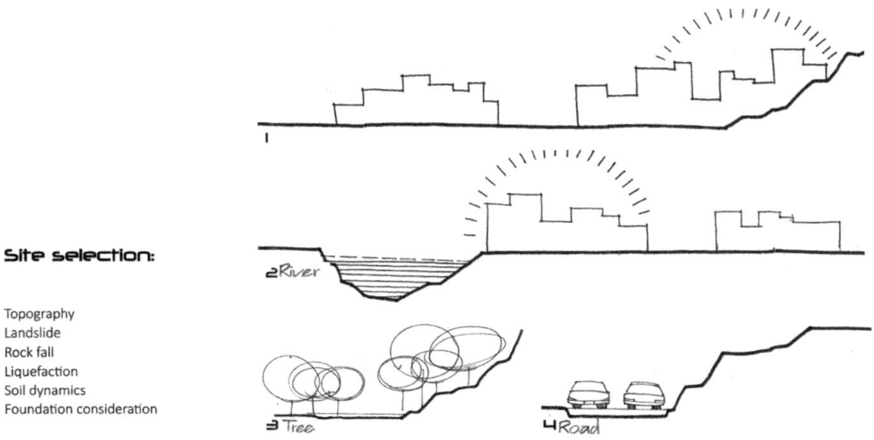

Fig. 7.3 Topography, landslide, rockfall, liquefaction, soil dynamics, and foundation considerations should be taken into account in any site selection decision

Fig. 7.4 Slopes and instable soils are high-risk areas and development in these areas should be prevented

7.6 There Should Be an Optimum Distribution of Activities Centers, While Avoiding Locating Sensitive Public Buildings in the Areas with Potential risk

See Figs. 7.5, 7.6, 7.7, 7.8, 7.9, 7.10, and 7.11.

- Recessed form at different levels. The design of any streets or public boundary contributes to a positive, attractive, and safe public realm.
- The boundary treatment balances appropriate views into any adjacent public realm while maintaining privacy for building occupants.
- Offer a defined edge between public, communal, and private open space.
- Extruding and recessed form and creating semi-private open spaces
- This open space creates an attractive street environment, while offering privacy and a sense of separation for the user.
- Also level change offers privacy to the internal spaces

7.6.1 *General Guidelines for Roads and Open Spaces*

- Creating open space along with strengthening road walls.
- Provide access to open spaces as the first priority, creating open space as the second priority.
- Provide access to existing open spaces as the first priority, and road improvement, as the second priority.
- Strengthening roads, as the first priority, and road improvement, as the second.
- Removing the non-structural damaging elements, as the first priority.
- Provide accessibility for the time of disaster, as the first priority.
- Urban renewal, as the first priority.
- Urban renewal, as the second priority, partial repair and improvement of physical type, as the first priority.

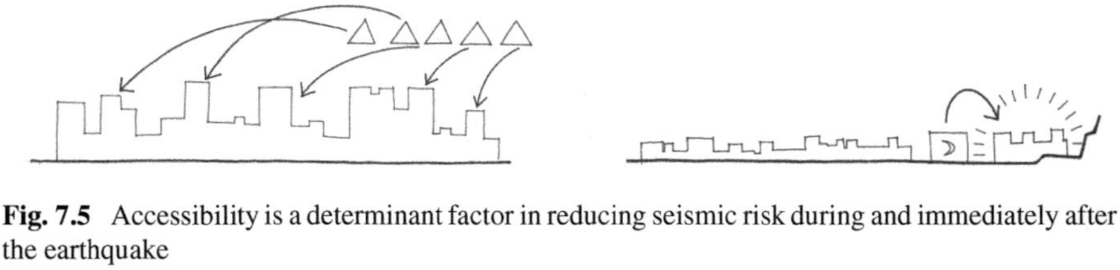

Fig. 7.5 Accessibility is a determinant factor in reducing seismic risk during and immediately after the earthquake

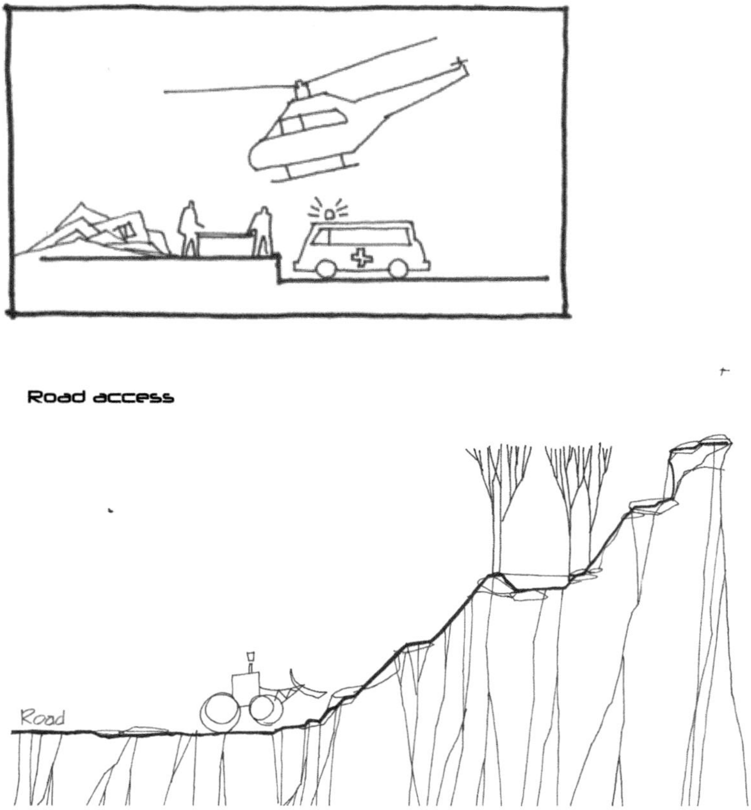

Fig. 7.6 Roads built on the slopes and instable soils are vulnerable to earthquake and landslide

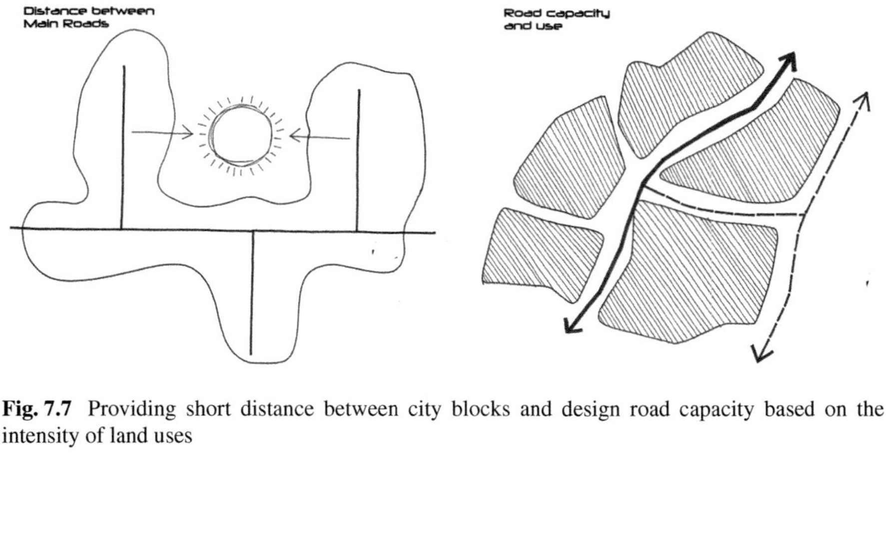

Fig. 7.7 Providing short distance between city blocks and design road capacity based on the intensity of land uses

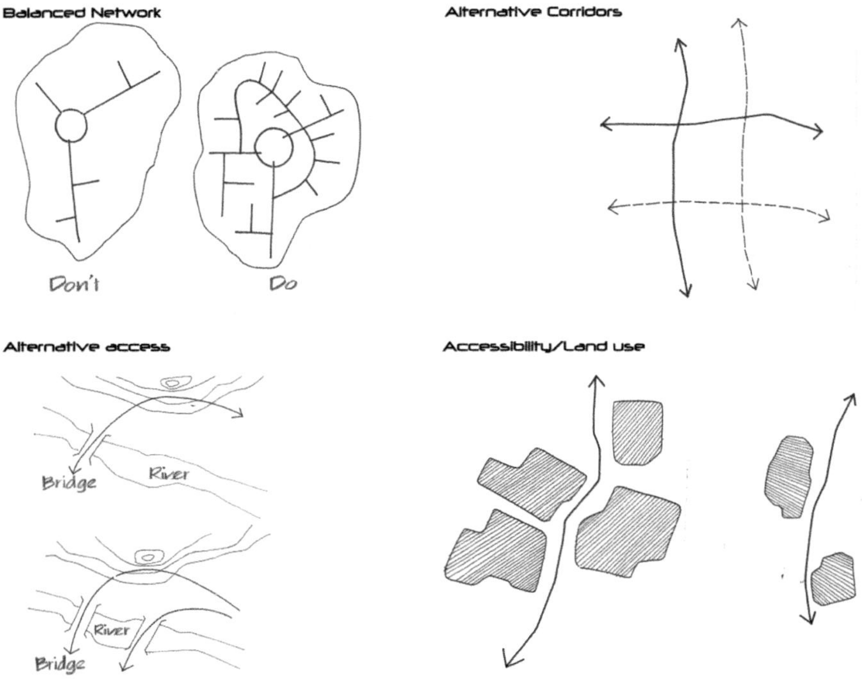

Fig. 7.8 Provide a balanced network, alternative corridors, and access

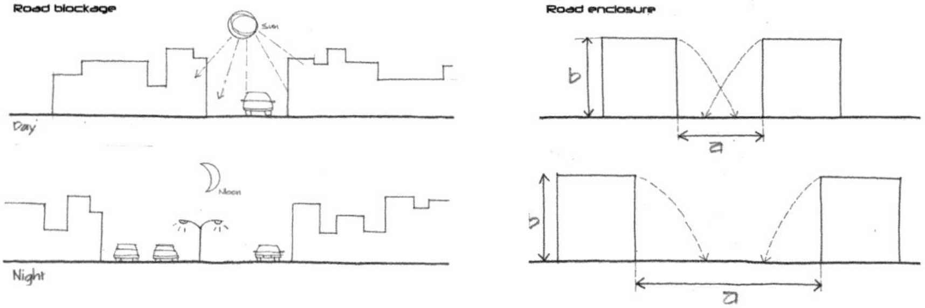

Fig. 7.9 Road blockage due to traffic density and possible building collapse require alternative access

- Repair and measures to facilitate access to existing and proposed open spaces outside the area and partial physical improvement.
- Strengthening as the first priority.
- Strengthening as the first priority and urban renewal along with improvement of physical type, as the second priority.
- Strengthening and partial improvement of physical type, as the third priority.
- The area has priority to provide measures for better access during disaster.

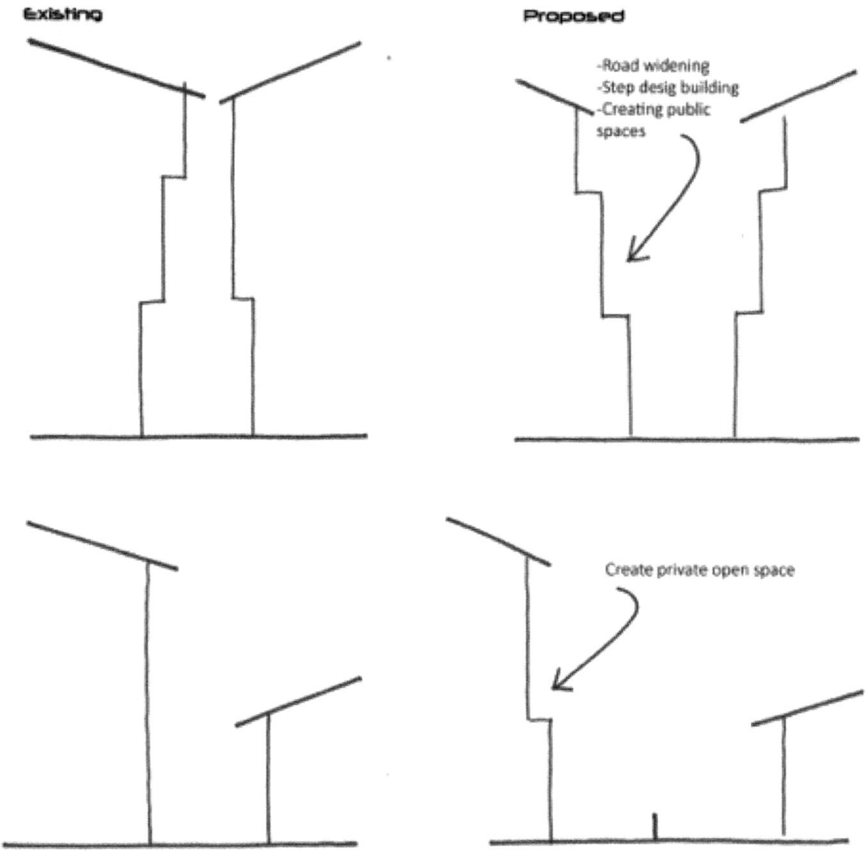

Fig. 7.10 Vertical setback is recommended in the dense old section of central districts of cities

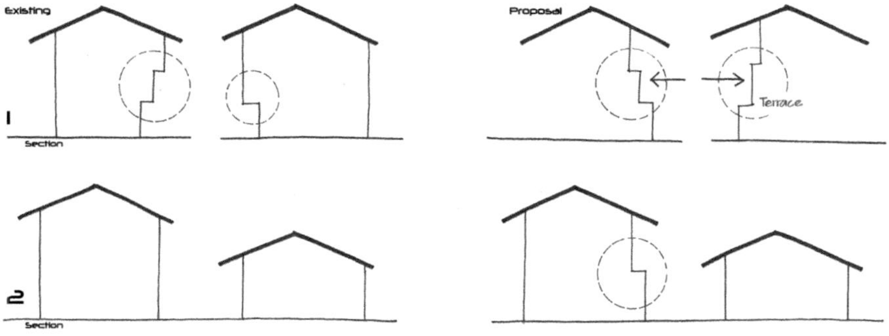

Fig. 7.11 Improving access network

- The area is in need of appropriate measures to prevent breaking fire and spreading to other areas. Physical type improvement to prevent fire from spreading.

7.7 Decentralization

See Figs. 7.12, 7.13, 7.14, 7.15, 7.16, 7.17, and 7.18.

Fig. 7.12 The idea of decentralized concentration may be used at the city scale to prevent high-density development in the inner city areas and, therefore, reduce earthquake risk

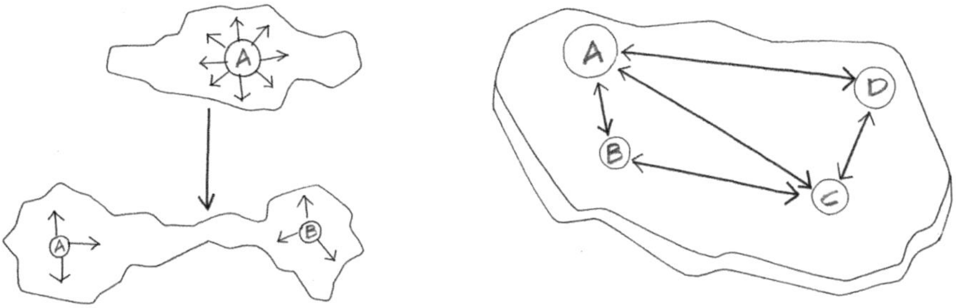

Fig. 7.13 Decentralization and hierarchy of activity centers in the city

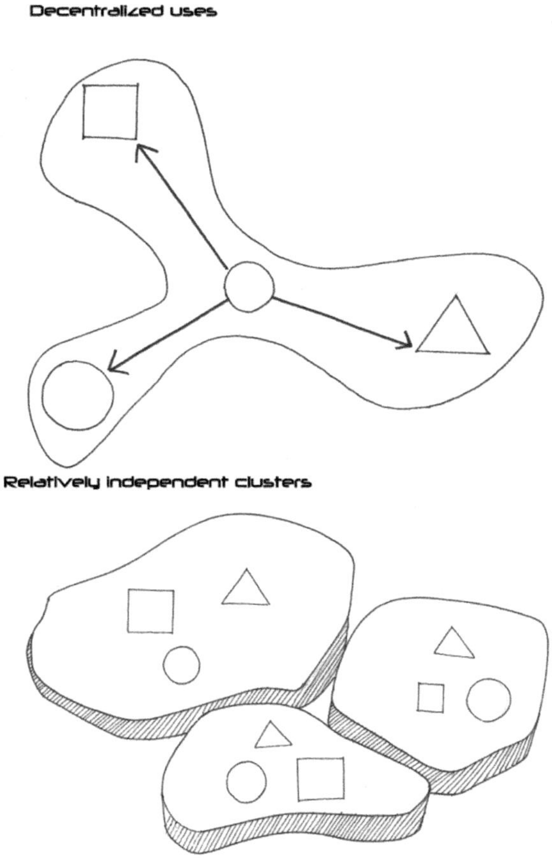

Fig. 7.14 Decentralizing activity centers by creating self-sufficient districts or neighborhoods will reduce seismic risk during and after disaster

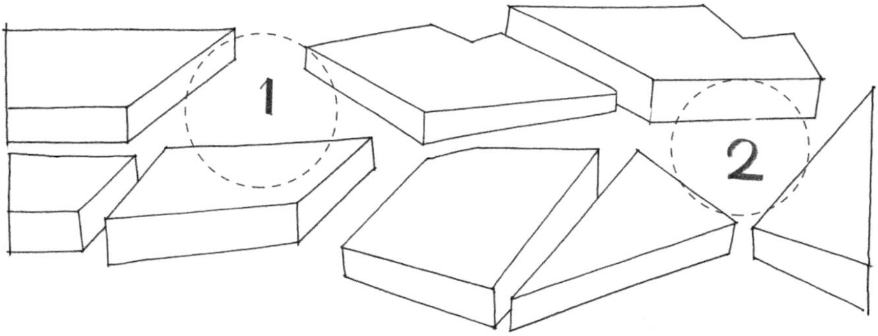

Fig. 7.15 Decentralization of open spaces will provide easy and safe access of residents in the surrounding areas to public spaces

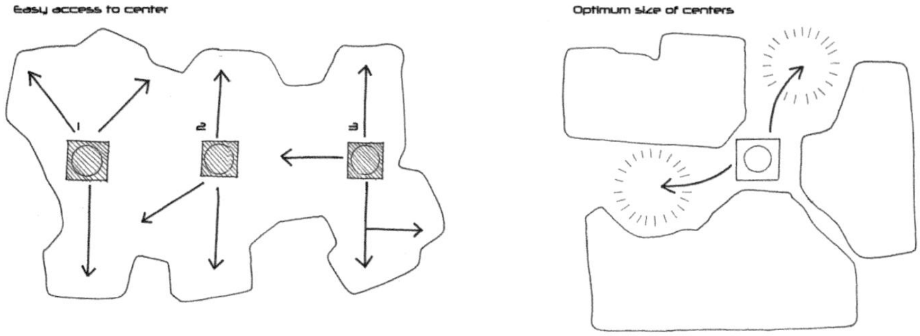

Fig. 7.16 Provide easy and safe access to activity centers

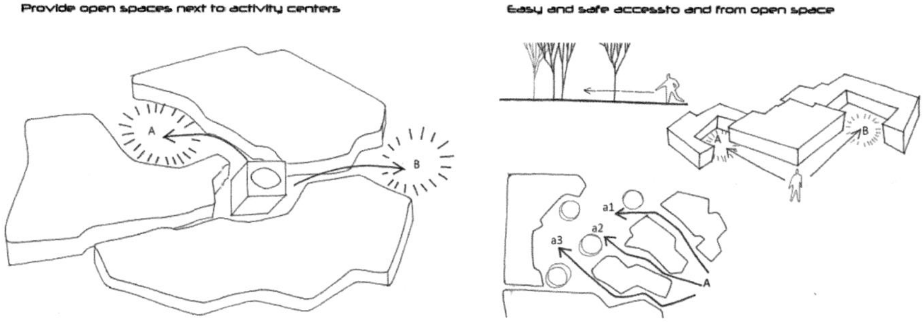

Fig. 7.17 Provide alternative access to public services

Fig. 7.18 Using seismic standards when decentralizing public utilities and services

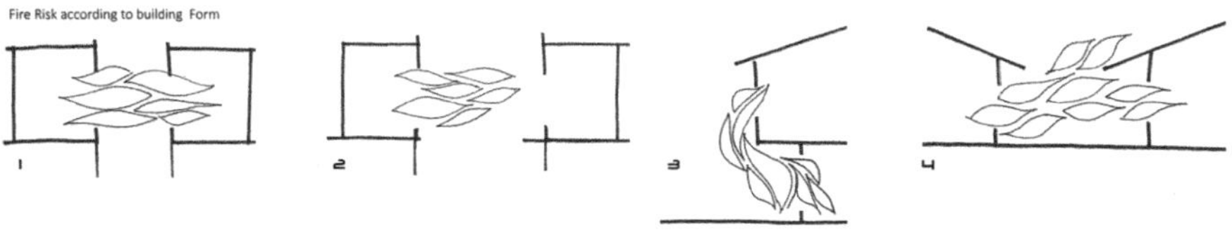

Fig. 7.19 How the fire spread to other buildings or higher levels of the same building

7.8 Protecting Buildings and Public Spaces from Fire

See Figs. 7.19 and 7.20.

- The design of buildings using multilayer edge poses a risk for fire safety.
- Fire spread via cavities, or fire spread from the interior of the building spreading to the exterior of the building via openings, balconies, windows, and doors.
- Vertical setback and decreasing windows size.
- Reducing the openings for fire protection.
- Using open space between buildings as a factor of fire protection.
- Designing for fire protection may include dividing a building into isolated modules (Fig. 7.21).

7.9 Preparing for Evacuation in Case of a Disaster

Simple building form, optimum population, and building density make reasonable evacuation time possible. The desirable characteristics of a building configuration are simplicity, regularity, symmetry, uniformity of mass and stiffness, and redundancy (Figs. 7.22 and 7.23; Table 7.7.1).

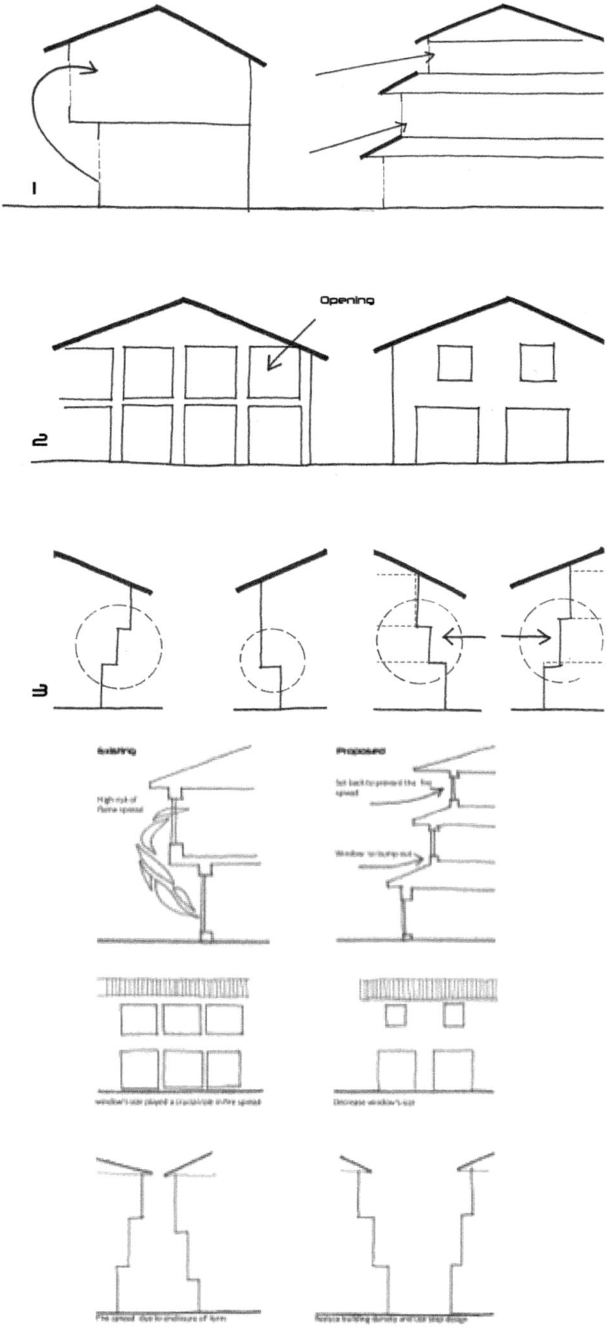

Fig. 7.20 The existing form of buildings may be modified to prevent fire from spreading

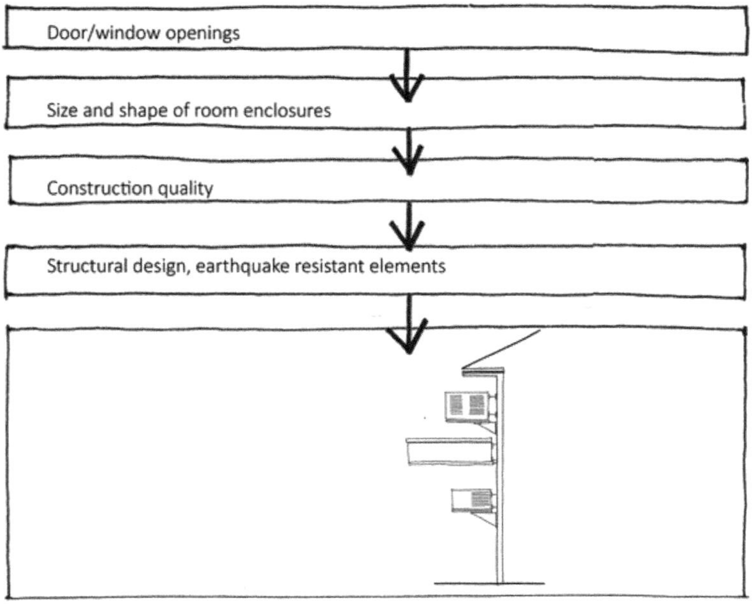

Fig. 7.21 Factors affecting vulnerability of buildings and adjacent areas

7.10 Assessing the Vulnerability of the Built-Up Area, Combining Various Factors

- External factors:
  - Land slope (L_1).
  - Soil type (L_2).
  - Dominant height of buildings (L_3).
  - Plan (L_4).
  - Dominant construction quality (L_5).
  - Population density in the district (L_8).
  - Texture complexity in the district (L_9).
- Internal factors:
  - Construction system of buildings (L_6).
  - Floor structure (L_7).

Qualitative data is then converted into quantity, and by using the following formula, damage ratio is calculated (Table 7.2):

$$(\mathrm{LR}) = (L_1 \times L_2 \times L_4 \times L_5 \times L_8 \times L_9 \times \frac{1}{4}[(0.666 \times L_6) + (0.334 \times L_7)]$$

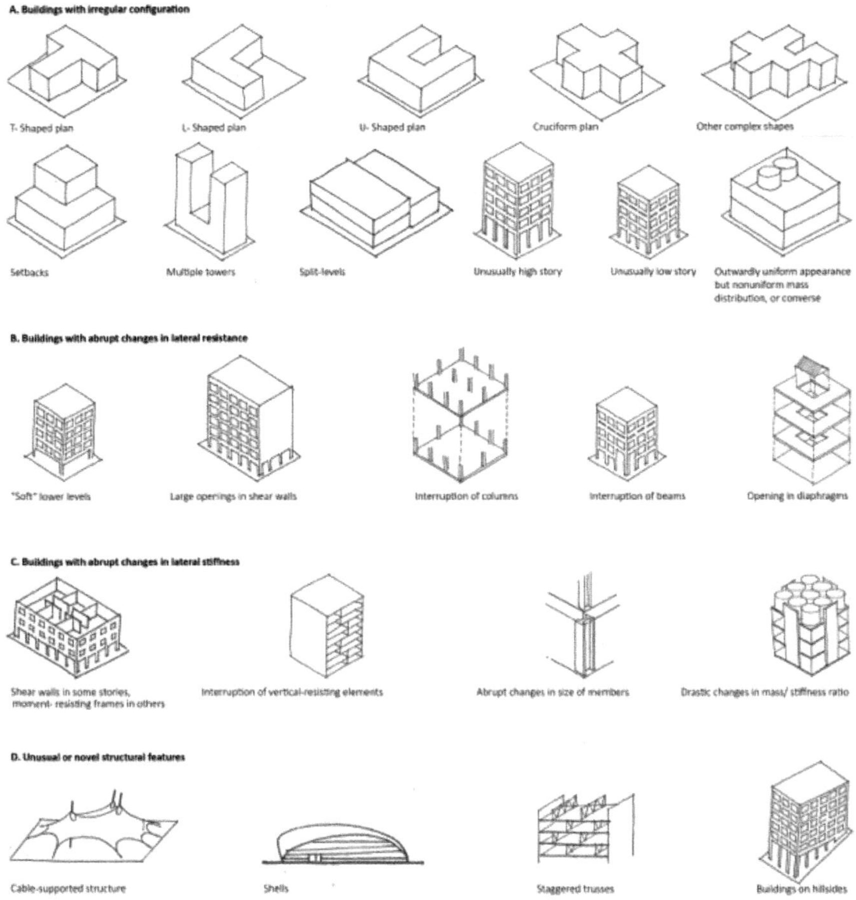

Fig. 7.22 Irregular structures or framing systems (SEAOC 1980)

7.11 Practical Guideline for Reconstruction of Passages is as Follows

H_1 and H_2= Height of buildings on both sides
D = passage width (recommended)

$$D - \frac{h1 + h2}{2} > 3$$

Three meters is the minimum width for walking and driving rescue people.

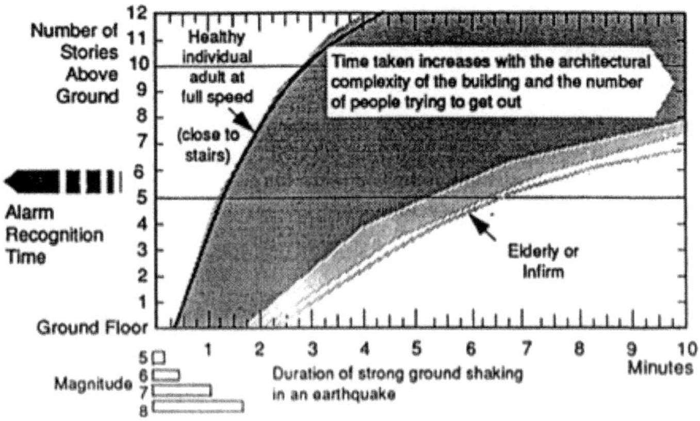

Fig. 7.23 Evacuation times from multi-story buildings (after Georgescu 1988)

Table 7.1 Relationship between the damage ratio, repair, and evacuation time

Damage ratio (LR)	Interpretation
LR ≤ 0.25	Low damage, usable but requires repairs, repairing is possible at the same time as evacuation
0.26 ≤ LR ≤ 0.5	Medium damage, repairing is possible after evacuation
0.51 ≤ LR0.75	Severe damage, evacuation is compulsory, reconstruction is required
0.76 ≤ LR	Building collapse, possibility of death

7.11.1 Reducing Potential Risk of Non-structural Elements

See Fig. 7.24.

7.12 Hospitals, Laboratories, Libraries, Sport Facilities and Education Centers, Avoiding Hazardous Areas/Activities

See Fig. 7.25.

7.13 Structural Modifications to Reduce Seismic Risk in the Vulnerable Historical Site Beam

See Figs. 7.26 and 7.27.

Table 7.2 Converting qualitative attributes of districts into quantitative data

No. of factors	Affecting factors		Damage Ratio
	Name of the factor	Description	
1	Land slope	0°–15° 16°–30° 30°≤	1 1.1 1.2
2	Soil type	A.F 2> A.F 2–4 A.F 4< A.F = Amplification Factor	1.2 1.3 2
3	Building height	One story Two story	1 1.3
4	Plan		1- 1.1
5	Construction quality	Good Average Poor	0.6 0.8 1
6	Construction system	Steel frame with brace Reinforced concrete frame, with brace Brick load-bearing wall Cement block load-bearing wall	1 2 4 3.5
7	Floor construction system	Steel beam with arched ceiling Wood beam and clay Reinforced concrete Beam with hollow block Scaffold with metal sheet	3 4 3 1
8	District population density	Less than 70 in hectares 7–150 " " 150–250 " " 250 and more	1 1.2 1.4 1.7
9	Texture complexity	– Open and regular with appropriate access – Open and regular with inappropriate access, organic open with appropriate access – Open, organic with inappropriate access, regular compact with appropriate access – Organic compact with appropriate access – Regular compact with inappropriate access – Organic compact with inappropriate access	0.6 0.8 1 1.2 1.6 1.8

- Using triangular anchors to prevent balconies from overturning.
- Frame base in the walls design provide both vertical and lateral support.
- Resistance components to prevent the top and bottom of the columns from rotating (Figs. 7.28, 7.29, 7.30, 7.31, 7.32, 7.33, and 7.34).

Effects of façade setbacks cannot be predicted by normal code equivalent-static analyses (second from top). Avoid changes of stillness with height problems with

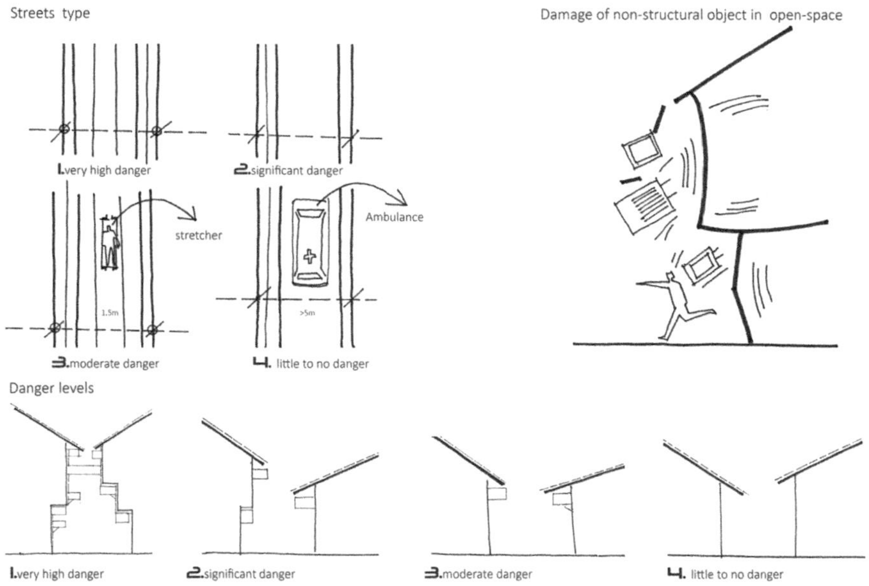

Fig. 7.24 Non-structural elements, such as coolers, signs, planters, even some kind of loose building materials on building facades may fall and aside from blocking the pathways, could also cause injuries or even death

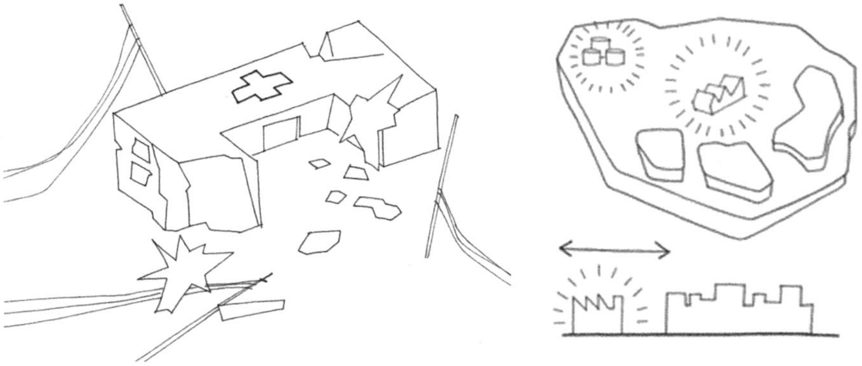

Fig. 7.25 Special attention should be paid to the location and resiliency of these activities because they may not only being damaged in case of an earthquake but also cause damage to adjacent buildings and facilities

analysis and detailing (two below). Remarks as above soft story demonstrably vulnerable (two below). Simple rules for vertical frames in aseismic buildings (two below).

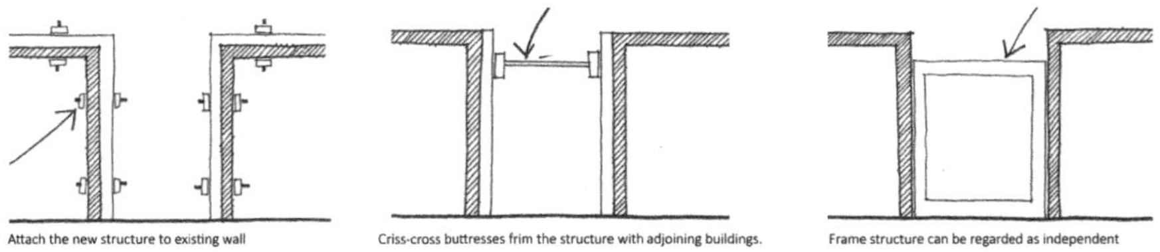

Fig. 7.26 Strengthening access in the historical sites, where no other solution is possible

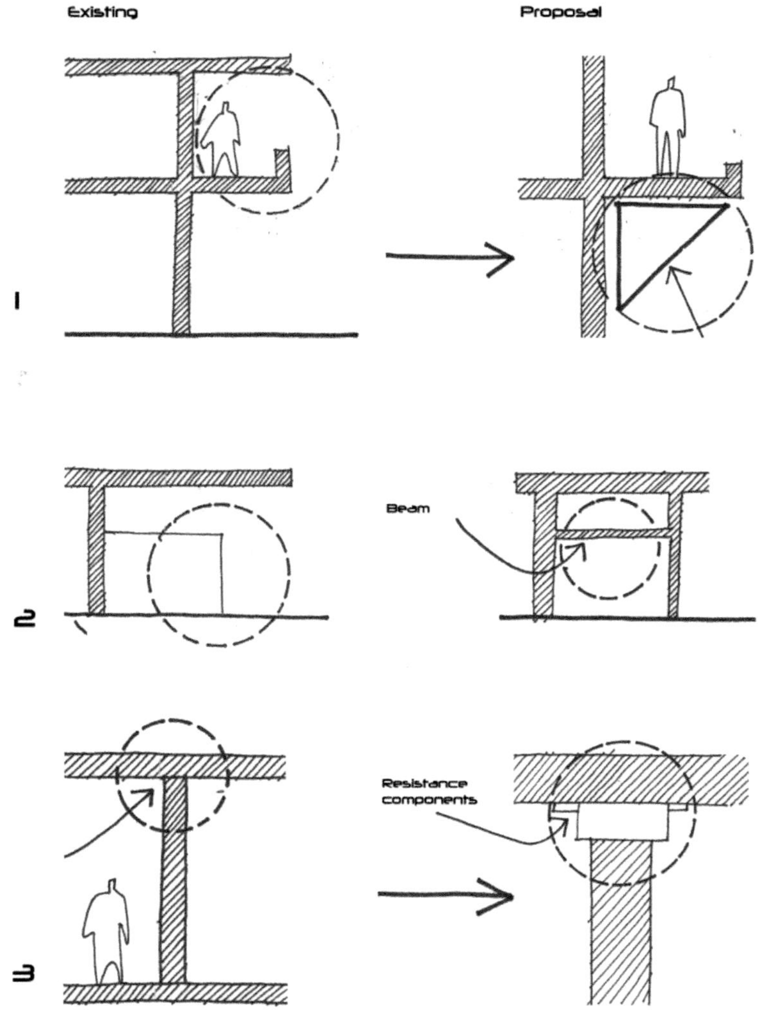

Fig. 7.27 Cracking and overturning of masonry parapets, large cantilever cornices, and balconies

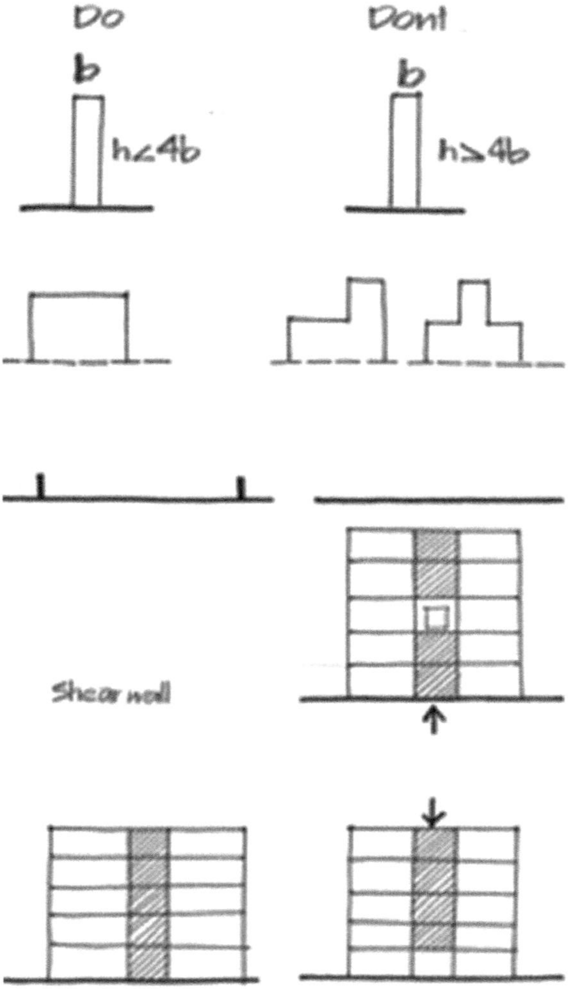

Fig. 7.28 Very slender buildings have excessive horizontal deflections (top)

7.14 Urban Design Quality and Seismic Risk Reduction

The overall purpose of urban design is to promote the quality of human settlements by using integrative rules and principles (see Bahrainy and Bakhtiar 2016). However, these rules and principles might not be complementary, but rather contradictory. In other words, disaster mitigation considerations may work against the quality issues in certain historical sites, where complexity of the built-up areas, narrow and winding alleys, and old buildings are valuable assets to these sites, but in the mean time they are quite vulnerable to earthquake. The situation becomes more critical when we realize that most of the historical cities around the world are located on the "seismic belt" with higher possibility of strong seismic activities. What is important from an

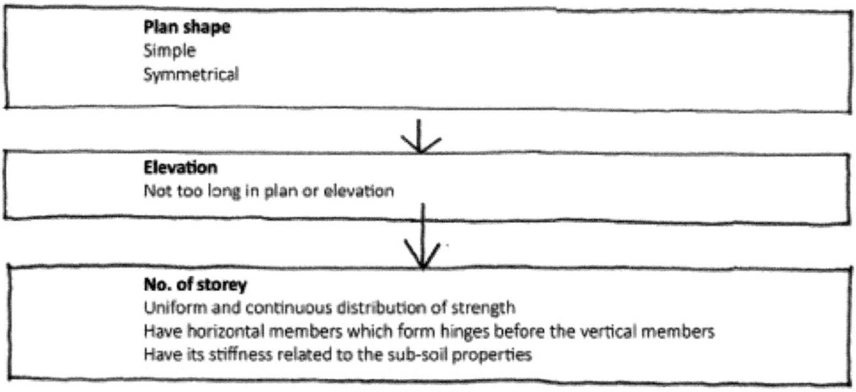

Fig. 7.29 Recommendations on plan shape, elevation, and number of story to reduce seismic risk

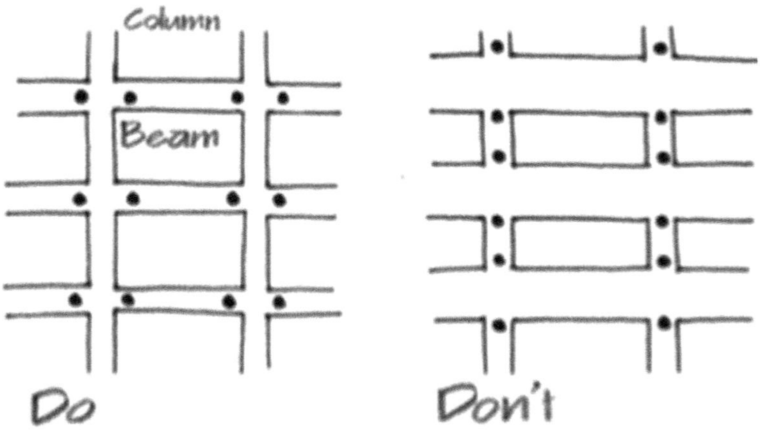

Fig. 7.30 Beams should fail before columns to maximize energy absorption before collapse Simple rule for relationship between column and beam strengths in aseismic frames

urban design point of view is to create a balance between the disaster-related criteria and the quality-based criteria. Following sets of criteria are suggested to be included in the urban design decisions in the seismic-prone regions (Figs. 7.35, 7.36, and 7.37).

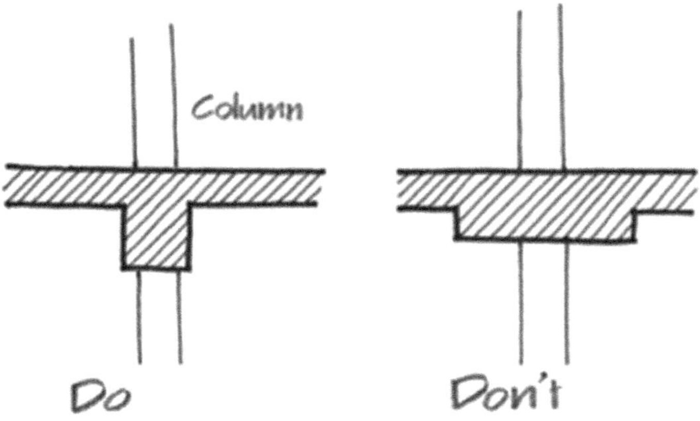

Fig. 7.31 Width of beam should not greatly exceed supporting columns continuity problems. Simple rule for width of beams and columns in aseismic rein forced concrete buildings

7.14.1 The General Design Concept in Relation to Reducing Seismic Risk

- Providing multi-function spaces, such as exhibitions, religious spaces, and small squares.
- Proposing main open spaces to be used during the disaster
- The main corridors to be used by the vehicular traffic during disaster
- Limiting the density to two stories along the passages with high vulnerability.
- Assigning crises-related land uses around the proposed open spaces, such as police and relief stations, mosques, pharmacies, and multi-functional spaces.
- Connection possibility between three centers.
- Providing multi-functional spaces around open spaces makes it possible to change their activities to disaster-related activities, such as rescue and relief and essential needs, during disaster.
- Vulnerable zones for fire need to be known and modifications made to make them safe.
- The area facing serious difficulties for accessibility during disaster because of blocked passages.
- Improvement of the passages and main pedestrian paths in the *Bazaar* area to promote spatial cohesion.
- Improvement and modification of passages to provide accessibility to the area during a disaster.
- Creating and improving of fringe roads to provide vehicular access to the warehouses and parking lots and secure the safety of local pedestrian passages.

Fig. 7.32 Strengthening, integrating, and reinforcing passageway walls to make accessibility possible during disaster and after

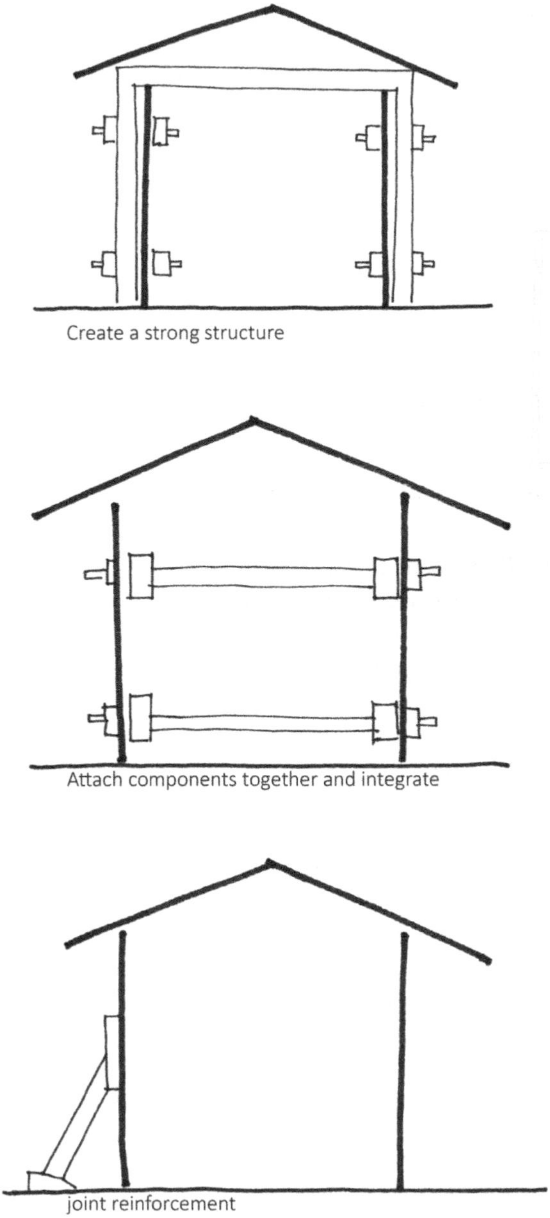

7.14.2 Some Design Examples, Based on Seismic Risk Reduction Goal

See Figs. 7.38, 7.39, 7.40, 7.41, 7.42, 7.43, and 7.44.

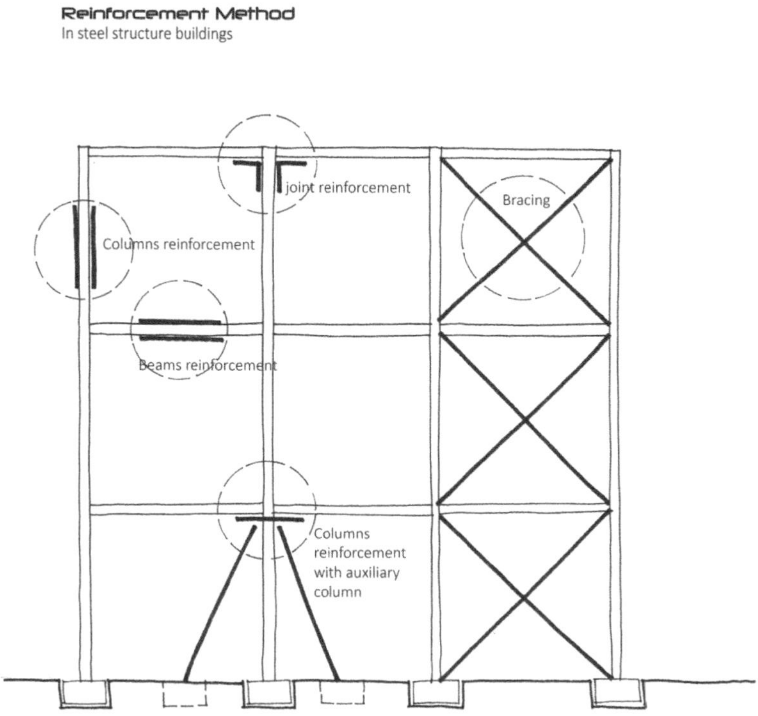

Fig. 7.33 Reinforcement method in steel structure: joints, columns, beams, and bracing

7.15 Vulnerability of Urban Textures

Each urban texture, based on its characteristics, has its potentials, as well as shortcomings, for the life and activities of human society. With regard to natural disasters, each will show different reactions. In the study of city's reaction against earthquake, the role of urban texture may be investigated in the following stages:

- Stage one: during earthquake and vulnerability due to earthquake and its consequences.
- Stage two: escape and refuge of residents to a safe place, during and after earthquake.
- Stage three: evacuation and rescue and relief operations.
- Stage four: temporary settlement of residents and assigning land uses.
- Stage five: clearance and reconstruction.

Following factors play significant role in the efficiency of urban texture—physical vulnerability, life loss, and post-disaster activities—against an earthquake:

- Building characteristics: building height, composition and form, building materials and structure, age, etc.

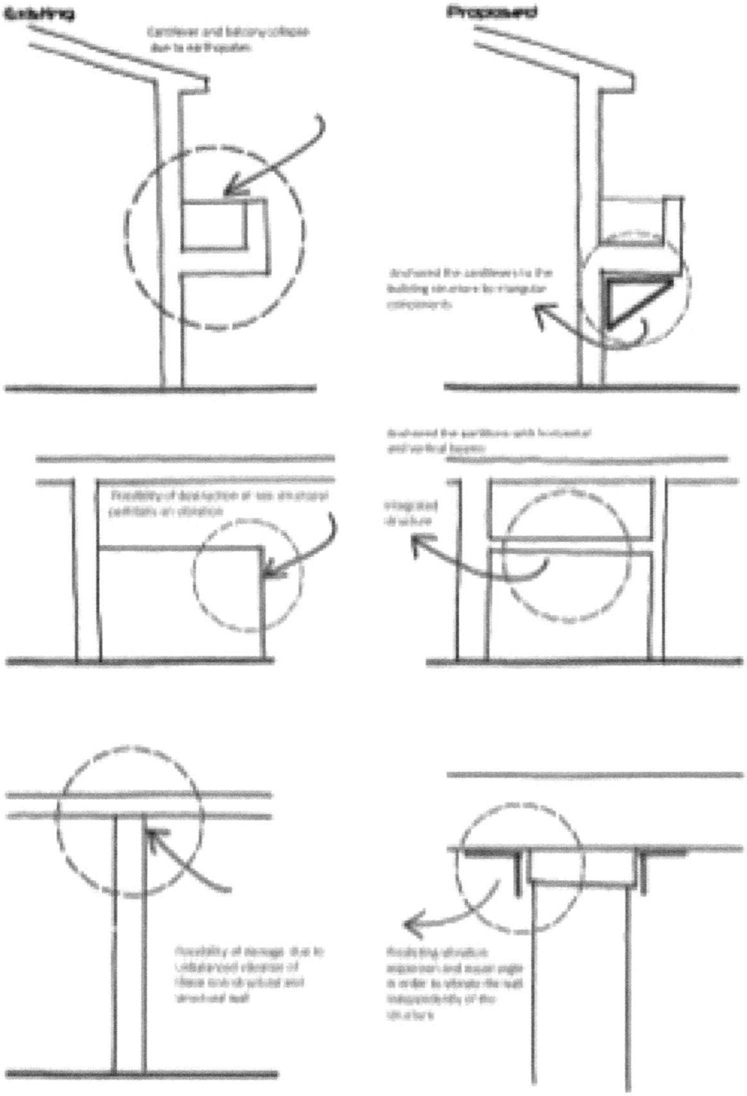

Fig. 7.34 Reducing potential risk through applying reinforcing techniques in vulnerable elements in steel structures

- Texture characteristics: land division, order and size, mass and void, compactness, continuity, road network, pattern, and size.
- Natural setting: topography, slope, natural elements, vegetation, and soil.
- Uses patterns: the period texture is used, density and intensity, the continuity of use, land-use type.

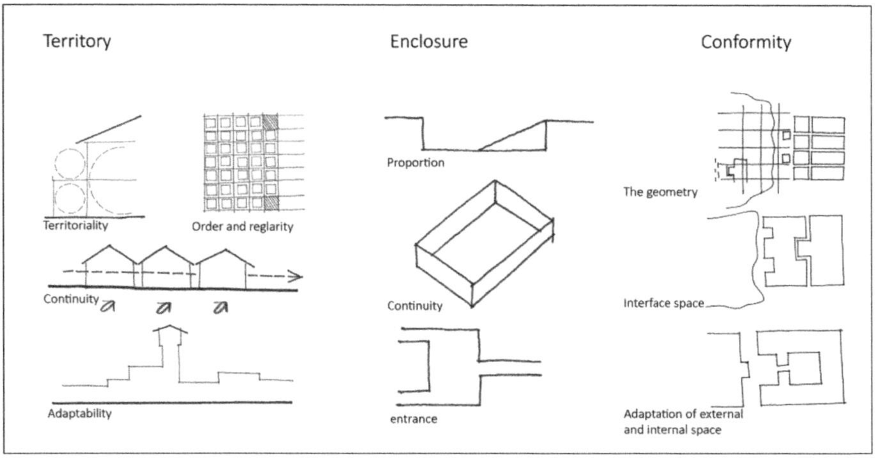

Fig. 7.35 Design ideas on territory, enclosure, and conformity to be applied, along with aseismic considerations, in the design/redesign of the seismic-prone regions

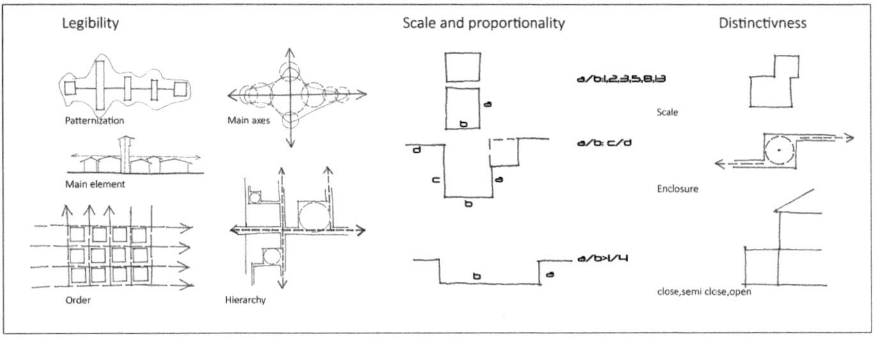

Fig. 7.36 Design ideas on legibility, scale and proportionality, and distinctiveness to be applied, along with aseismic considerations, in the design/redesign of the seismic-prone regions

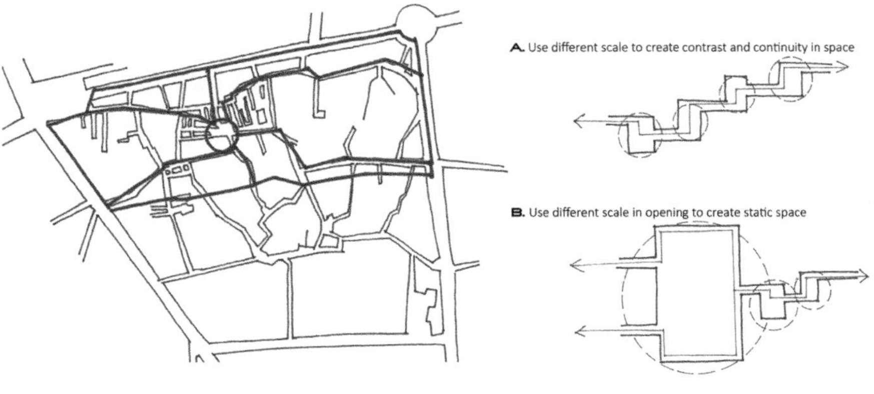

Fig. 7.37 General design ideas to create contrast, continuity, and stability in the city center

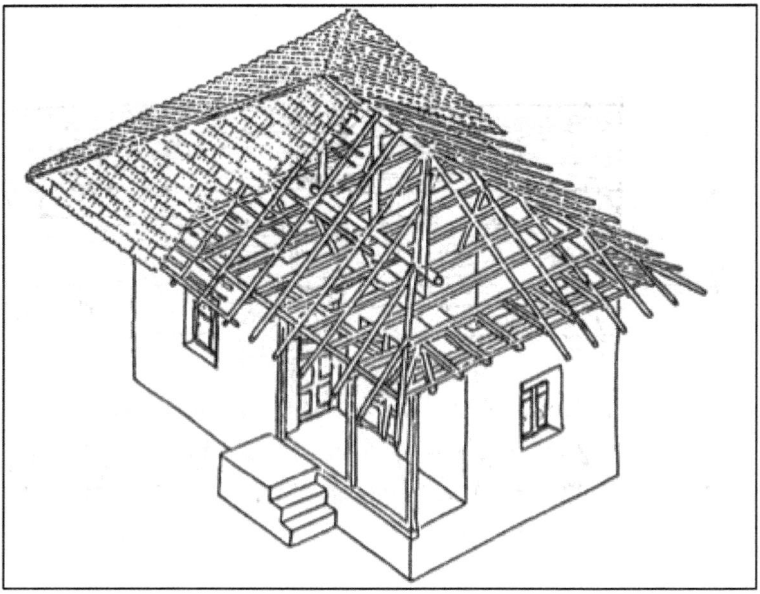

Fig. 7.38 A typical house along the Caspian shores (mud or brick wall, slope tile ceiling). The most seismic risk-resistant house

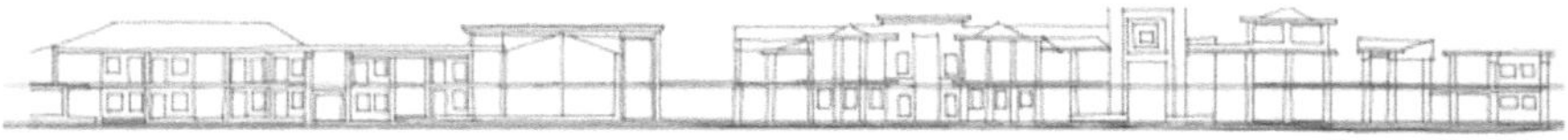

Fig. 7.39 Recommended design idea in a seismic-prone region

The role each element plays in vulnerability of buildings, and urban textures in different stages are shown in graphics to be easily understood and may be applied in practice (Hamidi et al. 1998).

7.15.1 Stage One: During the Earthquake, Vulnerability, and Its Consequences

The composition of mass and void, shape and size of the land, open spaces of adjacent lots, the adjacent buildings with open urban spaces (passages), balanced architecture, and the distance between buildings are analyzed bellow (Figs. 7.45 and 7.46):

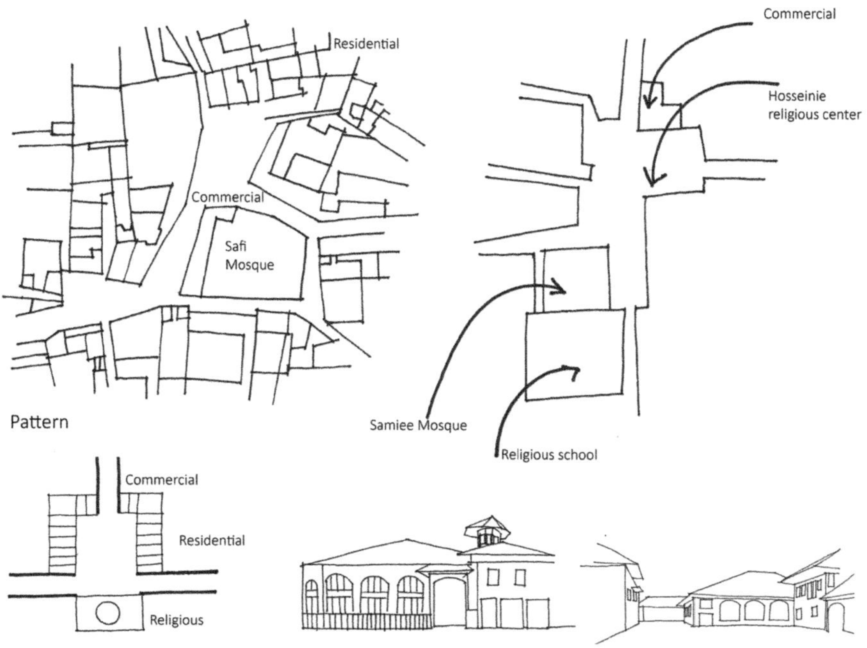

Fig. 7.40 Design idea for a neighborhood center in Rasht, based on seismic risk reduction goal

7.15.2 Stage Two: During Escape and Refuge

Building type, location of building on the lot, size and dimension of open space (court yard), building height (no. of story), use intensity, proximity of building with passages, direct access to passage, entrance obstacles to open space, the type of adjacent open space to be used for refuge, and proximity to green areas are described here (Figs. 7.47 and 7.48).

7.15.3 Stage Three: Evacuation, Rescue, Relief, and First Aid

Number and diversity of access to the lot(s) from the passage (pedestrian and vehicular), access to the open space from the lot, number of open spaces in the vicinity of the lot, proximity of building with passage, composition of mass and void, intensity and duration of space use, topography and slope, and dense vegetation will be analyzed bellow (Figs. 7.49 and 7.50).

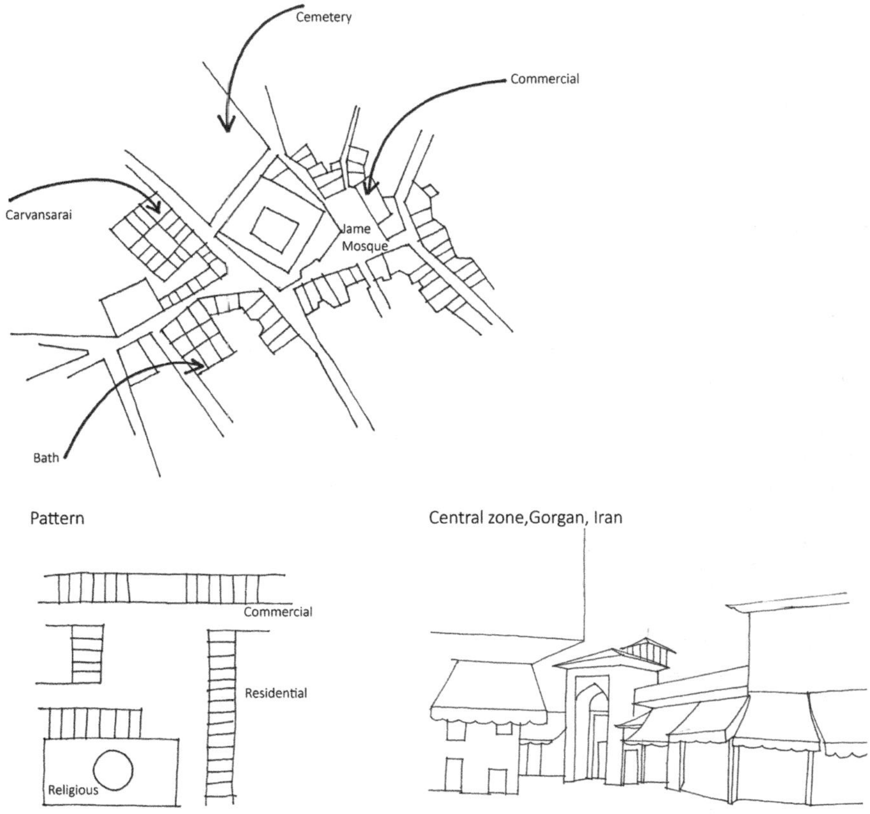

Fig. 7.41 Design idea for a city center in Gorgan, based on seismic risk reduction goal

7.15.4 Stage Four: Temporary Settlement

Construction characteristics within the lot, building type, building density, landscaping, and lot enclosure are the elements analyzed bellow (Fig. 7.51).

7.15.5 Stage Five: Clearance and Reconstruction

Size and dimensions of open space, building connection with passage, passage dimensions, land topography, and type of vegetation (Fig. 7.52).

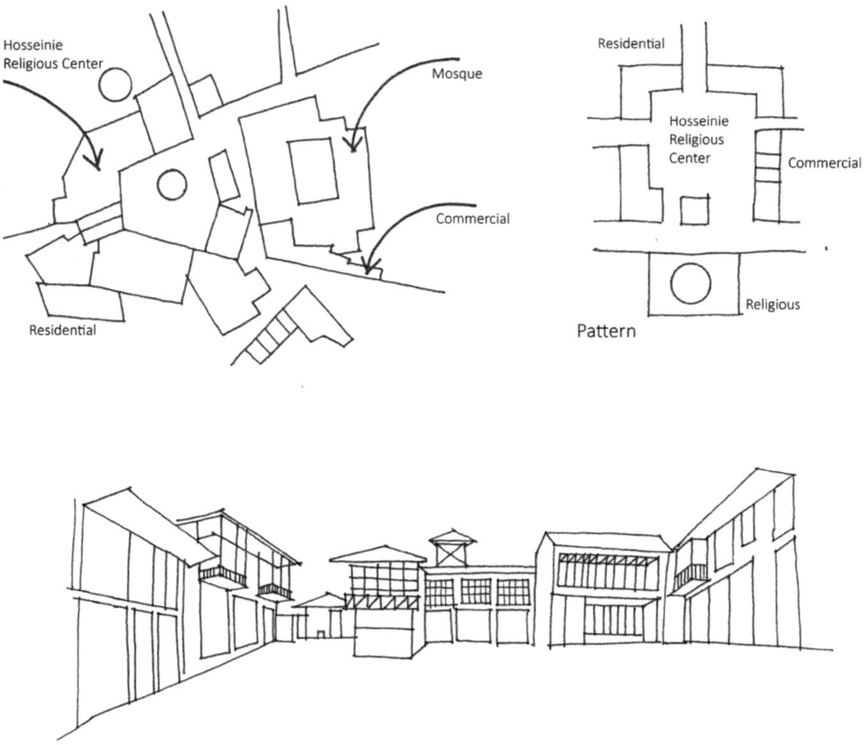

Fig. 7.42 Design idea for a neighborhood center, based on seismic risk reduction goal

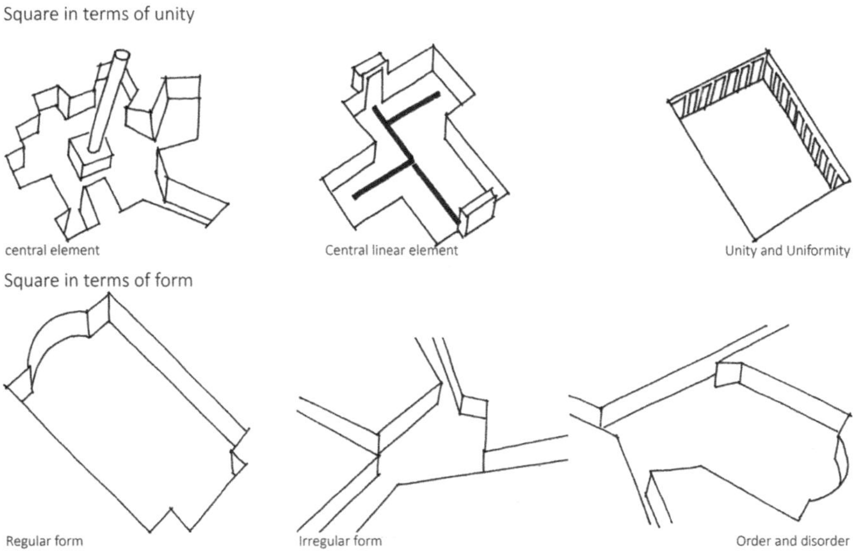

Fig. 7.43 Examples of square design, based on various principles

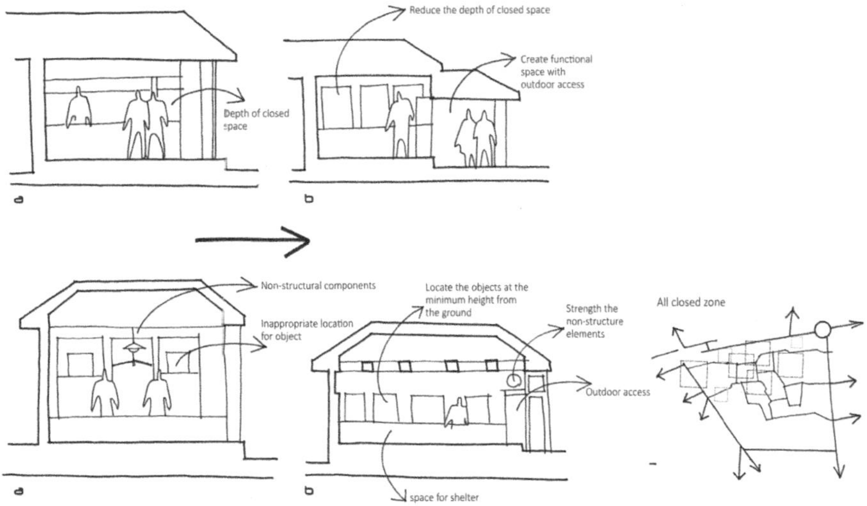

Fig. 7.44 Example of physical modification to lower seismic risk in a residential building

7.16 Conclusion

Here, in this chapter some guidelines are visually illustrated to help users better understand and apply the proposed ideas in practice. Information needed for the city scale design in the seismic-prone regions is divided into two groups: basic information (such as seismic zonation, geological condition, liquefaction). Deciding the right location for any specific urban development project could prevent a series of disasters in the future. Seismic zonation and especially microzonation is a required tool for this purpose. The second group is information needed for planning and design purposes (land uses, construction quality, building materials, etc.).

General principles suggested to reduce seismic risk include System redundancy, restorability, serviceability, and evacuation potential, etc. Specific principles include shorter distances between different land uses, network hierarchy, shorter distances between critical land uses during disaster, fast access to vital land uses, more possible control, more possibility of escape and evacuation through the network, adequate speed, shorter length and more safety of the network, higher ratio of open space over built-up areas, and more independent network for each district, etc.

Assessing the vulnerability of the built-up area is the first step for the analysis and risk reduction. Efforts should be made to decentralize all public services throughout the city and in the neighborhoods. Since fire is the major secondary factor threatening affected areas, therefore protecting buildings and public spaces from fire is necessary and at the same time preparing for evacuation in case of a disaster is also required.

Preparing earthquake scenario to include probable time of earthquake is the first step, on its basis which other plans, such as evacuation of different age groups and handicaps, and also, some structural modifications will be formulated. Secondary

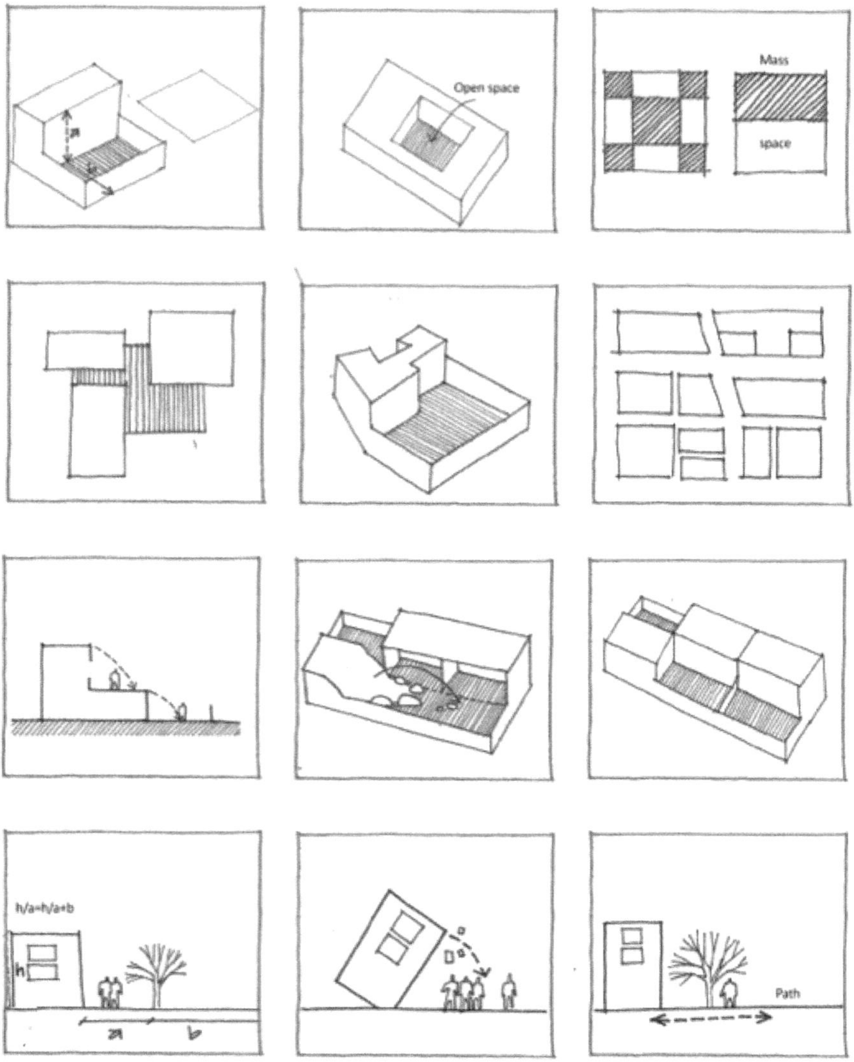

Fig. 7.45 The characteristics of buildings and urban texture in seismic vulnerability

damages, such as fire, explosions, flooding, subsidence, landslide, psychological trauma of shake, disorientation, uncertainty, and grief, are all concerned issues that are presented graphically to show the risks involved when disasters are not dealt with in the right way.

There should be an optimum distribution of activity centers, while avoiding locating sensitive public buildings in the areas with potential risk.

Due to the critical role roads and open spaces play during and after a disaster, some guidelines are presented here for roads and open spaces.

Fig. 7.46 Characteristics of buildings and urban texture in seismic vulnerability

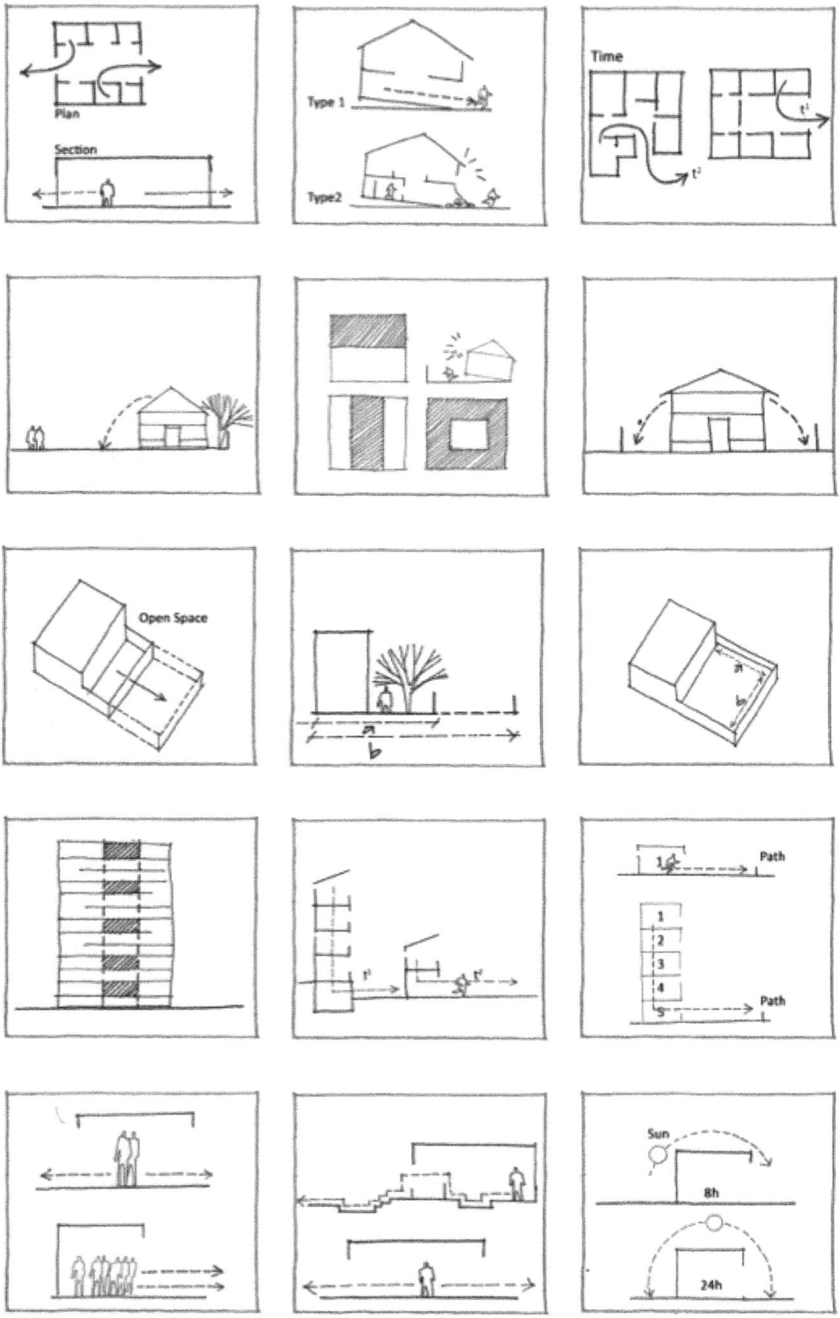

Fig. 7.47 Building and road characteristics during escape

Fig. 7.48 Building and road characteristics during escape and refuge

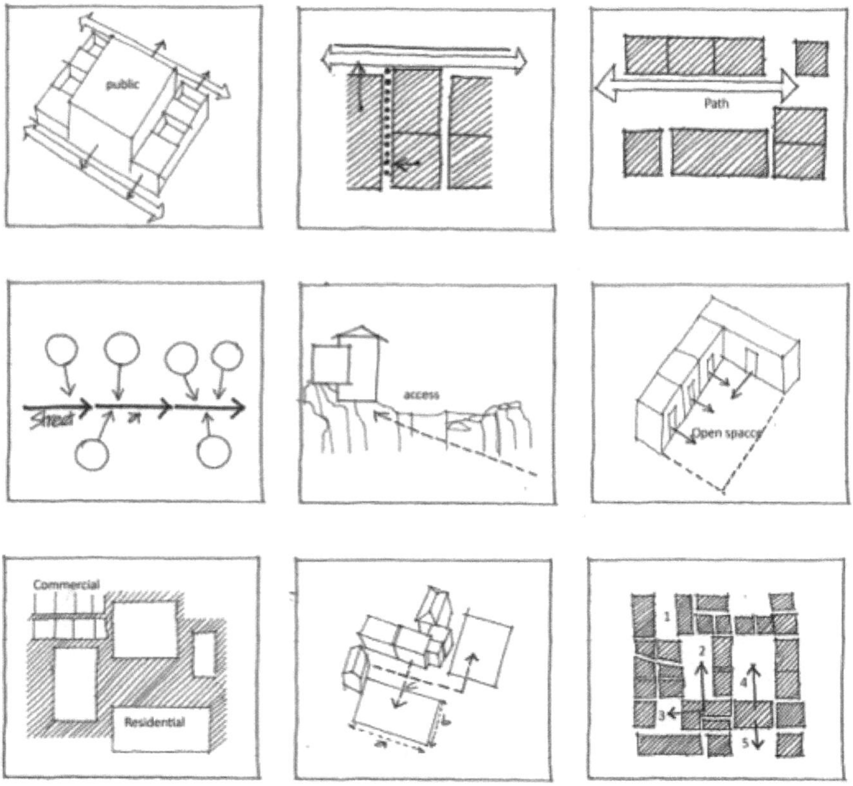

Fig. 7.49 Building configure characteristics of urban texture and use patterns during rescue and relief

Urban textures particularly in the old sections of cities are significantly vulnerable against seismic risk. Several measures may be taken to investigate the role of urban texture in reducing seismic risk:

The following general design concepts in relation to reducing seismic risk are presented here:

- Providing multi-function spaces.
- Proposing main open spaces to be used during the disaster.
- Assign main corridors to be used by the vehicular traffic during disaster.
- Limiting the density to two stories along the passages with high vulnerability.
- Assigning crises-related land uses around the proposed open spaces.
- Connection possibility between the centers.
- Providing multi-function spaces around open spaces makes it possible to change their activities to disaster-related activities.
- Vulnerable zones for fire need to be known and modifications made to make them safe.

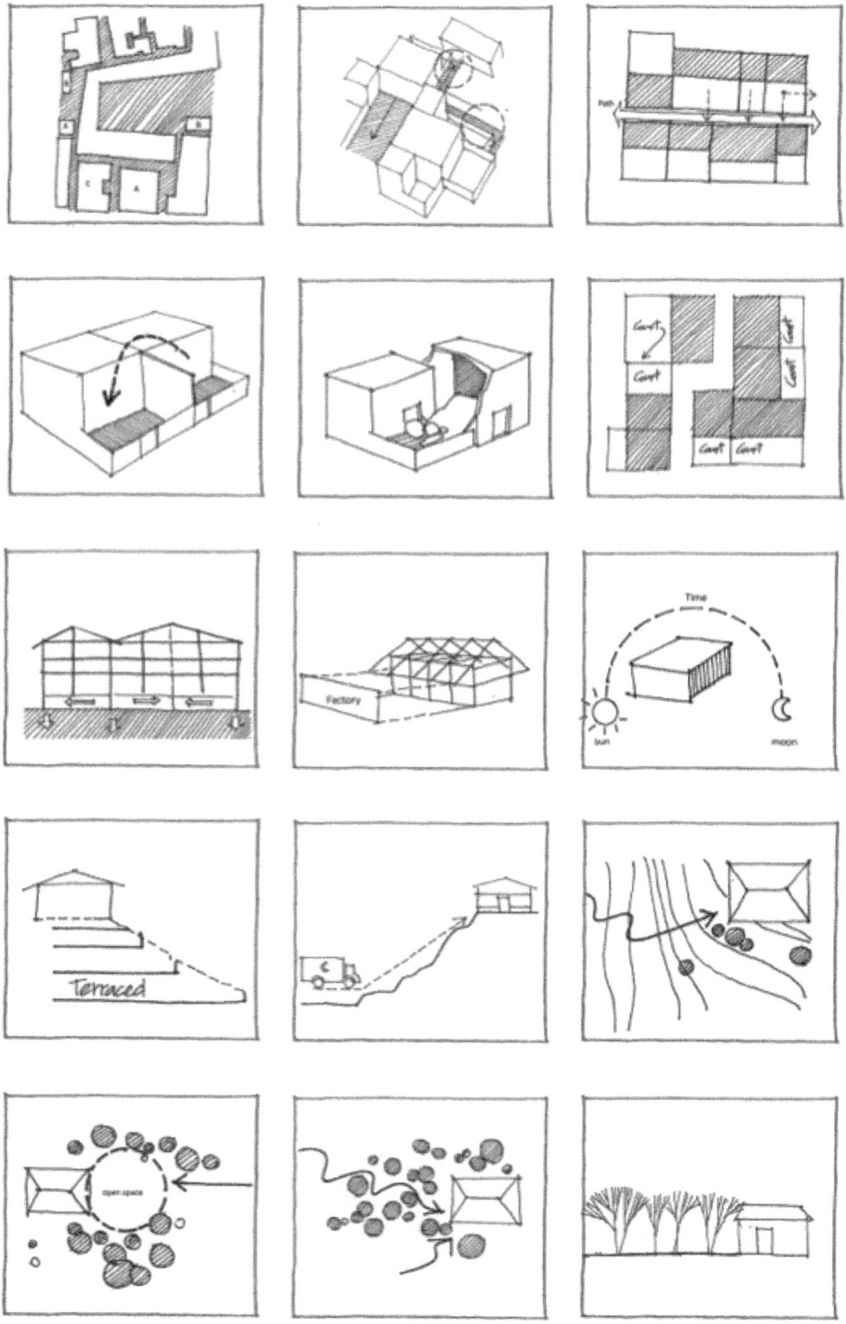

Fig. 7.50 The characteristics of urban texture and use patterns during rescue and relief, path, open space access roads

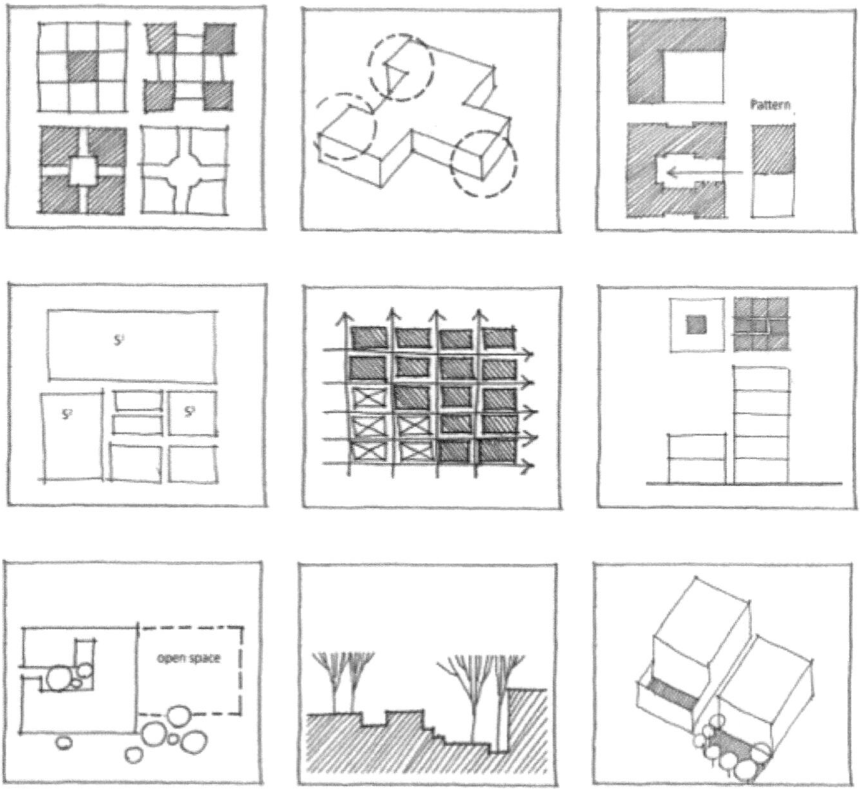

Fig. 7.51 Urban texture characteristics during temporary settlement

- Improvement of the passages and main pedestrian paths in the *Bazaar* area.
- Improvement and modification of passages.
- Creating and improving of fringe roads to provide vehicular access to the warehouses and parking lots and secure the safety of local pedestrian passages.

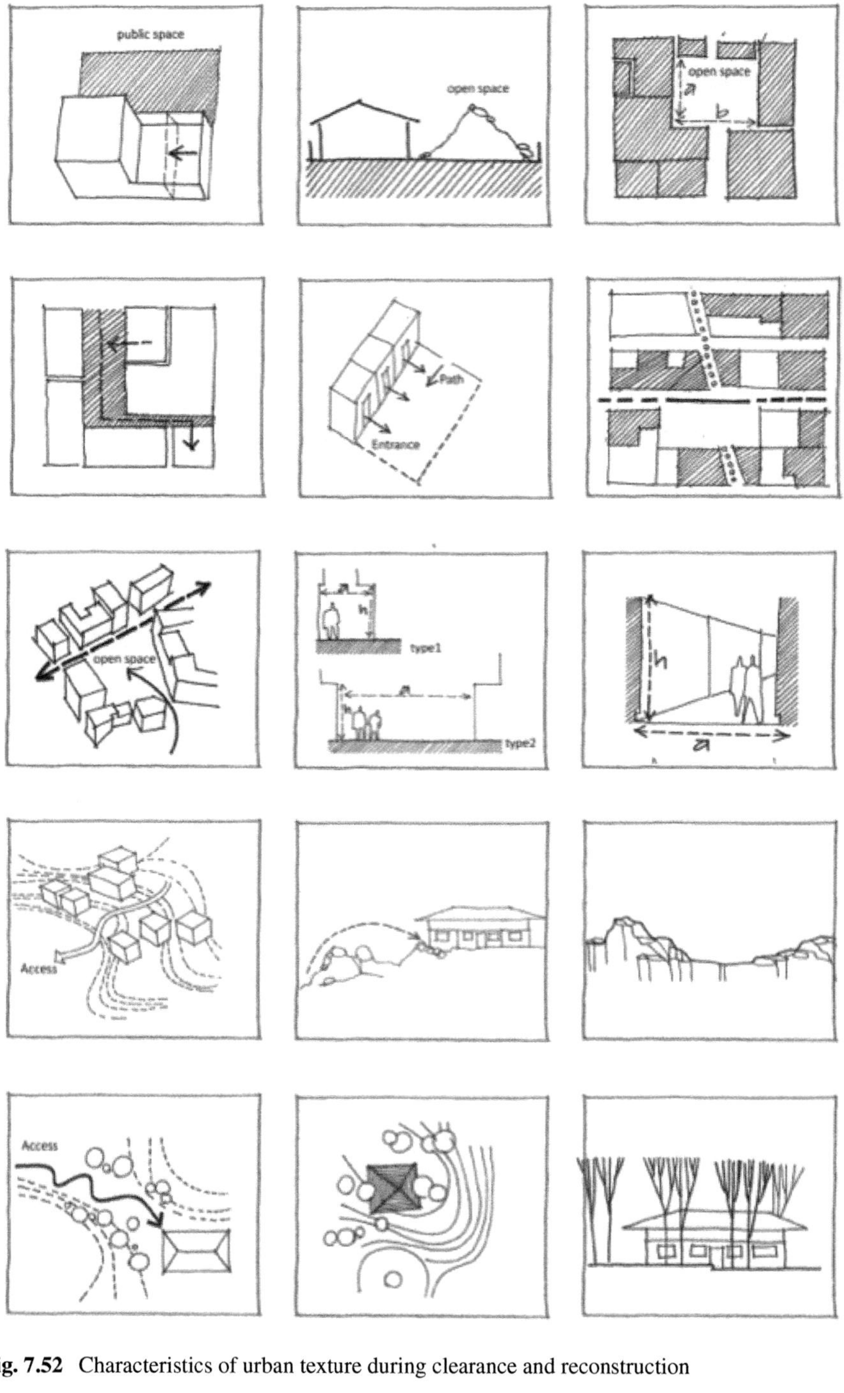

Fig. 7.52 Characteristics of urban texture during clearance and reconstruction

References

Bahrainy H, Bakhtiar A (2016) Toward an integrative theory of urban design. Springer
Bahrainy H (1998). Urban planning and design in seismic-prone region (the case of Rasht in Northern Iran). Am Soc Civ Eng Urban Plann Dev Div 124(4)
Bahrainy H (2005) What can urban design offer to make university campuses safer? In: Proc. of a symposium on Best practices in risk reduction for colleges and universities. University of Washington, Office of Emergency Management
Hamidi M et al (1998) Settlement size, form and pattern for earthquake hazard mitigation, National report, UNDP-UNCHS (Habitat) Project IRA/90/004, Housing Foundation of I.R.NDPC (translated in English), Tehran, July
SEAO (1980) Recommended lateral force requirements and commentary. Structural Engineering Association of California, San Francisco, p 1975

Chapter 8
Conclusion

Abstract The global number of yearly reported natural disasters has quadrupled since the 1960s, reaching more than 400 events and 200,000 affected people per year in the 2010s. Probably the most important factor is the increasing exposure of people to natural hazards, caused by human development patterns such as rapid urban growth and rising social inequalities, especially in the developing world. Moreover, as urbanization gathers pace, cities have become 'hot spots' for disasters. Our premise in this book is that urban design can play a significant and effective role in seismic risk reduction in urban areas. The June 20, 1990, *Manjil* Earthquake in North–West Iran was the most damaging event to date in the country. This experience and the valuable lessons learned provided an unprecedented context for research in urban design and related fields. On the basis of its findings, urban design recommendations are made for a region, city, city center and access networks and open spaces. Manipulating the physical environment encompasses a broad set of activities in a wide range of physical settings. Manipulation may include direct physical design and construction, regulatory measures to guide physical changes over time and economic and community development efforts. Some guidelines are visually illustrated to help users easily understand and apply the ideas in practice. At the end, some general principles such as system redundancy, restorability, serviceability and evacuation potentials are introduced to be applied in seismic-prone areas by urban designers.

Keywords Urban design · Seismic risk · Built environment · Environmental safety · Past experiences · Opportunities · Commonly used terms · June 20 · 1990 *Manjil* Earthquake · Deficiencies · Design ideas at three scales

The book starts with this very basic question of can urban design play any role in reducing seismic risk in the built environment. Then based on an accepted definition of urban design, which is the purposeful decisions and actions which aim at establishing functional and formal order in the physical environment, continues to claim that today's interpretation of order, however, has changed drastically, because the urban environment has gained unprecedented complexity. Our premise in this book is that urban design as a legitimate discipline and practice today depends on its ability and capacity to deal with the current issues society is facing and to offer

H. Bahrainy and A. Bakhtiar, *Urban Design in Seismic-Prone Regions*,
The Urban Book Series, https://doi.org/10.1007/978-3-031-08321-1_8

appropriate solutions to those problems, to the extent plausible and within the framework of the urban design tools and techniques. The global number of yearly reported natural disasters has been exponentially increasing due to the fact that the increasing number of people is exposed to natural disasters, which will result in lack of order in the built environment—urban form and space and urban activities. Urban design techniques and methods can be used to make changes and modifications in the built environment in order to establish desired order in the urban form and activities and provide cities with safety. Vulnerability and seismic risk analysis in urban design intend to mitigate seismic risk in cities.

A study on the quality of life indicators revealed that environmental safety is one of the major components of the quality of life in urban areas and the third critical element of the basic needs of any human settlement. Protecting human settlements against natural disasters is, therefore, a major requirement for achieving sustainability and resiliency in our urbanizing world.

By breaking down the disaster mitigation activities into active and passive activities, experience shows that developing countries rely heavily on passive activities to deal with problems in the post-disaster conditions of the affected areas. The type of solutions is presented here in this book are mostly preventive which will increase the prepared capacity and resiliency of the communities at risk.

Several experiences are reviewed here in this chapter to learn from the lessons gained, particularly with regard to the role urban design can play to reduce seismic risk in cities. They are The Kobe earthquake of 1995, *Ferdows* earthquake of 1968, and *Bam* earthquake of 2003. The *Ferdows* and *Bam* cases were discussed here to explain the important issue of risk management approach to the preservation of cultural heritage. Hurricane Katrina, although not a seismic disaster, but discussed here because of the critical role urban design could have played to prevent a disaster of that magnitude.

And finally, it is believed that natural disasters, in spite of the catastrophic impacts inflict on life and properties of communities, will provide golden opportunities to solve the chronic problems communities have been facing for a long time and local authorities have not been able to deal with them. They can take advantage of this event to rebuild the community on the basis of seismic-resistant guidelines and other desired goals.

A brief dentition and description of the commonly used terms and concepts in disaster, and particularly seismic risk management, are given in Chap. 2. Life losses and material damages of disasters may be attributed to three groups of elements and their mutual interactions: geophysical hazards, climatologically hazards, and biological hazards. These will affect the following elements: human systems, and building and construction systems. Some of the significant terms explained here in this chapter are liquefaction, landslide, rockfalls, tsunamis, acceleration lines, seismic zonation, mesoseismal, fault, active (capable) fault, disaster, hazards, disaster mitigation, exposure, vulnerability, preparedness, strengthening and retrofitting, epicenter, earthquake scenarios, intensity, intensity scale, isoseismal map, peak velocity, guidelines for earthquake disaster management, secondary damages, evacuation, and prevention.

The June 20, 1990, *Manjil* Earthquake in northwest Iran, which was the most damaging event to date in the country, was taken as a case study to analyze the impact of the earthquake on the built environment and look for ways and means to improve the conditions of the affected area in order to protect the area against future earthquakes. *Manjil-Rudbar* Earthquake, with magnitude Ms = 7.4 and estimated epicentral intensity of IX-X degrees of MM Scale, caused 13,911 deaths, 36,693 people treated for injuries, and over 8,000 people badly injured. It was strongly felt over an area of 600,000 km^2, including Tehran, Tabriz, and several other major cities in Iran.

First the spatial–physical conditions of the affected area were reviewed, focusing on the spatial structure and regional distribution of settlements. The natural condition has dictated the location and functioning of the towns, which has been resulted in linear situation of *Loshan*, *Manjil*, *Rudbar*, and *Rostamabad* towns along the extension of Tehran–Rasht highway. Rasht has acquired a central position for all those towns.

Social characteristics of the area include long history of settlements, tribal culture, religious and cultural variations, high population density in plains and lower density in mountains, wide differences in settlement population in number of plains in comparison with mountains. The main economic characteristics include dominance of agro-economics in plains, importance of industrial economy in industrial belts and centers, dominance of herding economy in mountains and economic variation of different parts of the region.

Landslides, rockfalls, and liquefaction are three of the most dramatic events that occurred because of this earthquake. One of the landslides triggered was the *Fatalak*. Liquefaction occurred on an area of 650 km^2 of the Caspian plain with north–east oriented strip from *Sangar* on the west almost to the cost of Caspian Sea. It caused extensive damage to the buildings in the districts of *Astaneh* and *Rudbaneh*, agricultural land and farms in the wider region as well as extensive deformations and damage to the regional and urban infrastructure.

The principal elements of the uniform methodology and procedure for post-earthquake damage evaluation are:

- Damage and usability classifications for buildings.
- Procedures for and organization of data collection.
- Earthquake damage data analysis and data bank organization.
- Estimation of economic losses, and human fatalities and injuries.
- Measures for reducing adverse earthquake consequences and for mitigating seismic risk.

Connected with these principal elements are earthquake damage evaluation, development of empirical vulnerability, or damage cost functions, and analysis of earthquake damage distribution.

Structural damages recorded as a result of June 20, 1990, earthquake may be classified as follows:

- Foundation failures.
- Cracks in walls.
- Wall separation at corners.
- Cracks around openings.
- Sliding of roof over walls.
- Collapse of walls.
- Collapse of buildings.

The following deficiencies were found in the design, detailing, and specifications:

- Wrong positioning of door/window opening.
- Large size openings separated by walls that are too slender.
- Insufficient wall thickness.
- Poor connection details.
- Columns that are not tied.
- Roof that is not anchored to the wall.
- Absence of bracings and tie beams.
- Poor detailing of connections.

Case histories which formed the basis of the observations were insufficient wall thickness, poor connection details, columns not tied, roof not anchored to wall, inadequate bracings, the absence of tie beams, and defective materials. Aside from these building deficiencies, major deficiencies were also found in the public form and space of the overall built environment, such as lack of public space, narrow alleys, high building density, lack of hierarchy in the transportation network, and lack of easy and safe evacuation.

Following a review of the lessons learned from the Manjil earthquake of June 20, 1990, and also the deficiencies found in construction activities and an analysis of seismic hazard a set of design criteria are formulated to protect buildings and built environment against earthquake. On the basis of these criteria, a series of essential guidelines are presented to be followed during the modification of existing areas, reconstruction of the damaged areas, or the design of new settlements.

A research project entitled: "Assistance in the Implementation of a Post-earthquake Rehabilitation Program" was defined, to not only contribute to the reconstruction of the seismic-affected areas, but also to develop a set of guidelines for earthquake risk reduction to be applied in possible seismic-prone areas.

In Chap. 4, urban design recommendations will be made for a region and city which are prone to seismic risk. Following chapters will deal with urban design solutions for the city center, Bazaar area, and access network and open spaces in a city. The city of *Rasht* will be used as example. Information needed for the regional and city scale study and design in a seismic-prone region include seismic and geological map and data, hazard assessment, risk and vulnerability assessment, built environment data, earthquake scenarios, urban and regional policies, an implementation guidelines and mechanisms.

The existing conditions in the cities are a result and a continuation of development planning processes which seldom account directly or indirectly for natural hazards. The favorable achievements of otherwise good planning can be, and often have been, wiped out in minutes because of failure to account for the disaster potential of natural hazards. Tendency to view such misfortunes as unavoidable tragedies, although to substantial degree in fact they are, is a testimony to poor planning that ignores that natural disasters are an integral feature of the colonized environment and that the extent of damages and losses caused are to a significant degree a function of decisions made (or not made) during the process of human settlements development.

Small scale is in fact the most effective and commonly used scale in which urban design, as an activity that manipulates physical environment to achieve certain goals, and particularly reducing seismic risks may be applied. Manipulating the physical environment encompasses a broad set of activities in a wide range of physical settings. "Manipulation" may include direct physical design and construction, regulatory measures to guide physical changes over time, economic and community development efforts, regional growth strategies, and institutional measures such as funding programs that ultimately result in physical actions.

Disasters cause formal and functional disorder in the built environment. Formal implies urban form and space and functional implies urban activities. Urban design, through its rules and regulations, intends to prevent disorder in case of a disaster. As discussed in chapter one of this book, one of the determinant factors of disorder or unsustainability of human settlements deals with basic needs, in which safety and security, play significant role.

To determine the vulnerability of the area and propose design ideas to reduce seismic risk, an analysis of the existing conditions is necessary, and this includes location of the area in the city, typology of the city structure (large scale), typology of the *Bazaar* structure, (macro-, medium, and microscales), area characteristics, and general patterns of the physical elements.

The existence of the old *Bazaar* and the concentration of uses and services in this area have led to draw residents from other parts of the city and therefore causing difficulties for the pedestrians as well as motor vehicles to use the area. Higher population density, compact, and fine grain texture and small plot sizes have intensified the problems. The dominant texture in the area is generally organic, and compact, with inadequate access into the area. Deteriorated buildings due to the old age, high density, and inadequate accessibility all have made the area the most vulnerable for all stages of the disaster management.

The main city corridors which are the same as the main *caravanserai* roads and roads for transporting goods and also canals for discharging surface water into the city rivers and pedestrians connection between neighborhoods and linear configuration of commercial spaces make the main structure of the city.

Four dominant types may be recognized in the structure of this area: central type (*caravanserai*), grid type, linear type and dense and integrated blocks. Risks involved in each type are fundamentally different.

Design ideas to reduce seismic risk at small scale begins with modifications in the city structure.

Rasht Bazaar which is the physical reflection of the commercial identity of the city of *Rasht* and plays a significant role at the regional scale, and it also plays its role as the only active and vibrant center of various commercial, religious, and to some extent administrative activities. Aside from securing the needs of residents of various parts of the city, it is also a place for exploring, choosing and purchasing goods, and destination for many daily trips for work, visit, walking, and leisure.

The process of vulnerability analysis and design with the goal of vulnerability reduction due to seismic risk at the small case includes the following phases:

- Recognition and determination of damaging and damaged elements.
- Preparing earthquake scenario and determining the vulnerability zoning on its bases of which design priorities could be known.
- Zoning.
- Specification—design implications.
- Design alternatives to mitigate vulnerability at the *Bazaar* scale.
- Reusing scenarios to evaluate design alternatives and choosing the optimum solution.
- Complete the selected alternative.
- Loss and damage estimate based on isoseismic map of the city center: (1) Very severe damage, (2) severe damage, (3) medium damage, (4) low damage.

Building vulnerability is determined on the basis of the following factors and using the specific guidelines: construction technique, roof structure, number of stories, construction quality, openings, overhangs, façade building materials; and also site characteristics such as soil, and land slope. As a general rule, the higher the number of population in a surface, an increase in loss life possible.

Vulnerability due to the lack of access to safe open spaces is a major concern. According to some criteria, the distance is run by a normal person would be 50 m in 30 s. But other factors such as damaged walls and population density would increase vulnerability.

Higher vertical density increases the chances of fire spreading, while more open spaces will decrease spreading fire, also difference of building heights in adjacent buildings increases the possibility of spreading fire from lower buildings to the higher ones, and more openings will increase the possibility of fire spreading.

Non-structural elements including signs, powerlines, sloping roof covers, consoles, façade materials, and utility elements such as coolers may cause life loss and injuries but also passages blockage.

As stated before, the city of Rasht and particularly its center and the *Bazaar* areas, in spite of vulnerability against seismic risk, have significant cultural, economic, and social values which deserve to be taken into account in any planning and design activity. One of the primary goals in any intervention in the city structure is to provide physical cohesion and unity, not only in the *Bazaar* area, but within the whole city. For practical applications, some operational objectives for roads and open spaces and also for the closed spaces are presented.

The measures introduced above to reduce seismic risk may be applied to a smaller (architectural) scale, and for this purpose, a central open space (square) is selected for analysis, on its basis of which design suggestions were made.

Open spaces and access networks, as the most significant elements of urban space, play a critical role in every phase of earthquake disasters, from escape, refuge, relief, rescue, first aid, emergency, and temporary settlements.

Urban space is defined here as the container of the average daily circuit which is composed of urban activity systems. The aggregate of the daily circuit of urban activity systems is the average of a person's daily, weekly, or yearly routine of activities in the urban environment. In normal situations, these activities include such things as recreational activities, shopping, driving to work, going to church, walking to school, visiting, etc. Repetitive patterns of such activities are called urban activity patterns, examples of which are shopping patterns, home-to-work commuting patterns, recreational patterns, and so on. When a disaster strikes these activities will be interrupted and urban spaces play quite different functions.

Application of seismic risk mitigation strategies in dealing with urban spaces in the event of a disaster can have more tangible results in smaller scales. Areas of the cities with high risk or vulnerability such as city cores, historical sections, mountain slopes, or areas with land instability potentials are examples for sector scale application of seismic risk mitigation. One should, however, bear in mind that changing the existing fabric of a city, particularly in the dense and traditional core of the cities, to secure the safety and well-being of residents, is a complicated and difficult task. Attempts should be made to make immediate and short-term modifications in such dense, built-up areas realistically and limited to the most crucial elements, i.e., those with significant positive consequences. This will ensure smooth implementation of recommendations at this level.

From an urban design point of view, project scale is the lowest possible level one can deal with, with short-term and specific objectives and easy-to-implement proposals. It could be applied to the old section of the Central Business District of traditional cities where the interwoven fabric, deteriorated structures, overpopulated passages, high density of buildings, and substandard safety against earthquake and fire make them the most vulnerable sections of the cities.

The goals of the area design to reduce vulnerability against seismic risk are as follows:

- The need for creating the main open space in the central core, to hold the commercial activities and, at the same time to reduce population density.
- The need for the modification and improvement of the main passages to provide accessibility during disaster.
- Creation or improvement of emergency access network and facilities in the vicinity to be used in case of disaster.
- Demolition and reconstruction of highly vulnerable buildings due to seismic risk.
- Creating open and semi-open spaces in order to connect closed spaces to make speedy evacuation during an earthquake possible.

- Change and modification of dangerous buildings, such as mosques and schools, with high density of population.
- Providing spaces with commercial activities to be flexible so that their uses could be changed in case of emergency.

Design ideas at the block scale to reduce seismic risk in the Bazaar area are presented in three alternatives. Specific solutions are also suggested to reduce seismic vulnerability. Selected cases are used for modification of passages.

Selected sites are analyzed on the basis of building height, roadblock due to building and wall destruction, vulnerability degree, and land uses, on their basis which appropriates open spaces are proposed.

Vulnerability of non-structural elements in the passages and open spaces and vulnerability against fire are also investigated, and necessary modifications are made.

What can urban design do to practically reduce the vulnerability of the built environment against seismic risk? In Chap. 7, some guidelines are visually illustrated to help users easily understand and apply the proposed ideas in practice. Information needed for the city scale design in the seismic-prone regions is divided into two groups: basic information (such as seismic zonation, geological condition, and liquefaction) Deciding the right location for any specific urban development project could prevent a series of disasters in the future. Seismic zonation and especially microzonation are a required tool for this purpose. The second group is information needed for planning and design purposes (land uses, construction quality, building materials, etc.).

General principles suggested to reduce seismic risk include system redundancy, restorability, serviceability, evacuation potential, etc., specific principles include shorter distances between different land uses, network hierarchy, shorter distances between critical land uses during disaster, fast access to important land uses, more possible control, more possibility of escape and evacuation through the network, adequate speed, shorter length and more safety of the network, higher ratio of open space over built-up areas, and more independent network for each district, etc.

Assessing the vulnerability of the built-up area is the first step for the analysis and risk reduction. Efforts should be made to decentralize all public services throughout the city and in the neighborhoods. There should be an optimum distribution of activities centers, while avoiding locating sensitive public buildings in the areas with potential risk.

Since fire is the major secondary factor threatening affected areas, therefore protecting buildings and public spaces from fire is necessary and at the same time preparing for evacuation in case of a disaster is also required.

Preparing earthquake scenario to include probable time of earthquake (although property damage might not change, but life loss and injuries depend to a large extent on the time earthquake occurs) is the first step, on its basis which other plans will be formulated, such as evacuation of different age groups and handicaps, and also some structural modifications. Secondary damages, such as fire, explosions, flooding, subsidence, landslide, psychological trauma of shake, disorientation, uncertainty, and grief are all concerned issues that are presented graphically to show the risks involved when disasters are not dealt with in the right way.

Urban design can be used as regularity measures to reduce seismic risk. Regulatory measures can direct their activities toward reducing seismic risk in human settlements and reduce life loss and property damage in case a disaster strikes. Most of urban design activities focus on regularity measures which include mitigation, preparedness, and emergency response and recover measures.

Due to the critical role roads and open spaces play during and after a disaster, some guidelines are presented here for roads and open spaces, as follows:

- Creating open space along with strengthening road walls.
- Provide access to open spaces as the first priority, creating open space as the second priority.
- Provide access to existing open spaces as the first priority, and roads improvement, as the second priority.
- Apply recommended practical guidelines for reconstruction of passages.
- Reducing potential risk of non-structural elements.
- Avoiding hazardous areas/activities.
- Structural modifications needed to reduce seismic risk in the vulnerable historical sites.

Urban textures particularly in the old sections of cities are significantly vulnerable against seismic risk. The following stages may be taken to investigate the role of urban texture in reducing seismic risk:

- Stage one: during earthquake and vulnerability due to earthquake and its consequences.
- Stage two: escape and refuge of residents to a safe place, during and after earthquake.
- Stage three: evacuation and rescue and relief operations.
- Stage four: temporary settlement of residents and assigning land uses.
- Stage five: clearance and reconstruction.

The following general design concepts in relation to reducing seismic risk are presented here:

- Providing multi-function spaces, such as exhibitions, religious spaces, and small squares.
- Proposing main open spaces to be used during the disaster.
- The main corridors to be used by the vehicular traffic during disaster.
- Limiting the density to two stories along the passages with high vulnerability.
- Assigning crises-related land-uses around the proposed open spaces, such as police and relief stations, mosques, pharmacies, and multi-functional spaces.
- Connection possibility between three centers.
- Providing multi-functional spaces around open spaces makes it possible to change their activities to disaster-related activities, such as rescue and relief and essential needs, during disaster.

- Vulnerable zones for fire need to be known and modifications made to make them safe.
- The area facing serious difficulties for accessibility during disaster because of blocked passages.
- Improvement of the passages and main pedestrian paths in the *Bazaar* area to promote spatial cohesion.
- Improvement and modification of passages to provide accessibility to the area during a disaster.
- Creating and improving of fringe roads to provide vehicular access to the warehouses and parking lots and secure the safety of local pedestrian passages.

And now let us turn to a different type of disaster which shook the world for almost 3 years (see Epilogue). We have been familiar with numerous kinds of natural disasters and the ways and means to deal with each group of them in the past. Most well known of them are: geological and hydrometeorological hazards. Although biological hazards have also been with humans all along time under endemics, the recent global spread of COVID-19 and its variants became so predominant that made it pandemic, meaning prevalent throughout the whole world. So for the first time in human history, biological hazards took a significant position within the list of natural hazards. The literature is quite extensive with regard to the known disasters such as Earthquakes, floods, droughts, tsunami, storms, avalanches, tornados, extreme colds, volcanoes, endemics, and hurricanes. But as for the new biological disaster—COVID-19, there are still much more unknowns, compared to what is known, particularly on its impact on human behavior and its implications on urban form, space, and activities, which are the subject matter of urban design.

The future will not be the same as before. Paradigm crisis has occurred. The old knowledge base of urban planning and design is incapable to solve the emerging problems—the new pandemic crises. We need to evaluate and revise the past trends in all areas; values and goals have to be reconsidered and new priorities have to be set. In the case of our specific interest, the content of urban design and planning needs to be fundamentally restructured (in theory, as well as in education and practice). Urban design and planning procedures and mechanism and the criteria on their basis of which decisions are made should be completely revised.

No other disaster, man-made or otherwise, in the history of mankind has been able to affect every aspect of life in our planet. Scientists, politicians, leaders, and the public were quite shocked by their inabilities to grasp the phenomena and find solutions to put it under control. All technological advances proved rather useless, at least in the short run. All societies, poor as well as rich, developing as well as developed, West and East, and people of all ages, genders, etc., faced the crises. The whole mankind fell in a stage of shock.

The question is: Can the concept of sustainable development offer reasonable solution to this worldwide problem? Over half a century ago the concept of sustainable development was introduced as a new paradigm to replace the classic development. Proponents of the ideal were so optimistic that this will fundamentally change the way of thinking and action in all areas, particularly in urban design and planning.

The concept started from environmental issues but soon spread to other areas such as economic, social, and technological aspects to be comprehensive and ideal. This was in fact the real reason why after so many years and diverse efforts by experts at various levels to implement the ideas and put them into action no tangible success has been reached. So that the founder of the idea in urban planning, Campbell (1996), maintains that after three decades of hope, no tangible results are gained. This is especially true at the city and neighborhood scales. Sustainable development claims too much that could be realized in real world.

The outbreak of COVID-19, for the first time and certainly not the last, showed that the sustainability concept is of no use to deal with this new problem. Recent experience shows that what is critical at this time is how to become resilient against an unknown phenomenon. The phenomenon has been so strong, contagious, and universal that is not comparable to any other disaster in history. It is a turning point in human life. Tackling the problem in a rather short time required strength and flexibility, mentally, socially, economically, and health-wise.

So in recent years, the idea of resiliency is gradually replacing sustainability. Resiliency is considered to be realistic, pragmatic, and to best fit the real-life situations. It is incremental, segmental, and easy to use and measure its performance in practice.

There are many unknown factors involved in the outbreak and spread of COVID-19, including where and how it actually started, how it was spread, how does it work, how does it change, and probably the most significant question to be raised is when is it going to end.

The point here is that from now on, and definitely for the post-COVID-19, the basic needs and its components should be given the first priority in any decision-making processes, at any level.

The question now is: What can urban planners/designers do to reduce the impact of a crisis on communities, and how should public services be designed to cope with surges in demand?

As it was stated earlier, the future will not be the same as before. Paradigm crisis has occurred. The old knowledge base of urban planning is incapable to solve the emerging problems—the new pandemic crises. We need to evaluate and revise the past trends in all areas; values and goals have to be reconsidered and new priorities have to be set.

The substance and the priorities of planning and design should be also changed. Resiliency, and not sustainability, should be the determining factor in evaluating the plans. In fact, recent event revealed that sustainability concept is too idealistic to be useful to deal with deep and global crises. It is more tangible and practical for communities to be resilient and prepared to cope with such anomalies. We will suggest here that current anomalies will eventually lead to major breakthroughs or a paradigm shift.

More challenges and justifications for the new post-COVID-19 paradigm are presented. From master planning entire communities to interior design, COVID-19 is reshaping the way developers, architects, and designers are looking at our homes and the environments around them. These changes may be divided into two groups of procedural and substantive areas:

Procedural areas

- Revising priorities.
- The need for alternatives.
- Information dissemination.
- Justice, equality.
- Decentralization.

Substantive areas

- Regional mobility, fostering connections.
- Big cities, the role of size.
- Public space.
- A different approach to open space.
- Mega malls, commercial centers, etc.
- Reclaiming sidewalk.
- Shared Streets
- Density and Its relevance in the context of public health and resilience.
- Low- and medium-density development.
- Housing, housing patterns, neighborhood idea, and urban village concept.
- Sociability.
- Walking and biking.
- Behavior patterns, changing habits, and lifestyle.

Reference

Campbell S (1996) Green Cities, Growing Cities, Just Cities? Urban Planning and the Contradictions of Sustainable Development. J American Plann Assoc 62(3): 296–312. https://doi.org/10.1080/01944369608975696

Epilogue: Natural Disasters, Sustainability, Resilience, COVID-19 Pandemic, and Urban Design

Abstract We have been familiar with numerous kinds of natural disasters and the ways and means to deal with each group of them in the past. Most well-known of them are: geological and hydrometeorological hazards. Although biological hazards have also been with humans all along time under endemics, the recent global spread of COVID-19 and its variants became so predominant that made it pandemic, meaning prevalent throughout the whole world. So for the first time in human history, biological hazards took a significant position within the list of natural hazards. The literature is quite extensive with regard to the known disasters such as earthquakes, floods, droughts, tsunami, storms, avalanches, tornados, extreme colds, volcanoes, endemics, and hurricanes. But as for the new biological disaster—COVID-19, there are still much more unknowns, compared to knowns, particularly on its impact on human behavior and its implications on urban form, space, and activities, which are the subject matter of urban design. The future will not be the same as before. Paradigm crisis has occurred. The old knowledge base of urban planning and design is incapable to solve the emerging problems—the new pandemic crises. We need to evaluate and revise the past trends in all areas; values and goals have to be reconsidered and new priorities have to be set. In the case of our specific interest, the content of urban design and planning need to be fundamentally restructured, in theory, as well as in education and practice. Urban design and planning procedures and mechanism and the criteria on their basis on which decisions are made should be completely revised.

Keywords Natural disasters, Biological hazards, COVID-19, Sustainability, Resilience, Urban design, Paradigm crises, New paradigm, The content of justifications and challenges

H. Bahrainy and A. Bakhtiar, *Urban Design in Seismic-Prone Regions*, The Urban Book Series, https://doi.org/10.1007/978-3-031-08321-1

Table E.1 Various natural disasters in the world

Geological Hazards	Earthquakes Volcanic eruptions Tsunamis Landslides (incl. all kinds of mass movements, cavity collapses, and ground failures)
Hydro metrological hazards	Floods Coastal erosion and flooding Wind storms (incl. cyclones, hurricanes, and typhoons) Extreme temperatures Storm surges Droughts Wildfires
Biological Hazards	Endemics Pandemics

E.1 Introduction

We have been familiar with numerous kinds of natural disasters and the ways and means to deal with each group of them in the past. Most well-known of them are: geological and hydrometeorological hazards. Although biological hazards have also been with humans all along time under endemics, the recent global spread of COVID-19 and its variants became so predominant that made it pandemic, meaning prevalent throughout the whole world. So for the first time in human history, biological hazards took a significant position within the list of natural hazards. The literature is quite extensive with regard to the known disasters such as earthquakes, floods, droughts, tsunami, storms, avalanches, tornados, extreme colds, volcanoes, endemics, and hurricanes. But as for the new biological disaster—COVID-19, there are still much more unknowns, compared to knowns, particularly on its impact on human behavior and its implications on urban form, space, and activities, which are the subject matter of urban design (Table E.1).

In the last months of 2019 and first months of 2020, the world witnessed an unprecedented event. It started as an epidemic in the city of Wuhan, China, and in a very short time spread to other parts of the world and turned into a global phenomenon as a pandemic. No other disaster, man-made or otherwise, in the history of mankind has been able to affect every aspect of life in our planet. Scientists, politicians, leaders, and the public were quite shocked by their inabilities to grasp the phenomena and find solutions to put it under control. All technological advances proved rather useless, at least in the short run. All societies, poor as well as rich, developing as well as developed, West and East, and people of all ages, genders, etc., faced the crises. The whole mankind fell into a stage of shock (Table E.2).

Table E.2 Normal disaster characteristics with those of COVID-19

Known natural disasters	Predictability	Duration	Geographic area	Return period	Intensity	Type of impact	Control
Earthquake	No	Very short	Limited	Not known	Mild to severe	Human, property, etc.	Preparation, mitigation
Floods	Yes	Weeks	Regional	Known	Average	Property, Human	Prevention
Tsunami	No	short	Limited	Not known	Mild to severe	Human, property, etc.	Preparation
Hurricane	Yes	Days	Limited	Seasonal	Average	Property, Human	No
Drought	Slightly	Long	Regional	Not known	Average	Human, property	No
Tornado	To some extent	Days	Limited	Seasonal	Average	Property, Human	No
Epidemics	To some extent	Limited	Limited	Seasonal	Mild to severe	Human	Relatively
COVID-19	**No**	**Not known**	**Global**	**Not known**	**Severe**	**Human, economic, social, psychological**	**Not quite**

Comparison of normal disasters with COVID-19

E.2 The Role of Sustainability Concept

For over half a century, the concept of sustainable development was introduced as a new paradigm to replace the classic development. Proponents of the ideal were so optimistic that this will fundamentally change the way of thinking and action in all areas, particularly in urban design and planning. The concept started from environmental issues but soon spread to other areas such as economic, social, and technological aspects to be comprehensive and ideal. This was in fact the real reason why after so many years and diverse efforts by experts at various levels to implement the ideas and put them into action no tangible success has been reached. So that the founder of the idea in urban planning, Campbell (1996), maintains that after three decades of hope, no tangible results are gained. This is especially true at the city and neighborhood scales. Sustainable development claims too much that could be realized in real world.

The outbreak of COVID-19, for the first time and certainly not the last, showed that the sustainability concept is of no use to deal with this new problem. Recent experience shows that what is critical at this time is how to become resilient against an unknown phenomenon. The phenomenon has been so strong, contagious, and universal that is not comparable to any other disaster in history. It is a turning point in human life. Tackling the problem in a rather short time required strength and flexibility, mentally, socially, economically, and healthwise.

So in recent years, the idea of resiliency is gradually replacing sustainability. Resiliency is considered to be realistic, pragmatic, and to best fit the real-life situations. It is incremental, segmental, easy to use, and measure its performance in practice.

What is common among all these disasters is that they are usually specific to certain geographic area and/or certain group of people. They mostly have specific return periods, which are to a large extent known to specialists, and start and end at certain points in time. Although exact prediction of certain disasters is not known, due to some unknown factors, for example, in the case of earthquake, there is ample information as to how to reduce the risk, how to be prepared, and how to mitigate the consequences. But as Fig. E.2 shows, there are significant differences between development trend of normal disasters compared to that of COVID-19. This is particularly significant with regard to the disasters duration (Fig. E.1).

There are many unknown factors involved in the outbreak and spread of COVID-19, including where and how it actually started, how it was spread, how does it work, how does it change, and probably the most significant question to be raised is when is it going to end. Figure E.1 shows the difference between known disaster trends and the COVID-19. What is critical in this figure is the indefinite length of the curve in the case of COVID-19. In other word, we still do not know of the exact behavior of the new variants, especially when it is going to end and if it is going to return and how and in what form?

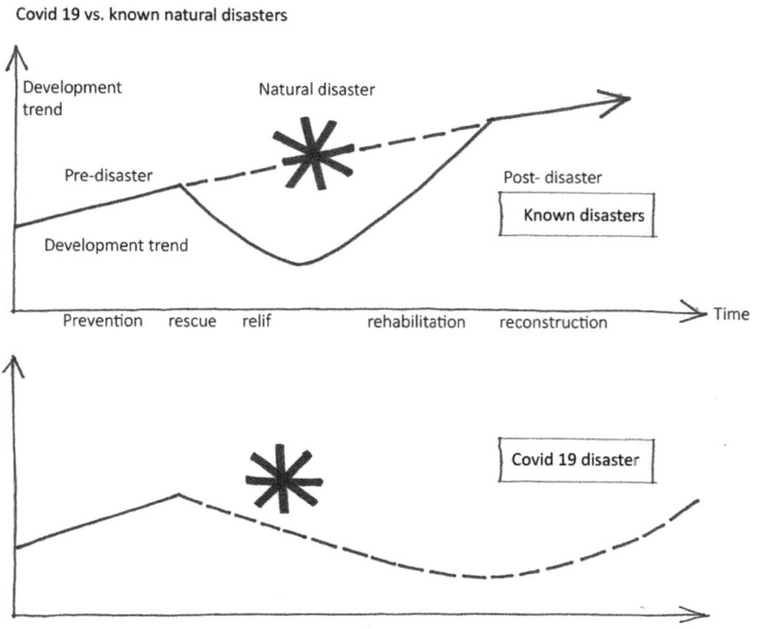

Fig. E.1 Comparison of the three stages of known natural disasters versus COVID-19

E.2.1 Natural Disasters, Sustainability, Resilience, and COVID-19

As mentioned earlier in Chap. 1, quality of life in human settlements consists of three main indicators: (1) Basic needs, (2) socio-economic needs, and (3) the cultural needs. The weight of these indicators, however, is not the same. Basic needs which include natural environment, health, welfare, and safety and security have, by far, much more value compared to the rest indicators. These three factors, in fact, make the vitality of any settlement. The point here is that from now on, and definitely for the post-COVID-19, the basic needs and its components should be given the first priority in any decision-making processes, at any level. This has not been the case in the existing paradigm of urban design and planning. Rather, physical appearance, beautification, and other non-essential aspects of cities have been given priority over the basic needs.

COVID-19 is an unprecedented natural disaster the world has ever witnessed. It started as an epidemic and in a very short time spread to other parts of the world and turned into a global phenomenon as a pandemic. No other disaster, man-made or otherwise, in the history of mankind has been able to inflict such a great impact on every aspect of life on the planet, so that as of today the number of cases and deaths are:

April 25, 2022, 01:10 GMT, Coronavirus Cases: 509,542,737, Deaths: 6,243,199.

https://www.worldometers.info/coronavirus/

Economies collapsed, social structures disintegrated, normal lifestyles stopped, mobility at the macro- and microscales were put on halt, sports, entertainments, demonstrations, religious activities were all canceled; streets, public spaces, shopping centers, parks, schools, and universities were all closed. It was like civilization was put to a halt, in order to change its trend, based on new set of criteria. "Big" as ideal criteria for better life failed, greater cities paid a higher price, while smaller ones were more successful in having control over the problem. So once again we realized that "small is beautiful." May be now it is time to rethink of "urban village" as a solution for future settlements.

COVID-19 is the most serious, most destructive event, with dangerous adverse impacts on all aspects of life, everywhere. No doubt a return to the pre-COVID-19 era is not probable. So it should be regarded as a turning point for all future activities.

In the past, diseases have shaped cities. But how can we shape our cities to better fight diseases?

The last three years have seen life disrupted in every city in the world. The variety and imagination of responses to this crisis have given us as urban planners/designers a huge source of insight. As with any experiment, the findings may take time to review and adopt. But already they point to practical ways in which we can and should be improving the planning and design of our cities. Here are three challenges reshaping urban planning: How can we improve our access to the real-time information that helps cities survive a crisis?

E.3 The Role of Urban Planner/Designer?

What can urban planners/designers do to reduce the impact of a crisis on communities, and how should public services be designed to cope with surges in demand?

Jon (2020) claims that "it becomes clear that there is not much time left for a 'radical shift' in restructuring our relationship with nature. They emphasize the element of uncertainty and complexity in today's challenges. We simply do not know, at this present moment in time, what will actually happen in the future. Especially for the urban planning discipline, which is fundamentally based on the projective spirit of making our communities a better place. what we are truly interested in is how the practice of planning—which is, ultimately, human action—can project a more sustainable, co-constitutive human/nature relationship."

E.3.1 Paradigm Crises

As it was stated earlier, the future will not be the same as before. Paradigm crisis has occurred. The old knowledge base of urban planning is incapable to solve the emerging problems—the new pandemic crises. We need to evaluate and revise the past trends in all areas; values and goals have to be reconsidered and new priorities have to be set. In the case of our specific interest, the content of urban design and planning needs to be fundamentally restructured; in theory, as well as in education and practice. Urban design and planning procedures and mechanism and the criteria on their basis on which decisions are made should be completely revised.

The substance and the priorities of planning and design should be also changed. Resiliency, and not sustainability, should be the determining factor in evaluating the plans. In fact recent event revealed that sustainability concept is too idealistic to be useful during such a deep and global crisis. It is more tangible and practical for communities to be resilient and prepared to cope with such anomalies. We will suggest here that current anomalies will eventually lead to major breakthroughs, or a paradigm shift.

Anomaly which is reflected in "violation of the current paradigm" will be modified by the scientific community in order to explain the anomalies and find a replacement, which could be regarded as a paradigm shift, or a turning point. The period of crises, which is generated when the existing paradigm cannot accommodate the anomaly, the paradigm is intensely scrutinized, and the parameters of research are broadened, leading eventually to paradigm substitution. Here I am suggesting a new paradigm to replace the existing one (Kuhn 1970, 1977).

Of course, this is not the first time that a paradigm shift is suggested in urban planning to replace the old one, because the old one is unable to solve the emerging problems. First, modernism, supported by the First Industrial Revolution, by introducing new dimensions for time and space revolutionized traditional life, then the Information Revolution, later the Sustainability Concept, and finally, the introduction of Critical Theory all claimed paradigm shift was occurring in urban planning. But the fact is that in spite of undeniable influences each of those developments had on all aspects of life, but none was as deep and global with significant and long-lasting social, economic, mental, and physical consequences as of COVID-19. That is why we regard this event as the "decisive paradigm shift."

Jon (2020) believes that while the pandemic should be considered a wake-up call for us to drastically rethink our relationship with nature, planning discipline cannot resign itself from its power and responsibility to make a difference in human and nonhuman lives. He further claims that it becomes clear that there is not much time left for a "radical shift" in restructuring our relationship with nature.

As it was stated before, new paradigm will emerge when the old one is not capable of answering the new emerging questions and, therefore, incapable of solving the new and future problems. Recent experience shows that the existing mechanism of decision making: autocratic, top-down (as urban planning/design procedural elements) prove inadequate to deal with the current situation. The old values and priorities need to be revised. The issues of intervention, control, and regulations seem to be more inevitable. As for the substantive elements, the problem is so subtle, that even the layman can understand it. Look, for example, at the issue of the size. Mega Cities such as New York, Wuhan, London, Tehran, and Chicago have been suffering the most during this recent crises, because it has been impossible to have any kind of control over such complex and giant phenomena. It is interesting that for the first time in human history, humans had to de-socialize. For years, urban planning and design have been constantly insisting on the importance of socializing and making public spaces attractive for people to make those spaces viable. Also, many regarded globalization as an ideal goal for cities throughout the world. But for the first time people in the cities are advised not to go to public places, not to travel, either local or international, or not to use public transportation, and rather use private car. As a result, the economics collapsed, social structures disintegrated, normal lifestyles stopped, mobility at the macro- and microscales were put on halt, sports, entertainments, demonstrations, were all canceled.

E.3.2 Emerging New Paradigm

- Knowledge base: Existing evolutionary knowledge has to be revised, because it is incapable of solving the new emerging problems.
- New paradigm: New revolutionary knowledge is needed to respond to the new needs (Galloway and Mahayni 1977).

E.3.2.1 The Process of Paradigm Development and Its Requirements

- We, in the fields of urban planning and design, are now in the process of creating new knowledge to deal with the emerging problems—COVID-19 pandemic. To be a qualified paradigm, some requirements should be met.
- Paradigm shift requirements
  - New shared rules, critical theory + resiliency
  - New shared language, " " "
  - New shared goals, " " "
  - New shared commitments, and
  - New shared exemplars of successful practice (Kuhn 1977).

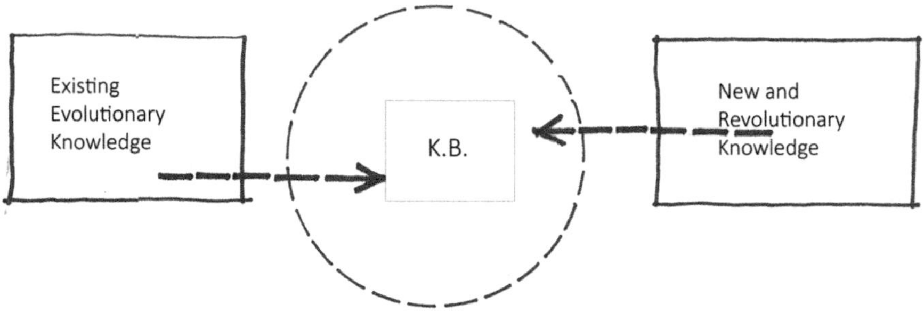

Fig. E.2 Existing knowledge base and the new knowledge of urban planning/design

The Content of the New Paradigm

The new paradigm will include two complementary elements—procedural and substantive:

- Procedural element comes from: **Critical theory**
- Substantive element comes from: **Resiliency.**

The contents of the emerging new paradigm introduce a new agenda (based on contingency, flexibility, and preparedness) for disaster management in general, and COVID-19 disaster management in particular:

- **Procedural**
  - objective, scientific, quantitative
  - subjective, artistic, qualitative
  - intuitive, interpretive, hermeneutic
- **Substantive**
  - urban (public)form, to be revised,
  - urban (public)space, to be revised, focusing on public domain
  - urban (public)activities, to be revised
- **Evaluation**
  - through EIA and Disaster Impact Assessment (DIA) to assess the adverse impacts of globalization, development, development pattern, comprehensive plans, standardization, urban renewal, compact city, centralization, malls, shopping centers, freeways, etc. (Table E.3).

E.3.3 More Justifications for the New Paradigm

There are also others who suggested such a paradigm shift. Horne (2021) for example focuses on the impacts of the pandemic on the planning profession itself. In short, he believes that the pandemic has precipitated a "wobbly" paradigm shift in planning, and in many other fields. This unbalanced paradigm shift creates opportunities to devise and catalyze good and lasting changes to our profession and to society, as long as we are consistent and persistent.

He describes this paradigm shift as a move from endless iteration to meaningful systemic change, and to be clear, my definition of systemic change includes both structural (i.e., organizational) and cultural (i.e., attitudinal) transformation.

The planning community, he adds, needs to step up and solidify our wobbly paradigm shift. Systemic change can create great and lasting societal gains, and that is something that planners are well-suited, qualified, and experienced to support.

Table E.3 Comparison of the procedural and substantive elements of urban design in the old versus new paradigm

Procedural elements of the old versus new paradigm

	Old paradigm: Theories, methods, concepts, etc.	New paradigm: theories, methods, concepts, etc.
Procedural	Political, autocratic, top-down decision making, technocratic, unethical, injustice, bureaucratic, private interest-oriented, lack of public participation, no collaboration, no effectuation, closed system, no feedback, physical aesthetics oriented, no concern for social and psychological aspects, globalization	Down-up and open decision-making participatory, NET' S collaborative, feedback and effectuation, public awareness, mutual learning contingency planning, flexibility, two kinds of thinking: divergence and convergence, evaluation (against part of decision making. Self-governance. The essence of critical theory can provide the procedural element of the new paradigm

(continued)

Table E.3 (continued)

Procedural elements of the old versus new paradigm		
Substantive elements of the old versus new paradigm		
Substantive	Modernism/post-modernism, extensive mobility, centralization, big cities, agglomeration, public spaces, socializing Beautification, high density, no concern for vitality, capitalism, mega malls, big shopping centers, large parking lot/structures, commuting pattern, car-dependent trips, fossil fuel consumption, disintegration, machine-based lifestyle, commercialism and consumerism, globalization	Post-COVID-19, Priorities will change: public health, natural environment, and safety and security. Limited and conditional mobility, redundancy, promoting virtual systems decentralization, smallness, urban village concept, lower density, no long-distance commuting, self-sufficient, a revised lifestyle, disregard traditional blueprints, the concept of resiliency may provide the substantive element of the new paradigm

Architecture journalist Lubell (2020) describes how the COVID-19 pandemic, like pandemics of the past, will trigger a shift in how visionaries of the built environment—planners, urban designers, and architects—imagine place and space in response to the disruption of the moment. He says "It's clear that the coronavirus will have—and is already having—a similarly profound effect on today's built world. It's shaking loose notions of what is 'normal' in a field still employing many of the same techniques it did a century ago." He, then, claims six methods can play prominent role in the age of COVID-19. Lubell suggests a new look at: modular construction, adaptive reuse, lightweight architecture, the healthy building, telecommuting and small-city living, the town square, reconsidered (Lubell 2020).

Since cities and towns are facing a new reality today, one of the founders of CNU has called for new designs that violate some of the long-held design doctrines of New Urbanism—but have the same social intention.

According to critics, the pandemic has brought to light several structural issues that cities are quickly trying to respond to. Are these reactionary design and policy plans effective for long-term recovery? These thought-provoking proposals advocate for a reimagining of what a city is and propose alternative ways for creating a resilient built environment. From making city infrastructures more flexible through design.

The pandemic jeopardizes many of the core concepts of New Urbanism, according to architect and urbanist Andres Duany. People are now suspicious of density, transit, face-to-face retail, social "third places," open communities, and in-person community engagement.

Jon raised this question: "The end of urban density?" And then recommends "rethinking and rebuilding the relationship between nature and human intervention" Jon (2020).

Jachnow (2020) has raised this question: "What will be the Post-Pandemic Urban Path?" And then continues "What will be the 'new normal' for urban planning and development after COVID-19?" Of course there are some who believe that little will change in the making of cities.

This is not about returning to normal. This is about adapting our downtowns and public spaces to become more resilient and more inclusive. We are rethinking, rebuilding, and reconnecting more than just the physical public realm.

Now, new goals are promoted for the post-pandemic city, which includes the functional division of space, safe mobility within its physical environment, and other, already existing mantras of healthy cities. However, though all these desired developments are promoted worldwide, they often remain unattainable visions, especially for the fast-growing cities in Africa, Asia, and Latin America. Many urban areas might be as far from being healthy cities as they are from being sustainable, spatially just, or economically resilient.

Unlike the airy, pristine emptiness of modernism, the space needed for quarantine is primarily defensive, with taped lines and Plexiglas walls segmenting the outside world into zones of socially distanced safety. Wide-open spaces are best avoided. Barriers are our friends. Stores and offices will have to be reformatted in order to reopen, our spatial routines fundamentally changed. And, at home, we might find ourselves longing for a few more walls and dark corners (Jachnow 2020).

Bozikovic (2021) suggests that "For the sake of our health, our cities need to be denser and more walkable. But too often, restrictive zoning and reluctant neighbors stand in the way." But there are many that oppose higher density for cities and believe that in fact, high density is one of the determinant factors in fast spread of the virus and less effective control (Jon 2020; Gehl: https://covid19.gehlpeople.com/ 2020).

During COVID-19, the need for responsive and creative solutions to sustain our local communities and economies has become more apparent than ever. But how might tactical solutions have a lasting impact on our urban spaces? How does our physical response to COVID-19 reshape urban space in ways that remain socially and emotionally connected? How do the decisions we collectively make today move us toward a more resilient and inclusive urban public realm?

Unlike a fire, flood, earthquake, tornado, or hurricane, COVID-19 has prevented us from coming together physically in response to its aftermath. A keen sense of urgency is propelling the need for adaptability moving forward; champions, innovators, and organizers are emerging to provide flexible engagement tools as an alternative to the traditional means of public *charettes* and open houses in order to gauge diverse input and feedback. Every community needs to find a place that IS right for them and their constituents, but only a thoughtful, inclusive process will contribute toward building enduring trust and public support. While the solutions offered today may be temporary, once in place they may have staying power. The lessons we learn today may underscore future needs for a more resilient and inclusive framework of public streets and spaces, paving the way for an enhanced quality of life while ensuring small business sustainability and success (Weidl et al. 2020).

E.3.3.1 Challenges Under New Paradigm

From master planning entire communities to interior design, COVID-19 is reshaping the way developers, architects, and designers are looking at our homes and the environments around them.

In the meantime, we as urban designers need to remind ourselves that aren't we placing too much confidence in how urban design may influence societal life? In others words, we are facing with serious shortcomings in our tools to effectively achieve our desired goals such as dealing with the pandemic. Possible changes may be suggested in the framework of procedural and substantive areas.

E.3.3.2 Procedural Areas

Revising Priorities

The COVID-19 pandemic has raised new questions, forcing us to reassess the way we design our cities. For example, the crisis has given us an opportunity to rethink the relationship between urban design and public health. The ability to assess and mitigate the effects that development has on health should become a new field of expertise, to help prepare cities to respond more rapidly and efficiently in future. Because physical activity is strongly associated with numerous chronic conditions, and walking is the most popular form of physical activity among adults, health-related policies and research should be given priority and put strong emphases on how the built environment will have impact on walking (https://www.arup.com/perspectives/three-design-lessons-for-the-post-pandemic-city, see also: Bahrainy et al. 2015; Bahrainy 1995).

The Need for Alternatives

One of the significant results of following an urban design process is leading to various alternatives (Bahrainy 2018). This is particularly important in time of crises such as pandemic COVID-19. Many studies suggest to bring home, work, and shopping into proximity, encourages pedestrian traffic, promotes development around mass-transit nodes, and mixes types of housing (Bahrainy et al. 2015).

Information Dissemination and City-Wide Digital Transformation

Information dissemination is critical in any disaster management for public involvement in decision-making processes, education and training, mass media, technology transfer, and warning system. The pandemic has proved urgent need for all kinds of information to be available to the decision-makers at all level and also the public. With the advance of digital technology the information for policies on health, sanitation, public services, limitation and the rules and regulations can be delivered in digital form to make it easily understood by all the people.

Grassroots methods of organizing have proven to be effective in creating a sense of community during this remote time and designers are finding ways to establish these ideas on a citywide scale. These proposals emphasize the residents' role in community recovery, the need to re-establish trust in the built environment, and the important steps designers should take to effectively connect with the people they serve.

In this regard, the value and usefulness of globalization, as a global channel of information transfer, particularly from the developed countries to the developing ones has been seriously questioned, partly because the spread of COVID-19 has been attributed to the global transportation and the mass transfer of people and goods. (https://www.arup.com/perspectives/three-design-lessons-for-the-post-pandemic-city).

Justice, Equality

People in low-income neighborhoods often rely more heavily on accessible public spaces. Studies have shown that the percentage of green space in people's living environment has a positive association with the perceived general health of residents. The relation is stronger for lower socioeconomic groups. The public realm can, therefore, offer a path forward to create social cohesions, heal communities, and advance health equity. Open spaces and greenery are not enough though, neighborhoods need to also have easy access to good schools, health care, nutrition, transportation, and affordable housing (Sarkin 2020).

So in the post-COVID-19 world, we have the opportunity, and responsibility to give marginalized groups equal consideration in our planning and design decision-making processes as we reshape our communication networks and our wider cities.

Decentralization

Past experiences show that decentralization plays a determinant role in decreasing community risks to almost all disasters, including the spread of the disease. So decentralization of human settlements, especially public services, such as waste management, health care, and safety and security organizations, as well as public land uses, such as public open spaces should be regarded as the integral part of decision making in the urban planning and design process. In the time of crises centralization works against resilience, which is so critical for disaster management policies. Lack of due attention to resiliency will, therefore, increase community's vulnerability against disaster (see also Chap. 7).

The participatory process is more important than ever before as we work toward a more equitable future. For public space to be relevant, we need to take the time to deeply understand the relationship between people's way of life, history, memory, and built environment. This includes focusing on the public health benefits of space, giving voice to marginalized communities, and seaming together our cities to repair past spatial injustices. Gender, age, and income gaps should be overcome, which will eventually lead to public awareness and involvement.

E.3.3.3 Substantive Areas

Regional Mobility, Fostering Connections

After months of isolation, the importance of developing safe social infrastructure is key to rebuilding healthy cities. The pandemic has drastically shifted the way we socialize and create networks between each other. We feature proposals that investigate emergency response methods on a local scale, providing elderly populations support in recovery, and the fast-acting power of mutual-aid groups.

Big Cities, the Role of Size

Perhaps foremost, will we see a return to ideas about decentralization and de-densification promoted most notably by Ebenezer Howard at the turn of the last century—when industrialization and immigration defined cities like London, Chicago, and New York and created genuinely harmful overcrowding and attendant epidemics?

No doubt we will see this, and we may see people traumatized from this experience retreat from city life. We have already seen a call to reduce "our dependence on megacities" (as if megacities were planned) and density described as the "enemy." But we have yet to see how rural and suburban communities, although perhaps later in the pandemic timeline due to their relative remoteness, will fare in the weeks to come.

Thus, it is difficult to imagine that decentralization will work now, even for present reasons, any more than it did when Howard (1902) conceived of his so-called Garden Cities, which were unable to be self-sufficient or create jobs, as he had hoped. As much as it was the case in the mid-twentieth century, when Jacobs (1961) wrote *The Death and Life of Great American Cities*, Americans are not good at regional planning—which was part of the reason she wrote her great book in the first place.

Public Space

Public space is essential infrastructure in the time of the COVID-19 pandemic. The value of public space is being brought to light during this pandemic, together with a renewed appreciation. In addition to the public health and environmental benefits, it can reduce socioeconomic segregation, build trust and reduce social isolation (Sarkin 2020). COVID-19 has demonstrated the importance of public space in cities. Throughout the COVID-19 pandemic, we have all had similar internal conversations: "Should I go out for a walk?", "Can I go to the store?", "Maybe I'll stay inside a few more days." The decision to go out into the city or stay inside is hugely influenced by how your city is designed. The key determining factor to being in public safety is to maintain one to two meters distance from other people. Yet streets, sidewalks and transit options don't always allow people to stay physically distant (Schuff 2020).

So much has transpired in 2020, but one thing has not changed; our innate desire to gather. Our urban areas serve a vital role in our communal identity, providing a place to come together, to celebrate, exchange ideas, and protest injustices. We're in a unique moment in time; a moment of self-reflection, an opportunity to examine traditional strategies and strive for more inclusive process that embraces diverse perspectives while amplifying their voices.

Even those working from home in their apartments seem to have a strong longing for "the urban"—the street life, the park life, the vibrancy and life that is the city: public, semi-public, and private spaces (businesses, shops, restaurants) which in the end provide you things you can order to your door. Is there now a growing recognition of its non-materialistic values?

We observe the tragic irony that quarantine is far from safe. An increase of domestic violence against women during the pandemic and lockdown situations is reported from around the world. These are already areas where "third spaces" like outdoor public space have already proven particularly important for inhabitants. Why not consider urban public spaces as "essential services" for women and people in vulnerable domestic situations.

A Different Approach to Open Space

COVID-19 has caused us to develop a more nuanced appreciation of open space's social and therapeutic potential, such as access to better air quality and more physical activity, among other benefits.

Safety and security now take on an entirely different meaning, one that goes well beyond Oscar Newman's (1973) Defensible Space theory. Urban planning must now push forward the idea that safety in the public realm is an investment in public health, thereby requiring us to think differently about open space in the face of this crisis. Cities and their local governments must embrace and manage security in our public realms to address this crisis and future pandemics.

Mega Malls, Commercial Centers, Etc.

COVID-19 has forced architects and urban designers to add a new set of criteria for the design of public spaces such as malls and commercial centers. These include health and safety of customers, through social distancing.

Reclaiming Sidewalk

Irrespective of the variance in social-distancing guidelines in place across countries, citizens have found a common issue: pavements don't provide enough space for pedestrians to observe the rules. Many see this as a once-in-a-generation opportunity to rethink a space that has, in many places, been relegated to just another part of transport infrastructure. While some of the emergency measures will be scaled back as the infection curves flatten, others will remain in place for the foreseeable future, and as such we need to harness this crisis to bring about positive change in cities. In particular, we want to see the street transformed into a space for the community, a public arena rather than a through route (ARUP, https://www.arup.com/perspectives/three-design-lessons-for-the-post-pandemic-city).

This will require a reevaluation of the role and quality of on-street amenities, principally street furniture (Figs. E.3 and E.4).

Alongside pavement-widening schemes, many local governments are relaxing licensing laws in order to allow restaurants to occupy street space.

Fig. E.3 How challenging it is for Seattle restaurants to extend out into the street (photo by the authors)

Fig. E.4 Closing the street to vehicular traffic in order to provide more safe space for pedestrians, U District, Seattle (photo by authors)

In Copenhagen, a city built around wide walking paths, bike lanes along all primary roads, and a network of public spaces and parks, we found that people have not had difficulty social distancing. And while the city retail center became deserted, with an 80% reduction in pedestrian movement, local neighborhoods saw a 21% increase in people eating, socializing, or partaking in other recreational activities. Here, it became immediately clear that where some areas witnessed a drastic reduction in public life, local neighborhood meeting places had been thriving. In an online survey, respondents claimed that their local public spaces were vital for their mental health and they appreciated the lower number of cars on the road (Schuff 2020).

Public Spaces/Squares

Urban squares, as the most important public spaces, are a reflection of urban planning cultures, and also to some extent even a reflection of overall nations' cultures. Squares are microcosms of urban life, offering excitement and repose, markets and public ceremonies, a place to meet friends and watch the world go by. They have been shaped by popular whims, by topography and architectural fashion. City squares seem to be a good alternative for occupying public space, since they are more accessible and not so crowded, thus healthier and more democratic (Soltanzadeh et al. 2007).

15-Minute City Planning Framework

We are also seeing how some cities are implementing policies and frameworks to guide future development. Paris, for example, is aiming for a "15-min city" with most daily needs a short walk, bike ride, or public transit stop away (Sarkin 2020).

Shared Streets

Streets are the most important public spaces, as far as land-use size and distribution within a city are concerned. They play critical role in various functions of the cities. For example, they became the sites of escape for residents under lockdown. Accessibility was a determinant fact to let people move around to secure their needs and to use parks and beaches. Through the expansion of businesses into streets for the sake of social distancing, outdoor space has gained a new meaning, which has to be taken into account in the design of any new public streets.

Density and Its Relevance in the Context of Public Health and Resilience

Brumfield and Cubillos (2020) believe that the idea of density is under intense scrutiny in this era of physical distancing, despite the many positive outcomes associated with density in the urban environment. They further claim that this is why the central question for city planners is: How can cities make themselves stronger by reconsidering the nature of density and its vital relationship to public health, wellness, and resilience?

Planning and design activities are, by their basic definitions future-oriented, involving two determinant factors: complexity and uncertainty. COVID-19 is now further adding to these two characteristics. In these circumstances, there is no any other way but to look for more flexible and adaptable responses to our planning and design activities. All relevant variables including private and public spaces, public activities, and forms need to be redefined and their relationships changed, according to the emerging conditions.

Low- and Medium-Density Development

Is higher density and compact cities have become obsolete due to COVID-19? Do we need to reconsider the relationship between natural environment and built environment? And choose between sustainable development and economic development (Bahrainy 2011).

Mehta (2020) claims that "social distancing measures during COVID-19 have altered the use of space. With the closure of places of work, learning, leisure, consumption, and more, the pandemic has limited our territories and public life. Yet, residents living in mid- to low-density places are experiencing their neighborhoods differently. They are repurposing residential streets, sidewalks, parking lots, and other spaces and transforming neighborhood space for active living, play, and sociability. In many neighborhoods, social distancing is generating a new *sociable space*. Can we build on our ingenuity to reclaim neighborhood spaces for public life, and the physical and psychological health of our communities?".

Many writers, city leaders, residents, and government agencies are creating a link between a city's density and its vulnerability to the spread of pandemics. Perceptions that low-density areas are safer could draw people away from cities. This has been the reaction after pandemics in the past. During the modernist movement, for example, following closely after the Spanish Flu of 1918, there were similar concerns about density and disease. As a result, utopian cities designed by modernist architects such as Le Corbusier's "City for Three Million People," focused on space, light, and air. The drawings for these new cities—which influenced many of the principles of modern urban planning—are often devoid of people and depict desolate open public space.

The reaction of many planners, architects, and urban dwellers in more recent times—influenced by Jane Jacobs (1961) among others—is that dense compact neighborhoods and vibrant public spaces create social cohesion and foster vibrant urban life. This is in addition to the fact that the correlation between density and vulnerability to the spread of the virus is not quite known. For example, the geographic breakdown of the virus shows that COVID-19 hit hardest not in dense areas but the lower density, but with low-income populations, immigrants, frontline workers, and people of color.

Housing, Housing Patterns, Neighborhood Idea, and Urban Village Concept

One of the most inventive ways we now see density realized is through the polycentric model, in which self-sufficient districts are distributed across cities and function like urban villages. Such models have the potential to improve the quality of life, promote walking, and free up space for other uses, such as parks and gardens (Frey 1999; Bahrainy and Valadkhani 2019).

Sociability

Social distancing measures during COVID-19 have altered the use of space. With the closure of places of work, learning, leisure, consumption, and more, the pandemic has limited our territories and public life. Yet, residents living in mid- to low-density places are experiencing their neighborhoods differently. They are repurposing residential streets, sidewalks, parking lots, and other spaces and transforming neighborhood space for active living, play, and sociability (Mehta 2020; Daneshgar et al. 2011).

Walking and Biking

It has been decades since reducing dependence on car has been an integral part of any successful urban design project. There are several reasons for this. Walking and biking will be the most efficient and convenient mode of transportation and will contribute to the health and well-being of people, and it is also suitable for recreation. So in normal situations streets can be redesigned to provide more space for pedestrians, bikers, and public open space.

But COVID-19 has provided a unique opportunity to rethink transportation and mobility furthermore. We can emerge from the pandemic with a stronger and safer approach to mobility and improved open space systems. Reallocating space previously used by cars—especially in neighborhoods without walkable access to parks and essential services—would go a long way in improving the public realm (Bahrainy et al. 2015).

Behavior Patterns, Changing Habits, and Lifestyle

By studying the movement and activities of people during this time, we are creating an evidence base for recovery plans and responses that react to people's natural behavior and allow for physical distancing to be the easy choice for people to make. The economy has opened but restrictions are still in place, and many of the habits that have emerged are likely here to stay, including an increase in walking, staying locally, and using public spaces for more diverse activities (Bahrainy 2015).

Urban planning and design must respond to these changing habits and build on the fact that public and accessible space is no longer a "nice to have" but a necessity for all people, especially during disaster.

We need to change our habits. We need to invest much more in local community infrastructure, mindful of their life-sustaining natural and social ecologies, whether the neighborhoods are urban, suburban, or rural. We saw, very quickly, how important and valued public spaces, both streets designed for people and parks, have been to people during this crisis.

E.3.4 And Finally Here Are More Changes Suggested by Various Writers in Order to Deal with the Post-corona Planning and Design of Our Cities

- A changing office work environment.
- More flexible public transit.
- A renewed appreciation for just plain old walking.
- A new kind of city.
- Mixed uses, land uses, balanced distribution of, e.g., health facilities, etc.
- Limits and constraints (natural resources: water, energy).
- Revision of old theories/methods and techniques.
- Leapfrog pattern/centralization/decentralization.
- The concept of privacy.
- Family structure (cohesion).
- Inward looking concept.
- Market support/intervention.
- City governance.
- Recreation and sports.
- Promote parkland over parking.
- Redundancy.
- Flexibility.

Conclusion

We have been familiar with numerous kinds of natural disasters and the ways and means to deal with each group in the past. Aside from geological and hydrometeorological hazards, biological hazards have also been with humans all along time under endemics, but the recent global spread of COVID-19 and its variants became so predominant that made it pandemic, meaning prevalent throughout the whole world. So for the first time in human history, biological hazards took a significant position within the list of natural hazards. With regard to the new biological disaster—COVID-19, there are still much more unknowns, compared to what is known, particularly on its impact on human behavior and its implications on urban form, space, and activities, which are the subject matter of urban design.

No other disaster, man-made or otherwise, in the history of mankind has been able to affect every aspect of life in our planet. Scientists, politicians, leaders, and the public were quite shocked by their inabilities to grasp the phenomena and find solutions to put it under control. All technological advances proved rather useless, at least in the short run. All societies, poor as well as rich, developing as well as developed, West and East, and people of all ages, genders, etc., faced the crises. The whole mankind fell into a stage of shock.

What is certain is that the future will not be the same as before. We claim that a paradigm crisis has occurred in some fields, particularly urban planning and design. The old knowledge base of urban planning and design is incapable to solve the emerging problems—the new pandemic crises. We need to evaluate and revise the past trends in all areas; values and goals have to be reconsidered and new priorities have to be set. The content of urban design and planning needs to be fundamentally restructured (in theory, as well as in education and practice).

The question may be raised here is that can the concept of sustainable development offer reasonable solution to this worldwide problem? Some believe that after so many years and diverse efforts by experts at various levels to implement the ideas and put them into action, no tangible success has been yet reached. No doubt, sustainable development claims too much that could be realized in real world.

The outbreak of COVID-19 showed that the sustainability concept is of no use to deal with this new problem. Recent experience shows that what is critical at this time is how to become resilient against an unknown phenomenon. Tackling the problem in a rather short time required strength and flexibility, mentally, socially, economically, and healthwise.

So in recent years, the idea of resiliency is gradually replacing sustainability. Resiliency is considered to be realistic, pragmatic and to best fit the real-life situations. It is incremental, segmental, and easy to use and measure its performance in practice.

The point here is that from now on, and definitely for the post-COVID-19, the basic needs and its components should be given the first priority in any decision-making processes, at any level.

But what can urban planners/designers do to reduce the impact of a crisis on communities, and how should public services be designed to cope with surges in demand?

Our premise is that a paradigm crisis has occurred in urban planning and design. The old knowledge base is incapable of solving the emerging problems—the new pandemic crises. We need to evaluate and revise the past trends in all areas; values and goals have to be reconsidered and new priorities have to be set.

The substance and the priorities of planning and design should be also changed. Resiliency, and not sustainability, should be the determining factor in evaluating the plans. We will suggest here that current anomalies will eventually lead to major breakthroughs, or a paradigm shift.

The emerging literature on COVID-19 provides extensive justifications, and thereby significant challenges, for this new paradigm, from master planning the entire communities to interior design. COVID-19 is reshaping the way developers, architects, and designers are looking at our homes and the environments around them. Suggested changes may be divided into two groups of procedural areas, such as revising priorities, need for alternatives, information dissemination, justice, equality, and decentralization. Substantive areas include regional mobility**, fostering connections**, the problem of big cities, the role of size, public Space, a different approach to open space, the problem of mega malls, commercial centers, etc., low- and medium-density development, **housing**, housing patterns, neighborhood idea and urban village concept, sociability, walking and biking and behavior patterns and **changing habits and** lifestyle are among other suggested subjects.

We have been familiar with numerous kinds of natural disasters and the ways and means to deal with each group in the past. Aside from geological and hydrometeorological hazards, biological hazards have also been with humans all along time under endemics, but the recent global spread of COVID-19 and its variants became so predominant that made it pandemic, meaning prevalent throughout the whole world. So for the first time in human history, biological hazards took a significant position within the list of natural hazards. With regard to the new biological disaster—COVID-19, there are still much more unknowns, compared to what is known, particularly on its impact on human behavior and its implications on urban form, space, and activities, which are the subject matter of urban design.

No other disaster, man-made or otherwise, in the history of mankind has been able to affect every aspect of life in our planet. Scientists, politicians, leaders, and the public were quite shocked by their inabilities to grasp the phenomena and find solutions to put it under control. All technological advances proved rather useless, at least in the short run. All societies, poor as well as rich, developing as well as developed, West and East, and people of all ages, genders, etc., faced the crises. The whole mankind fell into a stage of shock.

What is certain is that the future will not be the same as before. We claim that a paradigm crisis has occurred in some fields, particularly urban planning and design. The old knowledge base of urban planning and design is incapable to solve the emerging problems—the new pandemic crises. We need to evaluate and revise the past trends in all areas; values and goals have to be reconsidered and new priorities have to be set. The content of urban design and planning need to be fundamentally restructured (in theory, as well as in education and practice).

The question may be raised here is that can the concept of sustainable development offer reasonable solution to this worldwide problem? Some believe that after so many years and diverse efforts by experts at various levels to implement the ideas and put them into action, no tangible success has been yet reached. No doubt, sustainable development claims too much that could be realized in real world.

The outbreak of COVID-19 showed that the sustainability concept is of no use to deal with this new problem. Recent experience shows that what is critical at this time is how to become resilient against an unknown phenomenon. Tackling the problem in a rather short time required strength and flexibility, mentally, socially, economically, and healthwise.

So in recent years, the idea of resiliency is gradually replacing sustainability. Resiliency is considered to be realistic, pragmatic and to best fit the real-life situations. It is incremental, segmental, and easy to use and measure its performance in practice.

The point here is that from now on, and definitely for the post-COVID-19, the basic needs and its components should be given the first priority in any decision-making processes, at any level.

But what can urban planners/designers do to reduce the impact of a crisis on communities, and how should public services be designed to cope with surges in demand?

We claim that a paradigm crisis has occurred in urban planning and design. The old knowledge base is incapable of solving the emerging problems—the new pandemic crises. We need to evaluate and revise the past trends in all areas; values and goals have to be reconsidered and new priorities have to be set.

The substance and the priorities of planning and design should be also changed. Resiliency, and not sustainability, should be the determining factor in evaluating the plans. We will suggest here that current anomalies will eventually lead to major breakthroughs, or a paradigm shift.

The emerging literature on COVID-19 provides extensive justifications, and thereby significant challenges, for this new paradigm, from master planning the entire communities to interior design. COVID-19 is reshaping the way developers, architects, and designers are looking at our homes and the environments around them. Suggested changes may be divided into two groups of procedural areas, such as revising priorities, need for alternatives, information dissemination, justice, equality, and decentralization. Substantive areas include regional mobility**, fostering connections**, the problem of big cities, the role of size, public space, a different approach to open space, the problem of mega malls, commercial centers, etc., low- and medium-density development, **housing**, housing patterns, neighborhood idea and urban village concept, sociability, walking and biking and behavior patterns and **changing habits and** lifestyle are among other suggested subjects.

References

ARUP (2022) Three design lessons for the post pandemic city. https://www.arup.com/perspectives/three-design-lessons-for-the-post-pandemic-city

Bahrainy H (1995) Healthy City Concept, Criteria for urban planning, a WHO research project, Issue 17, No. 17, Winter

Bahrainy H (2011) Comparison of development and sustainable development: a theoretical analysis. In: Aminzadeh B (ed) Sustainable urban development, pp 7–21. University of Tehran Press

Bahrainy H (2015) Urban space analysis in relation to users' behavior pattern. University of Tehran Press (in Persian)

Bahrainy H (2018) Urban design process. University of Tehran Press (in Persian)

Bahrainy H, Khosravi H (2012) The impact of urban design features and qualities on walkability and health in under-construction environments, the Case of *Hashtgerd* New Town in Iran. J Cities

Bahrainy H, Valadkhani H (2019) Urban village: toward a sustainable settlement. University of Tehran Press (in Persian)

Bahrainy H, Khosravi H, Aliakbari F, Khosravi F (2015) The impact of built environment on walkability: case study: North-West of Shiraz. Armanshahr Archit Urban Dev 7(14):49–62 (Spring Summer)

Bozikovic A (2021, March) Yes, in my backyard: How urban planning must shift to meet our post pandemic challenges. THE GLOBE AND MAIL

Campbell S (1996) Green cities, growing cities, just cities? Urban planning and the contradictions of sustainable development. J Am Plann Assoc 62(3):296–312. https://doi.org/10.1080//01944369608975696

COVID Live-Coronavirus Statistics-Worldometer. https://www.worldometers.info

Daneshgar G, Bahrainy H, Einifar A (2011) Analysis of sociability of physical environment due to the nature perception in the built environment. J Fine Arts Archit Urban Plann 45 (Spring)

Duany A (2020) Three Corners plan for post-pandemic future. https://www.tcpalm.com/story/news/local/shaping-our-future/growth/2020/07/22/vero-beach-andres-duany-change-three-corners-plan-post-pandemic/5485539002/

Frey H (1999) Designing the city, toward a more sustainable urban form. E& FN SPON

Galloway TG, Mahayni RG (1977) Planning theory in retrospect: the process of paradigm change. J Am Inst Plann 43(1):62–71

Gehl: https://covid19.gehlpeople.com/

Horne RRPP (2021) Excellence through humanity: planning's wobbly paradigm shift—part one. MOVING FROM THE DECLINING RETURNS OF ITERATIONTO MEANINGFUL SYSTEMIC CHANGE, Planner-in-Residence School of Planning, University of Waterloo: April

H. Bahrainy and A. Bakhtiar, *Urban Design in Seismic-Prone Regions*, The Urban Book Series, https://doi.org/10.1007/978-3-031-08321-1

Howard E (1902) Garden cities of tomorrow, Sonnenschein *co
https://independent.academia.edu/reshmarameshan?swp=tc-au-17494346
https://www.worldometers.info/coronavirus/
Jachnow A (2020) What will be the Post-Pandemic Urban Path?
Jacobs J (1961) The death and life of great American cities. Random house, New York
Jon I (2020) A manifest to for planning after the corona virus: towards planning of care. Plann Theory 19(3):329–345
Kuhn T (1970) The structure of scientific revolutions, 2d edn. University of Chicago Press, Chicago
Kuhn T (1977) Second thoughts on paradigms, in Suppe, F. 1977, The Structure of scientific theories, 2nd edn. Urbana
Lubell S (2020, April) COVID-19's impacts on urban design & the built environment. LA Times
Mehta V (2020) The new proxemics: COVID-19, social distancing, and sociable space. J Urban Des 25(6)
Newman O (1973) Defensible space; crime prevention through urban design. Macmillan Publishing
Sarkin G (2020) Cities at the front line: public space in the time of the COVID-19 Pandemic, June, smithgroup.com
Schuff S (2020) Gehl, Making cities for people
Soltanzadeh H, Bahrainy H, Varmaghani H (2007) Urban squares: the tale of two cities in west and east, Savannah, USA and Hamadan, Iran, unpublished research paper
Weidl B, Huff A, Hynes C (2020) Rethinking public space with empathy: returning downtown during & after COVID. Adapting our Urban Realm with Empathy

Index

H. Bahrainy and A. Bakhtiar, *Urban Design in Seismic-Prone Regions*, The Urban Book Series, https://doi.org/10.1007/978-3-031-08321-1

C

D

Q

R

S

T

GPSR Compliance

The European Union's (EU) General Product Safety Regulation (GPSR) is a set of rules that requires consumer products to be safe and our obligations to ensure this.

If you have any concerns about our products, you can contact us on ProductSafety@springernature.com

In case Publisher is established outside the EU, the EU authorized representative is:

Springer Nature Customer Service Center GmbH
Europaplatz 3
69115 Heidelberg, Germany

Batch number: 10837251

Printed by Printforce, the Netherlands